STATIC ELECTRIFICATION

BY

LEONARD B. LOEB

PH. D. PROFESSOR OF PHYSICS
UNIVERSITY OF CALIFORNIA
BERKELEY, CAL./USA

WITH 63 FIGURES

SPRINGER-VERLAG

BERLIN · GÖTTINGEN · HEIDELBERG

1958

ISBN 978-3-642-88245-6   ISBN 978-3-642-88243-2 (eBook)
DOI 10.1007/978-3-642-88243-2

Dedication

To Professor F. P. BOWDEN and his associates, whose beautiful pioneering investigations on the neglected classical problems of friction have done so much to clarify that important field and by its clarification has aided so materially in the understanding of the more baffling aspects of static electrification, this book is humbly dedicated.

October 5, 1957   LEONARD B. LOEB

Preface

In our preoccupation with the dramatic developments in the numerous fields of modern physics with their beautiful instrumentation and exciting revelations, we tend to forget our profound ignorance of some of the longest known phenomena of physics. Among these were, until the middle nineteen hundred and thirties, ferromagnetism, friction, lightning stroke, the common electric spark, and static electrification. The first two have now been pretty well clarified and the understanding of both of these phenomena have contributed greatly to our understanding of the structure of matter and surface physics. The lightning stroke and common spark are well on their way to clarification. Strangely despite the ever expanding importance of static electrification in industry affecting as it does, a wide diversity of processes either as a useful tool or adversely and extending even to the realms of meteorology, this field has awakened little curiosity and stimulated little investigation in recent years except in so far as the immediate industrial problems it invoked required an immediate and often make-shift remedy.

Trained in his early years as a chemist, and brought into contact with some aspects of colloidal chemistry involving electrokinetic potentials, cataphoresis, and spray electrification, the author had his curiosity aroused by a number of these strange phenomena. Entering physics as a life career coincident with the development of the early studies in atomic structure, in part through his teacher, R. A. MILLIKAN, the author became aware of the earlier confusions and controversies concerning electro-chemical potentials, the Volta potentials, the contact potentials and work functions and their role in electrification. Thus as the years have gone by while his chief interest focussed on the basic processes of gaseous electronics and the related atomic physics, the author never could resist the urge of his curiosity concerning various aspects of the several processes leading to static electrification. In consequence, over the years, in addition to following the progress in the fields of major interest, he and his students have, from time to time, engaged in studies of various aspects of the static phenomena.

These led to a series of researches beginning in the middle nineteen hundred and thirties on spray electrification of liquids which in the immediate prewar years and subsequent to World War II, ended in a rather long continued program involving first the static electrification of dusts and subsequently of contact charging of solids. With improving vacuum techniques, increasing knowledge of the solid state and especially inspired by BOWDEN and TABOR's excellent book on *Friction and Lubrication*, the studies took on a form such that much clarification concerning the processes at work was achieved. Through the generosity of the Office of Naval Research, beginning in 1947, grants became available so that it was possible to support some excellent half time graduate assistants while they worked for their doctorates whose curiosity also was aroused by this field. Thanks to this support, the study of P. E. WAGNER, in some measure, has brought the program to a degree of completion that makes a general summary of the work seem desirable at this time. However, in addition to the integration and reporting of the program completed in the author's laboratory, the crying need for a *general* and *critical summary* of *the whole field of static electrification processes* and with the encouragement of Professor BOWDEN, the author undertook the task which resulted in this book at the end of 1955.

Since the processes of static electrification, and there are many competing and confusing processes obscuring our understanding, have never been brought together under one head before, the present attempt no doubt will have its weaknesses and many flaws. It will, however, it is hoped, pave the way for more and perhaps more properly directed and controlled researches and thus will, it is hoped, lead to the publication of a bigger and better book when the time is ripe.

The book is written for all those who are interested in, or curious about the phenomena in their various aspects. It is aimed at explaining and clarifying the basic physical principles underlying the processes. It does not attempt to solve the many technical problems which industry requires. These, where solutions exist, are buried in confidential files and in any case, present individual problems that must each be solved separately, using knowledge of general principles.

The book is therefore written for the engineer or scientist whose professional applied work requires a basic understanding of these elements in order to control and apply them to the solution of his problems.

With the wide diversity of background of those interested, it has been necessary to develop the analysis of the various phenomena from the simplest principles and the ground upward. Thus the introductory portions usually begin with the simplest concepts and pictures and develop these further as the book progresses.

The book begins in Chapt. I dealing with *electrolytic processes* in *static charging*. In so doing, it begins with the concept of electrode potentials of a single metal in solution. Then it considers the application to charging mechanisms, where such charging is to be found and through this, leads to the more complicated processes involving electrokinetic potentials. These are first presented as explained by PERRIN and then extended to the Debye-Hückel analysis. The chapter ends with a presentation of the newly discovered freezing potentials of WORKMAN and REYNOLDS. In such a chapter, examples of the electrolytic type of charging action are drawn from the past literature in so far as the older studies were carried out with adequate controls to insure that this action occurred.

In order to understand the more involved processes of solid-solid contact charging, the second chapter presents a condensed and concise account of the electron concept of metals, the Fermi bands and the nature of the surface barrier. It shows how this is affected by external fields and adsorbed atoms. This sets a pattern for the analysis and thought processes on contact charging of metal-inorganic insulator surfaces and is an essential introduction to what follows in the later chapters. It also presents what evidence there is for static charging between metals by virtue of this effect.

Chapter three discusses the electrification by spraying and bubbling of liquids. It begins with an analysis of the significance of the work of COEHN concerning the potentials across liquid-gas or liquid-liquid surfaces leading to the oft quoted COEHN's law. While basic analysis gives a rational and simplified interpretation of the law, it shows also that the derivation of the law under COEHN's experimental conditions presents a true enigma. Then follows the work on cataphoresis of gas bubbles ending with the very important and significant findings of ALTY on the true nature of the electrical double layer at liquid gas interfaces which emphasizes the time element in the formation of the double layers so vital to understanding the differences in the spray electrification data. Against the background of ALTY's work, the studies of CHALMERS and PASQUILL, of FRUMKIN and of LENARD, but more especially of CHAPMAN on spray electrification of liquids is unified and becomes comprehensible. The chapter closes with the remarkable new spray electrification phenomenon observed by WOODCOCK and BLANCHARD on the positive charge carried aloft on spray from sea water bubbles caused by breaking waves. This phenomena may have very important meteorological implications certainly involving coastal haze, sometimes miscalled smog (which has another origin), and perhaps on the maintenance of the atmospheric potential gradient. In this chapter, the studies of DODD on symmetrical charging of sprayed liquid droplets is presented and its importance on the falsification of data on spray charging mechanisms is discussed.

Chapter four deals with the contact charging between solids. It emphasizes the more recent studies of the controlled contact charging between simple inorganic insulators e.g., single crystals and clean metals. The techniques for such studies are presented as well as results and conclusions. The question of the action of external fields, the mechanics of rolling contact and charge transfer, etc., are discussed in detail in terms of BOWDEN's work. The study of electrification of dusts and the nature of and conditions for, symmetrical and asymmetrical charging, are discussed in detail. The chapter closes with a discussion of charging by asymmetrical heating of homogeneous substances, charging by common ion exchange (the Henry model), by rupture of surface dipoles and the evidence therefore.

Chapter five starts with the electrostatic charging processes incident on ionization of gases. It indicates the nature of charging by flue gases, by unipolar discharges, and allied matters. There is a short section devoted to the principles of elimination of static in so far as this is possible and closes with a short section on thunderstorm electrification and protection against lightning stroke.

One thing this book does not do. While enough references are cited to illustrate the principles and matters considered, the voluminous older literature and findings are not quoted in extenso although sources are given. The reason for this is that the more one considers the earlier investigations in the light of the processes as we now know them, the more it is clear that controls were inadequate to yield significant data. This is no reflection on the many excellent earlier workers who could not have been endowed with the understanding which modern techniques and knowledge imply.

In the hope that this initial step in the integration of past findings and analysis of basic mechanisms of static electrification in the light of present day knowledge will be of inspiration to the curious and of help to industry, the author presents the book to the reader.

Some months after the manuscript of this book had gone to press, an article in a recent issue of *Zeitschrift für Angewandte Physik*, by Prof. Dr. P. BONING,[135] brought to the author's attention the work on Static Electrification of that investigator and indicated that he was author of a book on the subject published in 1938. Dr. BONING is an engineer who became interested in the subject of Static Electrification in connection with colloidal chemistry inspired in part by Prof. Dr. WOLFGANG OSTWALD. Unfortunately, the extensive studies carried on by Dr. BONING leading to his book and those that subsequently appeared, were published in journals that are not often read by physicists and vice versa, so that he appears to have been little acquainted with the work published in the Physical Journals. The publication of this book shortly before the war and its very nearly complete destruction during the war, resulted in its being little known. So far, after nearly six months' search, only two copies have been discovered in the United States' libraries and the author has not been able to obtain the loan of a copy. However, later papers by Dr. BONING have been sent the author and he has maintained a correspondence with Dr. BONING. In principle, Dr. BONING has, in consequence of his colloidal chemical viewpoint, regarded all surfaces as being the seat of electrical double layers, having ions of one sign or the other exposed on the surface. In some cases, the ions are inherent in the nature of the substances; otherwise, the double layers are caused by ions adsorbed on the surfaces, including, possibly, air ions. Thus, on contact between two surfaces, transfer of ions from one surface to the other is inevitable, leading to static electrification. Dealing with a wide range of complex organic solids, data appear to substantiate these views. In principle and broadly speaking, as brought out in the several chapters in this book, static electrification is, in all cases, caused by transfer of carriers on contact between what could be loosely called electrical double layers on surfaces, if this term includes the electron atmospheres in metals, the electrical stratification at polar liquid-gas interfaces, and the electrolytic double layers on moist surfaces. However, as amply exemplified by what follows in the present volume, such generalization ignores the wide variety of binding forces and their magnitudes, much of the present day knowledge of the solid state and above all, the nature of surface interactions in the solid state as exemplified by the frictional studies of BOWDEN, such that each case of charge transfer must be analyzed separately with adequate experimental controls to separate it and establish its specific mechanism. Thus it is believed that the author's book and its analyses of the varied physical mechanisms and semi-quantitative illustrations, fulfills a definite need in the field of static electrification. However, the important prior disclosures of Dr. BONING's book cannot be ignored and the author gladly lists under the reference numbers, all of those papers in regretting that he came upon these too late to include a discussion of them at appropriate points in this book.

Berkeley (Cal.), Nov. 25, 1957   LEONARD B. LOEB

Acknowledgements

The writing of this book has been made possible through the extended investigations of a group of interested students whose studies were made possible by grants from the Office of Naval Research, United States Navy, and the Ordnance Department of the U.S. Army. While the initial investigations of C. DYK, Dr. SEVILLE CHAPMAN, Dr. DAVID DEBEAU, and Dr. J. W. HANSEN were carried out before the existence of this office, the extended studies of Dr. WULF B. KUNKEL, Dr. E. E. DODD, Dr. J. W. PETERSON, and Dr. PETER E. WAGNER were carried out under ONR Contract with the assistance of the Ordnance Department of the United States Army. To this generous support and more to the excellent work of this group of able young men, the author expresses his deep gratitude and thanks.

The author's profound thanks are given to Mrs. BEATRICE GALBRETH, who has carried out all of the typing and preparation of the manuscript.

He is also indebted to Messrs. WARREN WIESENFELD, BEN OSHINA, and RICHARD FRANCIS for the drawings and to Mr. GENE WEBER for the photography needed.

Table of Contents

Introduction

It is paradoxical that many of the physical phenomena earliest known to mankind appear to be phenomena of such complexity that we are just today learning to understand them. Among these phenomena are ferromagnetism, sliding friction, the lightning stroke, the electric spark and static electrification processes leading to the last named. Of all these phenomena, the understanding of processes leading to accumulation of electrical charge by static means, in contrast to electro-dynamic processes, has, in general, shown the slowest progress. This general statement requires some comment for among the static processes we must list the electrolytic cell and the contact potential charging between dissimilar metals, Volta effect, about which a great deal is now known.

The slow progress of clarification of static electrification mechanisms stems from the fact that there are several processes, or better, types of static electrification process, which without careful experimental control could be responsible for any particular manifestations observed and may even simultaneously be active. Thus the discordant and erratic results which have often led to confusion in the past must be ascribed to failure of observers to recognize the possibility of the action of several mechanisms which resulted in a lack of control of conditions to insure study of the manifestation of one mechanism only. In 1945, the author indicated this cause for confusion in an article entitled "The Basic Mechanisms of Static Electrification"[1]. In a later article, P. S. H. HENRY[2] has made a similar attempt at classification. Since 1945, notable progress has been made in the understanding of several of the more obscure mechanisms and even hitherto unsuspected processes have been discovered.

It is the purpose of this book to summarize the processes and the progress as of the present time in order to acquaint those interested as well as the workers applying such processes in industry with the present state of knowledge.

It is essential first, to define the term as here used. *Static electrification* covers all processes for producing segregation of positive and negative electrical charges by mechanical actions which operate by contact or impact between solid surfaces, between solid and liquid surfaces, or in the rupture or separation of solid or liquid surfaces by gases or otherwise including also ionized gases. These involve such processes as frictional, contact, or tribo-electrification, spray electrification and electrification in dust, snow, or in thunderstorms.

These phenomena aside from arousing our curiosity as to their nature, have enormous industrial implications as well. They are responsible for the extremely destructive dust explosions in sugar mills, granaries, sulfur mills, and in coal handling. They are responsible for explosions in the handling, filtering, or "bucketing" of volatile, inflammable fluids, such as gasoline and cleaning fluids. Static electrification is basic to the action of the earlier high potential generators, although the present devices, such as the van de Graaf generator, operate by more direct electrical processes. Electro static charging is becoming an industrial problem, of considerable magnitude through the use of certain of the highly insulating new plastics, presenting, for example, a serious inconvenience in the newer record discs in consequence of charging and dust accumulation. The paper and textile industries run into serious difficulties with the accumulation of static charge. Projectiles in flight through the air acquire static charges. More serious is the precipitation static charging of aircraft in flight leading to corona discharge which is sufficiently electrically noisy to wipe out communications entirely. The charging of dusts can have useful purpose such as in the rapid reproduction of printed and X-ray pictures by the process called xerography. Finally, the role played by static processes in charging water droplets and dusts from a meteorological point of view cannot today be underestimated. In view of these and the countless other aspects of industrial use, it is of importance to present current knowledge and to clarify the nature of the phenomena.

It is next of interest to segregate the various basic mechanisms which lead to electrification in order to delineate more in detail the scope of this book. The basic processes listed are:

1. Electrolytic phenomena ranging from the processes basic to the galvanic cell to those processes leading to the formation of the so-called Helmholtz double layer at the surfaces of metals or other substances in contact with liquids, usually of high dielectric constant. It is a consequence of *electrolytic ion* transfer.

2. Contact, or Volta, electrification between clean dry metal to metal or metal to semi-conductor surfaces. It is a consequence of *electron* transfer between boundaries with differences in electron energy levels.

3. Spray electrification produced by the disruption of surface films of dielectric liquids and solutions by mechanical forces such as atomizing and bubbling of liquids and perhaps shattering of high velocity jets by solid surfaces. It is a consequence of *intrinsic electrical double layers* at dielectric liquid interfaces. In some aspects, it is difficult to separate from Helmholtz double layers at solid-liquid interfaces and it is like those subject to strong ionic influences.

4. What used to be called frictional or tribo-electrification and should in most instances more properly be called *contact* electrification occurs as a result of impacts leading to contact between two dry solid surfaces and subsequent separation. This usually involves contact between dissimilar surfaces, e.g., two insulating solids, or an insulator and a metal or two solids containing the same ion in different concentrations. This is the phenomenon which, at present, has received new clarification. It involves transfer of *electrons, or ions*, between the two surfaces as a result of surface forces some of them akin, perhaps, to the processes under (2) in metals. True frictional electrification could occur if carrier densities and energy differ on asymmetrical heating resulting from rubbing.

5. Homogeneous or symmetrical charge separation. This occurs in separating relatively small solid particles, or liquid droplets, from similar particles or larger portions of the same matter. Created are particles of equal opposite charges with no net charge accumulation of one sign, though charges positive and negative of individual particles may be high. It results from the statistics involved when relatively small numbers of charge crrriers are distributed at random on both sides of the ruptured boundary.

6. Effects by which ions and electrons created by electrical phenomena in gases in contact with solids or in flames, are segregated and separated by mechanical forces, by mass motions of gases or solids, with or without the presence of fields. Examples appear in processes such as the electrical segregation in flames in exhaust gases from flues and in the van der Graaf static generator which are well known and understood.

7. There are occasions where various types of electrical or mechanical forces produce polarization of crystals or dielectric substances. In this condition appropriately strong surface interaction by friction might abrade sections of the polarized surfaces leading to charge separation. Thus piezo distortion of quartz crystals, or the pyroelectric effect where heat induces strains leading to the piezo electrification of quartz or by pyroelectrification itself could lead to charging if the charge can be removed from the surface. Similarly, polarization of dielectrics in fields could yield charge separation if mechanical separation of the polarized layers can be achieved. At present, there is no evidence of this for solids, though it is not precluded. With thin aqueous layers and liquids, it has been observed.

In what follows, the seven different categories will be discussed in five chapters. The subdivisions of the seven different processes into only five chapters is necessitated by the fact that the treatment of spray electrification and of dust electrification under contact electrification of solids involves and is overlain by symmetrical charging in such a fashion that it cannot well be treatet as a separate mechanism without undue repetition. Again it is proper to treat piezo and pyro electric segregation processes as aspects of contact charging. Polarization charging, coming as a source of error and being in a measure inherent to certain spray electrification studies as well as in some contact electrification studies, will be treated with them although it will appear in Chap. V as well. Because of the overlapping indicated, the phenomena delineated will be presented as follows: Chap. I—*Electrolytic processes*. These include the Helmholtz double layer as well as the freezing potential phenomena. Chap. II—*The volta potential* and *contact charging of metals*. A clear picture of the processes involved is essential as a proper introduction to the study of surface contact electrification on metal—non metallic solid and non metallic solid—solid charging. Chap. III—*Spray electrification*. Here the peculiar Helmholtz double layers at gas liquid interfaces are for the first time clearly delineated, for in fact, the Helmholtz double layers are probably more clearly understood in this simple case than in that of solid-liquid interfaces. Here an understanding of the symmetrical charging is required in order to interpret the better data on spray electrification. Chap. IV—*Contact electrification of solids*. This deals both with dust and contact electrification studies under controlled conditions. These studies, as will be seen, throw much light on the processes at work leading to certain illustrative theories of mechanisms and quantitative verification of processes at work. In this area, great help comes from the excellent studies of BOWDEN and TABOR on contact and friction between solid surfaces. Chap. V—*Other processes*. This deals with numerous problems not otherwise touched upon. It gives the principles of charging by ionization of gases, induction charging the prevention or elimination of static charge accumulations, the charging processes in thunderstorms, the lighthning stroke and protection against it.

I. Static electrification by electrolytic processes

A. Basic principle in terms of galvanic action from metal surface

If one regards a metal in contact with a liquid of high dielectric constant such as water, in which acids, bases or salts, i.e., electrolytes are soluble and dissociated, the metal immersed in the liquid will tend to go into solution in the form of ions. Depending on the nature of the dissolved substances in the liquid and the nature of the metal, the metal will tend to go into solution as positive ions, or reacting with the solution, it may go into solution in the form of complex negative ions. The action is most strongly pronounced in water, although liquified NH_3 gas acts in an analogous fashion. When solution takes place, the process continues until under the action of the kinetic bombardment by the ions and of the field of the charge acquired by the charged metal, the metal takes on a specific potential leading to equilibrium between processes of solution and precipitation of ions on the surface. The potential depends on the nature of the metal and the concentration of ions, or dissolved salts, in the solution. Many metals like Zn or the alkali atoms are strongly metallic and tend to send their own positive ions into solution. Other metals, like Al will, in slightly acid solution, go in as positive metal ions, in this case as trivalent Al^{+++} ions, but in an alkaline NaOH solution will dissolve as Na_3AlO_3, sending the negative ions $H_2AlO_3^-$, $HAlO_3^=$, and $AlO_3^{\equiv}$ into solution. At a certain hydrogen ion concentration, e.g. pH, Al metal exposed to an aqueous solution is quite inert. This is called its iso-electric point at which it sends neither one ion nor the other into solution. Any neutral metal either at its iso-electric point, or that is inert to its environment while *not contributing* ions, may *receive* ions of the solution gaining charge until it is in equilibrium with its environment as regards liberation and resolution of the ions as influenced by the charge. For example, Cu in a dilute HCl solution or Pt in the presence of H^+ ions, will receive these, gaining a positive charge and leaving the liquid negative. It does not take a high concentration of dissolved acid to yield considerable charge.

The potentials acquired by the metals in equilibrium under a given set of circumstances are sharply defined and range in magnitude from very small values up to the order of, perhaps, a volt or two at the most. The values relative to a standard electrode such as the *normal calomel electrode* are given in standard Physical Chemical Tables such as in those of the Electro-motive Forces of Reversible Galvanic Cells shown in LANDOLT-BÖRNSTEIN's Tables[3]. An illustrative electro-motive force table for the elements relative to the H^+ ion from $\frac{1}{2}H_2$ are shown in Table 1.

If two dissimilar metals are immersed in a solution, the difference in potentials built up between each and the liquid remain static until the two metals are brought into metallic contact. At this point, reduction or equalization of charge of each allows more ions to go into solution and other ions to come out on the other metal until either the metal or the ambient ions are used up. Thus current is derived as a result of a chemical reaction, the heat of reaction determining the quantity of charge that passes and fixing the potential difference except for losses to entropy. This action of the galvanic cell is of no significance in the static generation; however, unless a metal surface is very clean and homogeneous, local cell actions between neighboring points on the surface of different potentials may lead to disturbing currents.

The simple character of the metal electrode in equilibrium with its layer of electrolyte and solution furnishes a point of departure in the consideration of static electrification in consequence of electrolytic potentials. For given such a layer there are two ways in which charge separation may be achieved. (1) A tangential force parallel to a charged surface could presumably remove the liquid layer, i.e. sweep the liquid along the surface into an insulated container carrying with it the ionic charge in the liquid. This could accomplished by a flow of the liquid through a section of metal tube insulated from the receiving vessel. Thus the ions in the volume of flowing solution are separated from the metal and charge is allowed to accumulate, the metal remaining grounded relative to the solution, initially at ground potential on entering the metal tube section. This will be termed *flow electrification*. (2) It is conceivable that a neutral surface of some sort that is hygroscopic might be placed in contact with the metal having its ionized film of water in equilibrium such that on separation the liquid film in whole or in part remained with the neutral surface, taking the ions in solution and leaving the metal charged. This might equally well be achieved by wiping the surface with a neutral absorbent of the film. This is the *equivalent* of many *processes to be designated* as *contact charging* in what follows. In such charging, the *rubbing* of the two surfaces together, as with cat's fur and sealing wax, serves *primarily* to *expose new areas of the surface to contact* rather than to generate frictional actions which might lead to electrification in other fashions.

Table 1. *Electro-motive force series of some elements relative to H_2 and electrode potentials in volts*

Li^+	$+2.96$	$\frac{1}{3} Bi^{+++}$	-0.226
Rb^+	$+2.93$	$\frac{1}{2} Cu^{++}$	-0.344
K^+	$+2.92$	$OH^- = \frac{1}{4} O_2 + \frac{1}{2} H_2O$	-0.397
$\frac{1}{2} Sr^{++}$	$+2.92$	Cu^+	-0.470
$\frac{1}{2} Ba^{++}$	$+2.90$	$I^- = \frac{1}{2} I_2$	-0.535
$\frac{1}{2} Ca^{++}$	$+2.87$	Ag^+	-0.798
Na^+	$+2.71$	Hg_2^{++}	-0.799
$\frac{1}{2} Mg^{++}$	$+2.70$	$\frac{1}{4} Pb^{++++}$	-0.80
$\frac{1}{3} Al^{+++}$	$+1.70$	$\frac{1}{4} Pt^{++++}$	-0.86
$\frac{1}{2} Mn^{++}$	$+1.10$	$Br^- = \frac{1}{2} Br_2$	-1.065
$\frac{1}{2} Zn^{++}$	$+0.762$	$Cl^- = \frac{1}{2} Cl_2$	-1.36
$\frac{1}{2} Cr^{++}$	$+0.557$	$\frac{1}{3} Au$	-1.36
		F^-	-1.90
$\frac{1}{2} Fe^{++}$	$+0.441$		
$\frac{1}{2} Cd^{++}$	$+0.401$		
$\frac{1}{2} Ni^{++}$	$+0.231$		
$\frac{1}{2} Sn^{++}$	$+0.136$		
$\frac{1}{2} Pb^{++}$	$+0.122$		
$\frac{1}{3} Fe^{+++}$	$+0.045$		
H^+	0.000		

In order to evaluate the electrification by such processes, it is essential that one form a picture of the actual situation existing at the metal-liquid interface. The picture to be presented is a rather crude one and more accurate descriptions will follow at appropriate places.

Let the chemical energy available, less the work needed for the removal of ions from inside the metal to a distance r_0 just outside the metal surface be represented by the quantity Ae, where A is an equivalent potential for the process in electron volts and e is the electronic charge. The value of A can be positive or negative except for the work term deducted to bring the ion from the interior to the distance r_0. If the metal gives out energy in going into solution A will be positive and the work term will be negative and may be larger than or slightly smaller than A. The work to remove the ion from r_0 at the surface to a distance r

in the liquid against the image force of the ion in the metal is given by

$$\int_{r_0}^{r} \frac{e^2}{D\,r^2}\,dr = \frac{e^2}{D}\left(\frac{1}{r_0} - \frac{1}{r}\right). \tag{1.1}$$

The motion of the ions from the metal will continue until the metal is charged to a potential V_- relative to the liquid. Let D be the dielectric constant of the liquid and let the ion concentrations at r be N_r and those in the metal be N_m. Then the quantity N_r is given by the Boltzmann equation as

$$N_r \doteq N_m\, e^{-e\left[-A + \frac{e}{D}\left(\frac{1}{r_0} + \frac{1}{r}\right) + V\right]/kT}, \tag{1.2}$$

with k the gas constant per atom and T the absolute temperature.

At
$$r = r_0, \quad N_{r_0} = N_m\, e^{-e\,[-A + V]/kT}$$

and at

$$r \doteq \infty, \quad N_{r\infty} = N_m\, e^{-e\left[-A + \frac{e}{r_0 D} + V\right]/kT}.$$

The quantity N_{r_0} represents a constant of the metal and of the ambient solution governed by A, V, N_m, and T. If N_{r_0} for the ion in question is altered by presence of similar ions placed into the solution from outside then V will decrease. Thus the basic galvanic potential V depends on A, N_m, N_{r_0} and T. Assuming this fixed and N_{r_0} a constant then

$$N_r = N_{r_0}\, e^{-\frac{e^2}{D\,kT}\left(\frac{1}{r_0} - \frac{1}{r}\right)}. \tag{1.3}$$

This indicates a decrease in ion concentration as one recedes from the surface at the distance r_0 at which the ions are tightly bound by their image forces. If now $\frac{e^2}{r_0 D kT} \gg 1$ then as r increases from r_0 at which $N_r = N_{r_0}$ towards infinity the ion concentration rapidly decreases. This signifies that the ions are rather firmly bound to the surface layer and those at some distance r outward in the solution are relatively few. If $\frac{e^2}{r_0 D kT} \sim 1$, then even at very large r there are of the order of 0.37 of the N_{r_0} ions available. For $r_0 \sim 5$ Å and assuming $D = 80$, $N_{r\infty}/N_{r_0} = e^{-1.7}$. Here there is little attenuation. The value of the dielectric constant $D = 80$ for water in bulk can hardly be applied to $r_0 \sim 5$ Å so that the effective D may be much less. Thus were $D \sim 16$

$$\frac{N_{r\infty}}{N_{r_0}} = e^{-8.5} = 2 \times 10^{-4}.$$

Thus only a small fraction of the ions would be in a layer at more than 15 Å where binding forces are such as to allow the ions to be removed by flow or wiping.

It is next of interest to calculate the amount of charge separated in such processes. Consider 1 cm² of geometrical surface of metal in a tube 1 cm long exposed to an electrolyte flowing at the velocity of 1 cm/sec. The actual surface of metal exposed to transfer owing to roughness of the order of 10^{-5}cm might be 3 cm².

If $r_0 \sim 5$ Å then the equivalent ions of opposite charge in the metal (i.e., image force ions) will be 10 Å distant and these comprise a condenser of capacity C,

$$C = \frac{A D}{8 \pi r_0} \sim 2 \times 10^8 \text{ cm}.$$

If this is charged to 1 volt $= 1/300$ esu it yields the charge $q \sim 2.6 \times 10^6$ esu or 5.2×10^{15} electrons, per 3 cm^2, and σ the charge density $\sim 1.7 \times 10^{15}$ electrons/cm^2.

This is equivalent to 1 ion per a square 2.4×10^{-8} cm on a side. This is a rather high charge density and again it must be ascribed to the high value of D chosen at 5 Å. If this corresponded to $D \sim 16$ then there are 0.75×10^{15} electrons/cm^2, or one ion per square 3.6×10^{-8} cm on a side. If all N_{r_0} ions were separated, the rate of charge generation would be of the order of 3.7×10^5 esu/sec or about 1.2×10^{-4} ampere. If separated as a film from 1 cm^2 geometrical surface there would be about 1.2×10^{-4} coulomb of charge.

In contrast observed rates of charging by flow of sufficiently insulating liquids to insure against loss to back flow give currents of the order of 10^{-7} ampere, and contact charge separations of the order of 5×10^{-9} coloumb/cm^2 have been observed.

In the case of flow segregation, the difference of a factor of 10^3 is attributable to lower values of N_{r_0} and thus *largely* to *attenuation by the binding* of a large fraction of the ions to the surface. The binding factor of the order of magnitude ~ 100 is substantiated by observations to be cited later. Again since as the small slug of ions is pushed out of the cylinder by incoming electrolyte there is a macroscopic field set up between the surface and the slug of ions, depending on the distribution of charges and the movement of the liquid, the fields set up by separation and the thickness and length of leakage path will provide for a *return flow current* so that charge separated is reduced. Specific resistances are as follows: 0.5×10^6 ohms for a cube 1 cm on a side for distilled water, 1.3×10^4 ohms for 10^{-2} N KCl and 1 ohm for N HCl. As potentials increase with the return path of the 1 cm long tube with 1 cm^2 surface area, it is clear that charge separation can hardly accumulate. In fact, experience shows that for flow electrification specific resistances of the liquid of the order of 10^9 ohms $\times$ cm will already begin to cause loss by current flow while dangerous accumulation of charge only occurs above 10^{10} and more ohms $\times$ cm for this process. This statement holds for larger tubes such as used above and does not apply to capillary tubes. Thus, for example, if the tubes are of the order of 2×10^{-5} cm in diameter, the resistance of the return flow path becomes considerable since the area of cross section is $A \sim 10^{-10}$ and even for 0.1 mm length of capillary the specific resistance is multiplied by 10^8 so that distilled water, 0.02 N KCl and N HCl would have return current flow paths of 5×10^{13}, 1.3×10^{12} and 10^3 ohms. Thus to counteract observed charging rates of 10^{-7} ampere potentials an separation across 0.1 mm would need to be 5×10^6, 1.3×10^5 and 10 volts. Such estimates are not but approximate. They serve to show that for small channels or tubes this process can be effective even with ordinary aqueous solutions while for tubes of the orders of cm^3 volume and cm length specific resistance must exceed 10^9.

For the low charge observed on separation of surfaces, especially if the separation is one of wiping off a thin charged moisture film, the situation is more complicated. First, as before N_{r_∞} may be a fraction of N_{r_0} as for liquid flow. The separation of surfaces especially on wiping off thin layers such as 10^{-5} cm thick does not suffer from too much back current flow for solutions of less than $10^{-3} N$ of salts. For thicker layers back conduction is an important factor. In the contact charging especially if performed in air the limitation to high density of accumulation is of a different sort. Assume for simplicity that the charge of 10^{-8} coulomb has been wiped off an isolated sphere of 1 cm² geometrical surface area. Such a sphere has a radius of roughly 0.3 cm and its capacity is thus 0.3 esu. The field produced by 10^{-8} coulomb or 30 esu is then 100 esu/cm or 30000 volts/cm. This exceeds the normal electrical breakdown strength in air at 760 mm. Thus the possible separable 10^{-7} coulomb/cm² estimated from flow studies will have been reduced from 10^{-7} to the order of 2×10^{-9} esu/cm by back electrical discharge in air at some point during separation of the surfaces.

In consequence of these factors it is possible to conclude that observed charge separations either in flow electrification or contact electrification will fall far short of the potential values of σ_0 corresponding to the charges at r_0 on the surface of the double layer as a result of the following factors:

a. Binding forces that may reduce N_r available to about 10^{-2} of the charge N_{r_0} at the outer part of the double layer.

b. Loss by return current through the liquid film if the resistance of the leakage path on separation, as potentials increase, is not in excess of 10^9 ohms or thereabouts, depending on the rate of accumulation.

c. Loss by electrical breakdown of the air or ambient gas. With liquid films gas must be present to afford the humidity essential to maintenance of films. The lower the gas pressure down to about 1 mm the greater the gaseous breakdown loss. Below about 60% of the saturation pressure of water vapor at the ambient temperature moisture films may not exist. Such needed pressures of water vapor are well above one or two mm of Hg, so that gas discharge will usually play a role.

B. Nature of systems leading to charge separation and their limitations

In dealing with the simple case of metals against solutions, the problem was simplified in that by the flow method with grounded metal separation was indefinite since N_m remains constant. With contact electrification, in this instance, say wiping, as long as the film of moisture reforms before the next contact, the source is replenishable provided ions diffuse fast enough. However, with contact of other surfaces this may not always be the case. Consider for example some organic material which has a limited number of H⁺ ions that can come off. For example an insoluble substance with fatty acids having long aliphatic chains built into a lattice but with COOH groups oriented outward. Then the H⁺ ions would dissociate and enter the aqueous film. This film might be all pretty well removed by the time the surface was twice or more wiped or perhaps a few cm³ of water surface had passed over the surface. In this event, *flow charging* would not be an efficacious mechanism and surface contact charging from the same surface would be limited in scope. One may then classify possible surfaces that can charge as primarily *contact chargers of limited scope*, or *flow chargers of indefinite scope*, which will also yield repeated contact charging.

1. Flow charging systems

a. Metals against solutions, at the proper p_H.

b. Adequately conducting amphoteric electrolytes opposite solutions of appropriate hydrogen ion concentration to cause solution so as to yield only one sign of ion.

c. Solutions of electrolytes behind a *semi-permeable* membrane which allows one sign of ion to go through and the other not, e.g., a collodion membrane with a solution of gelatine, either acid or alkaline, which permits ions of one sign to diffuse into the moisture film on the outside following the Donnan equilibriums, or ferrocyanide films against certain solutions of bivalent metals.

2. Limited contact charging

Applies to any type of surfaces with aqueous films in which the substance is relatively insoluble but can liberate one sign of ion or the other into the aqueous film. The surface need not be conducting. This applies to practically all surfaces which may be wet by thin aqueous films and are not at their isoelectric point relative to the condition of the film of electrolyte.

From the list above, it is noted that flow charging to achieve any high degree of static charge accumulation by electrolytic action is applicable to a relatively limited range of substances. On the other hand, contact charging, limited or otherwise, can be common to practically all substances that will accept surface films of moisture and have ions of one sign or the other that they can lose to the film when the film is wiped off by some other surface. Such substances need not be conductors.

The question of flow charging will be discussed somewhat more at a later point. The vital question involving contact charging depends on the existence of surface films of such extent that the electrolytic processes discussed can occur. This requires that these films be such that they can receive adequate ions and are capable of being removed on contact or being wiped off. The limit of optical detection of any surface structure is of the order of 10^{-5} cm in thickness. Thus aqueous films could be 1000 Å thick and not be visually detected. Presumably films as thick as 10^4 Å would not readily be seen if uniformly spread over surfaces. It is unlikely that films much under 50 to 100 Å thick would contribute much to a contact charging process. Here binding forces of both water molecules and ions to the surface become appreciable.

C. Do adequate aqueous surface films exist for electrolytic charging ?

It is now essential to determine how common films of moisture of 10^{-5} cm or so thickness are on surfaces and under what circumstances they are to be found. Colloidal chemists classify surfaces into two categories, *hydrophillic, that is* having an affinity for moisture and *hydrophobic*, i.e., moisture repellent. In the chemist and physicist's terminology surfaces are hygroscopic if they absorb moisture.

Certain salts are called deliquescent if they absorb moisture to the point of dissolving in view of condensation of atmospheric moisture, or efflorescent if they give up water of crystallization to the ambient atmosphere.

These terms indicate a wide range of observed interaction behaviour of water with surfaces. It is also clear that within the range of behaviour the individual surfaces will vary widely in their degree of adsorbtion with relative humidity and temperature.

Within our own individual observation there are a large range of acids, bases, and salts that even in air of 60% humidity at 20° C condense visible films of moisture. Such films are entirely too thick and too conducting to yield any appreciable contact charging owing to their conductivity. Thus NaCl, NaOH and many other substances will probably not charge under those conditions. In air of much lower humidity, however, the growth of films may not reach visible dimensions and could lead to charging. When one turns to surfaces of other substances, the situation is not as clear. These are the better insulators and the surfaces that show static charging. The question is for example whether the many new plastics, glass, quartz, metal, amber, sulfur and similar surfaces have aqueous films and thus lead to *contact electrolytic* or to *true contact charging*.

In the case of glass, quartz, metal and sulfur, there is some information. F. P. BOWDEN and W. R. THROSSEL[4] have relatively recently shown that for *clean glass* and *metal* surfaces in air up to 90% saturation in relative humidity at about 20° C the aqueous surface films are less than two molecules, 6×10^{-8} cm, thick. However, when such surfaces had layers of salt, or other contaminents, on them adsorbed films up to 1000 Å thick could form. These clean surfaces could be classed as hydrophobic. Into this class would also fall most of the rather highly insulating new plastics. The films of moisture observed by BOWDEN and THROSSELL were in many cases rather patchy and the surfaces were not uniformly soiled. W. B. KUNKEL[5] in the author's laboratory arrived at the same conclusions for powdered quartz and borosilicate glass by direct measurement. In these studies, the substances were baked out, dessicated, and weighed, and then exposed to room air of varying humidity. The gain in weight indicated that using the apparent geometrical surface, which is a minute fraction of the total surface exposed to air, on the average only 2 molecular layers were possible. This means that in reality, H_2O molecules were spotted at only certain few special sites on the whole exposed surface. This situation existed up to above 80% or more relative humidity at which point the powders became wet and stuck together in clumps. E. E. DODD[6] confirmed this with smooth small fire polished spheres of soda glass around 3×10^{-4} cm in diameter. KUNKEL also inferred that sulfur surfaces are very much like quartz and glass, but no quantitative data could be obtained. Yet it appears that finely divided sulfur exposed to moisture and sunlight on plant leaves oxidizes to H_2SO_4 in time and the latter is quite hygroscopic. The very uneven crystalline surface of sulfur however provides a very long leakage path so that sulfur remains a good insulator exposed to air for years.

However, that apparently dry small borosilicate glass spheres in air at relative humidities above 15% do have sufficient moisture on them *to influence electrostatic charging* was proven by PETERSON[7] in the writer's laboratory. These spheres rolled on slightly oxidized Ni surfaces. They were cleaned by washing in fat solvents, then alcohol, then water then placed in chromic oxide, rinsed in water in dilute HNO_3 and rinsed clean in distilled water, dried in vacuum and heated to 210° C. Thereafter they rolled on the Ni surface and their charging properties in an auxiliary electrical field were studied as a function of humidity. The investigation was undertaken to study an obervation of GILL and ALFREY [8] who had reported contact charging in external electrical fields by glass or other insulating particles. In fact, on the basis of the influence of fields on the charging, they ascribed all contact charging by glass or other insulators to electrical fields without however, specifying the fields.

PETERSON did, in fact, observe an *induced* charging of the borosilicate glass spheres rolling on the Ni metal. The action was such as one would expect of a locally conducting surface of the sphere in the field in which ions from a conducting layer were attracted to the sphere surface of sign opposite to that on the upper inducing plate while ions of the opposite sign were driven to the lower conducting surface. As the sphere rolled breaking contact charge separation occurred, the conductivity of the surface film being inadequate to discharge the newly acquired patch of surface charge, even though potential rose as it separated. In fact, polarization of the sphere in the field could produce no charging if electrical carriers did not migrate across the boundary glass-metal before contact was broken in rolling. There are alkali ions in the borosilicate glass sphere in small numbers. Conductivity experiments with this type of glass indicate that *these cannot migrate to the metal in the field*. The metal has electrons which can migrate to the glass if the lower metal surface is negative, but not when it is positive. Again borosilicate glass does not normally have free electrons which migrate from it to the metal when the lower plate is positive. In the absence of a conducting film containing ions then the borosilicate spheres could charge negatively by induction when the upper inducing field plate was positive but could *not* charge equally positive when that plate was negative.

PETERSON's measurements indicated that as observed by GILL and ALFREY in *room* air the spheres showed *induced charge* of *both signs* proportional to the inducing field, the magnitude of which as PETERSON showed was a function of relative humidity. The induced charging overlay the *natural* negative charge transfer from Ni to the spheres which occurred in a dry system and consisted of an electron transfer from Ni of which more later.

The effect of moisture was studied in some detail by admitting H_2O vapor at partial pressures from 10^{-5} to 3 mm and then admitting dry air to 760 mm. What was measured was the change in charge Δq upon reversal of the applied potential from $+1200$ to -1200 volts, thus eliminating the natural negative charging process of q_0 which took place simultaneously. This was done for various humidities ranging from dry air with outgassed chamber to various values of moisture admitted. The charge q_0 measured in the absence of moisture, abruptly reduced to one half at 15% and to zero at 60% relative humidity at 20° C by surface conduction. These borosilicate glass spheres had been cleaned in chromic acid and distilled water and later heated as noted. They could well have been contaminated with traces of metal oxide from repeated rolling though this was not visible. The quantity Δq was relatively constant as humidity increased but

$\Delta q/q_0$ showed as sudden rise at about 15% humidity and a much sharper rise at 60%. For the dry system, $\Delta q/q_0$ was of the order of 0.2. Borosilicate glass is known to be hygroscopic under some conditions and requires heating above 300° C to remove *all* moisture. Thus there is indication of the existence of appreciable films of moisture that influence contact electrification above 60% humidity on borosilicate glass surfaces if not even at 15% humidity.

It must be noted here, however, that the aqueous films on glass and/or Ni and external fields did not alter the character or the sign of the basic contact charging q_0. Thus the charge observed by PETERSON was a *composite surface charge* between *areas charged* by *electron transfer to the asperities plastically deforming the Ni surface in which there were no aqueous films* and an *induced electrolytic charging* for the *moist portions of* the *two surfaces not normally in contact* (see Chap. IV—Sect. F4b), *by ion transfer in the field*. With the high enough fields, the net induced charge could neutralize and reverse the sign of charge q_0 of the sphere as a whole. The details of PETERSON's study are reserved for another chapter where they are in their proper setting. The small induced charge transfer for the dried sphere baked out in situ at 210° C in dried air, or vacuum, still indicates some ion transfer in the field. As later studies show the sphere was probably contaminated by condensate from the metal and *not* dry. This was evidenced in the later studies by increased surface conductivity on bakeout.

These results however, definitely indicate that moisture films do occur on fairly well cleaned borosilicate glass at relative humidities above 15% and become thick at 60% at 22° C. These films contain ions of both kinds sufficient to lead to polarization charging of the sphere in areas normally unaffected by contact charging. This action overlies and may mask the normal contact charging if applied fields are high enough. It therefore answers the question as to the existence of moisture films on metal and glass surfaces under the conditions outlined and that they contain electrolytes.

They do not answer the question as to whether they are a few molecules thick in patches or 1000 Å thick. They do not indicate whether they arise as a result of a relatively unclean metal surface, or initially clean glass soiled by exudate and condensate from the metal.

These films at 15% and more so at 60% relative humidity increase surface conductivity of the borosilicate glass so that the normal contact charge q_0 is notably decreased. In fact, above, 60% humidity, what would pass as normally clean glass, is so conducting that normal charging ceases. This could imply surface layers of moisture $\sim 10^{-5}$ cm thick. When this occurs, there is no evidence of contact charging between borosilicate glass and Ni of an electrolytic nature. Thus here an aqueous film is established, destroys the normal contact charging between Ni and glass produced by electron transfer, and does no charging on its own.

D. Past evidence of electrolytic charging

Having thus indicated the nature and mechanisms of electrolytic static charging, indicated where, and for what systems they operate and finally having shown the circumstances under which the needed moisture films appear in so far as we know them, it is of some interest to attempt to find illustrations in past observations of such electrification. Here the task is exceedingly difficult in that lack of control in many earlier studies allow no decision to be made as to what actions may be involved.

1. Electrolytic flow electrification

Such an effect was observed by the writer in 1931, in some studies never published. A clean carefully insulated section of copper tube was connected to an electrometer while kerosene was allowed to flow from an insulated reservoir through the section of tube and then to a grounded reservoir. As long as the kerosene was *dry*, there was no charging observed. As the kerosene became slightly contaminated with moisture, a marked charge separation occurred by the flow across the metal surface. Here the kerosene was merely an insulating vehicle for the ionization charging of the presumably slightly acidulated water, (around 1 % by weight), dissolved in the kerosene. This observation is quite parallel to that presented in a report by W. FORDHAM COOPER[9], who studied charge generation by C_6H_6, paraffines, water and ethyl ether. He did not expect much charging from C_6H_6 and paraffines, but most *commercial grades* used were impure enough to yield charges. With added water, charges were produced, but mostly weakened by leakage. Ethyl ether which has a strong dipole moment, produces very dangerous charging as it is an insulator. He ascribes the action of commercial grade benzene and paraffines to included water and dissolved electrolytes. Charging becomes dangerous in such flow when the specific resistance of the liquid reaches 10^9 to 10^{10} ohms $\times$ cm. Thus 95 % ethyl alcohol or diacetone alcohol gave no charging owing to conduction, while Xylol yielded charging.

Measurements along this line were made in 1913 by F. DOLEZALEK[10]. He used Cu, brass, iron, Pb and Al insulated as in the author's study. Liquids were benzol and ether. The charge on benzol varied in sign with the metal. The degree of purity affected the sign of charging. More consistent results were obtained on metals that had been exposed to air for some time. This insured oxide and moisture layers. Dry ether showed strong charging and moist ether less. Later data by the same author in 1930 indicated that charging increased with moisture content in gasoline or hydrocarbons to an optimum concentration and thereafter declined. If the liquid was passed over filings of certain metals, the charging increased, but it decreased by others. The charging also varied with the velocity of flow and was reduced when the flow was too rapid. All this behaviour points to electrolytic action.

Flow static electrification has led to fires in cleaning establishments until low flash point liquids were prohibited. Soaps have been used to reduce the resistance of the cleaning fluids and thus reduce charge generation. S. S. MCKEOWN and V. WAUK[11], likewise measured charging in flowing petroleum through a metal supply pipe.

These observers obtained currents of the order of $1-80 \times 10^{-8}$ ampere from various grades of petroleum.

2. Electrolytic flow charging in impacts of liquid jets on surfaces

There is one other circumstance where electrolytic flow charging could play a rather important role. This is in impact, or spray, electrification. Its predominance, or importance, depends solely on whether there is a chance for return flow in the processes of contact, distortion and separation of the liquid surface. There is no question about the physical observation that spraying or shattering of various fluids in impact with metallic and even other surfaces produces strong electrification. In some cases, negative electrification in the finer particles suspended in the gas phases have been reported in the case of water, especially in LENARD'S studies.

In his original article on waterfall electrification, which is probably *largely spray* and *not electrolytic* in nature, LENARD[12] observed heavy negative electrification of the fine spray on impact of water drops on water surfaces. If the drops impinged on a zinc plate and particularly when the plate was ventilated by a light current of air, the *charging was two to three times greater*. This was definitely caused by ionic and chemical action superposed on the normal spray effect. It is akin to the troubles encountered in the author's laboratory in 1946, when metal surfaces were involved in bubble creation. The influence of surfaces was clearly noted in this work and paraffine surfaces gave the least charging as no ions were present. Thus electrolytic and spray effects both appear to be present under impact shattering of liquids.

The electrification of aircraft structures by rain has produced heavy charging of both signs depending on the treatment the surface of the plane has undergone[13]. Analogously, very heavy electrification is observed from spraying of gasoline and similar fluids, even in filtering them through sieves or handling them in such a fashion that spray or mist is produced.

One of the earliest observations of charging of droplets was that observed from the electrification produced by wet steam blowing through a small nozzle, or a crack, on a locomotive at Sedgehill, England, by Lord ARMSTRONG, in 1840. The steam jet phenomenon has been further studied, but the experiments carried on in many cases such as those of RUDGE and PORTSMAN, were too poorly conceived and controlled to lead to any suitable conclusions. It is clear that if the steam is dry, there is no charge and that charging appears coincident with water drops. It would seem, then, that the phenomenon is closely related to the spray electrification with probably very little electrolytic action, though since metal surfaces are involved, it could be present. With the effective atomization of water in such jets ,the true spray effect should be predominant.

On the basis of what has been indicated from the earlier studies, it is doubtful whether the impact electrification of water against solid surfaces can be ascribed to ionic effects in its entirety as it cannot readily be separated from the so-called spray electrification to be treated in Chap. III.

The two effects are akin in that they involve respectively charge distribution at solid-liquid and gas liquid interfaces. In the former, ascribed to electrochemical effects, the ions derive from the solid phase or from ions in solution in the liquid and well known electro-chemical interactions. (a) At gas-liquid interfaces, surface tension forces in the liquid create a charge segregation related to the dielectric constant of the liquid and polar chemical linkages in the interface. This segregation in all pure liquids appears to have the same polarity with negative charge at the gas interface. What the polarity is in liquid-inert surface interfaces is not surely known, but *presumably* it is negative if the liquid has the higher dielectric constant. This charge segregation differs from (b) the *electrochemical potentials* in that the latter depend primarily on the ionic nature of the surfaces and the ions in the liquid and are not uniformly of the same sign. These two types of double layers, (a) and (b), do have one feature in common in the presence of ions—the ions in solution can alter the potentials and, to some extent, vary the sign of electrification produced. The polarization segregation (a) at liquid-gas interfaces in the spray phenomenon studied largely by P. LENARD[12] and his students and named *waterfall* electrification by him, is characterized by a difference in sign for droplets of different size, the finest droplets being negative and the intermediate ones being positive. However, the sign of all but the smallest droplets can be altered by ionic impurities in the liquid, trivalent positive ions appearing to be especially effective in this regard. In the electrolytic type of processes, (b) the charge can be of either sign and will not vary too much with droplet size.

To date, no really carefully controlled studies have been made clearly to distinguish these two mechanisms where charge by spraying, impact, or flow has been observed.

In fact, many, if not most observers have not been conscious of the difference between the two phenomena which are even confused by many colloidal chemists. Primarily, the colloidal studies deal with ionic systems. It is strongly suspected that the flow electrification of the less conducting liquids is probably electrolytic due to small amounts of dissolved moisture. In the case of the impact electrification of water against surfaces, it can only be suspected that the effect is, in considerable measure, the spray electrification, but it can be masked by electrolytic effects.

3. Electrolytic effect in contact charging

The evidence for electrolytic effects in straight contact charging processes is meagre indeed. Most of it stems from older work which has not been adequately controlled. The studies of H. F. VIEWEG[14] point to such action in some measure. He studied largely pure substances. He observed that different crystal faces of the same substance give different potentials on separation. He found application of pressure between substances was desirable to get rid of surface film action. He found that humidity caused moisture films that always add a positive charge to a substance. This appears to have been caused by some spray electrification produced in scattering the moisture layer as the air acquired a negative charge. He used weak solutions of acids and alkalies applied to surfaces. Critical concentrations were found below and above which opposite charges were manifested. He stressed the necessity of working with dry substances. RICHARDS[15] points out that in such studies there is likely a superposition of true contact charging inherent in the substances and other effects such as caused by moisture film.

Further instances of electrolytic charging effects may appear in the early studies of DAVY[16] who showed that "dry" acids in contact with metal plates charge them positively, but that powders of "dry" alkaline substances charged them negatively. The term is relative as practically all such studies are carried out in room air. Here liquid films on the metals would have transferred H^+ and OH^- ions to the passive metals.

O. KNOBLAUCH[17] dispersed dust in room air by letting it slide from a Pt or other plate. Pt and paraffine plates were nearly always charged positively with acid substances and negative with alkaline substances. Sulfur was usually charged negatively, but by a few acids, it was charged positively. Glass was usually charged positively, but negatively by a few alkalies. He interpreted his results in terms of electrolytic action in thin layers of moisture on his surfaces. This would be in keeping with BOWDEN and THROSSEL's theory owing to his contaminated surfaces, resulting from his *flaming* surfaces to discharge them and thus condensing H_2O vapor if the surfaces were not heated rather hot. Even then flamed surfaces on cooling have water films in virtue of the condensed impurities. It is clear that evidence for electrolytic ionic effects in static charging appear under conditions where they might be expected. However, in most of these studies, controls were not adequate to make conclusions certain.

E. Electrical endosmosis, streaming potentials, and cataphoresis (verified aspects of electrolytic charge separation)

In discussing the mechanics of flow charging it was indicated that much charge separation was impossible if the separation occurred where the resistance of the electrolytic solution return path became so low that leakage rate equalled generation rate. In general, it will be observed that in all electrostatic contact charging processes, there is a limit to the amount of charge separation set by one mechanism or the other by *return flow*, be it by conductivity of an electrolyte, by gaseous discharge or even by field emission. In the case of electrolytic solutions, the resistances can rapidly become low and this *circumstance alone* is *responsible for the lack of dominance* of the *electrolytic processes* over most other mechanisms. Flow electrification in large tubes and for bulk movement was observed to be confined to liquids having a volume resistivity in excess of 10^9 ohms $\times$ cm, and became serious above 10^{11} ohms $\times$ cm. It was there pointed out that if the tubes in flow electrification amounted to capillaries of the dimensions of 10^{-5} cm radius even 10^{-2} N KCl would yield electrification. In consequence, it must not appear surprising that the physical and colloidal chemists have encountered a variety of phenomena having to do with this situation. The phenomena have been named as follows:

1. Electrical endosmosis. This is the flow of a liquid through capillary tubes in consequence of a potential difference placed across the ends of the capillaries. The flow continues until the hydrostatic pressure p is sufficient to overcome the force of the field on the plug of liquid with excess ions of one sign in the double layer in the interior of the capillary.

2. Streaming potential is the name given to the phenomenon that for larger tubes was called *flow electrification*. It is the electrical potential set up across the ends of a capillary tube by the flow of the fluid with excess ions of one sign in the double layer resulting from an externally impressed hydrostatic pressure. It is limited by the return flow current.

3. Cataphoresis is the inverse of electrical endosmosis. If instead of a capillary tube surface, the surface with its double layer and excess ions surrounds a small solid particle suspended in the liquid, an imposed electrical field will move the outer portion of the double layer towards one electrode and the solid particle towards the other. This is observed not only with solid particles but with small gas bubbles in liquids. Obviously since forces are small, the motion will result only for small particles or bubbles.

The theory of these processes was first developed by HELMHOLTZ[18] extended by PERRIN[19], v. SMOLUCHOWSKI[20], GOUY[21], DEBYE and HÜCKEL[22]. The elementary theory as given by PERRIN will first be developed.

Consider the *endosmosis* set up by the field X created by a potential V applied across a fine cylindrical tube of radius R und length l. This field sets in motion the mobile section of the charged double layer in the tube. Call the surface charge density of this mobile double layer at the solid-liquid interface σ expressed in esu per cm². The field strength X will be uniform and given by $X = V/l$. It produces a force σX per cm² of surface. This force is resisted by the viscous friction of the liquid. If the linear gradient of velocity is U/d with d the thickness of the double layer and the coefficient of viscosity is η the force per cm² is $\eta\, U/d$ such that at equilibrium $\sigma X = \eta\, U/d$. The electrophoretic mobility is $k = U/X = \sigma d/\eta$. The quantity σd is the electrical moment per cm² of the mobile double layer at the surface of the capillary.

The potential across the double layer is the famous *electrokinetic Helmholtz potential* ζ, often called the Zeta potential. It can be calculated by the ratio of the charge/cm² to the capacity of the double layer per cm² of surface. Thus

$$\zeta = \frac{\sigma}{D/4\pi d} = \frac{4\pi\sigma d}{D}. \tag{1.4}$$

This approxiation is permissible since d is small compared to R. Here also D is the dielectric constant of the liquid. In consequence

$$k = \zeta D/4\pi\eta. \tag{1.5}$$

The total volume outflow of liquid per second as a consequence of the field X is $v_X = k X \pi R^2 = k X g$ with $g = \pi R^2$. Whence it it is possible to write

$$v_x = k X g = \zeta D X g/4\pi\eta. \tag{1.6}$$

Assume the liquid to have a specific electrical conductivity λ, the electrical current I is then $I = X \lambda g$ so that X is found from I and λg by $X = I/\lambda g$. Whence the velocity of flow can be expressed as

$$v_x = \zeta D I/4\pi\eta\,\lambda. \tag{1.7}$$

By POISSEUILLE's law the velocity of flow under a static pressure difference P through the capillary is

$$v_p = \frac{\pi\; R^4 P}{8\; \eta l} = \frac{P}{W}, \tag{1.8}$$

with $W = 8\eta\, l/\pi R^4$, the resistance to flow. Equilibrium indicates that $v_x = v_p$ so that flow ceases with a static pressure P^* set up across the capillary by the field X or applied potential V. Thus P is given by

$$P = \frac{2\zeta D V}{\pi R^2} = \frac{\zeta D I W}{4\pi\eta\,\lambda}. \tag{1.9}$$

---

* This treatment is idealized for simplicity and neglects the hydrodynamic nicities resulting from the fluid plug boundary and the POISEUILLE's flow.

Now P can be measured for a given V and if D and R are known ζ may be evaluated. Furthermore v_p may be evaluated as a function of P in a separate measurement yielding W. The values of λ and η may be determined for the electrolyte and I may be measured.

2. The streaming potential

It is now of interest to evaluate the charge separation produced by the flow, or the streaming of the liquid through the capillary, in virtue of a hydrostatic pressure difference P across the ends of the capillary. This streaming separates off the mobile double layer and carries it along as a streaming current, I_s. The streaming current in time builds up a potential V_s across the capillary. This sends a *return current*, or *back discharge current* I_v through the capillary. At equilibrium $I_v = I_s$ and this should allow the value of V_s to be determined.

In the derivation of the POISSEUILLE'S flow law, the linear velocity of flow at a distance r from the axis, if the velocity v of flow at $r = R$ is 0 e.g., if there is no slip, is given by,

$$v_r = \frac{P}{4\eta l} (R^2 - r^2).\tag{1.10}$$

For the double layer of mobile ions, the thickness is d such that the layer contributing to the current I_s from the axis is $r = R - d$. Thus

$$v_d = \frac{P}{4\eta l} (R^2 - R^2 - 2Rd - d^2) \sim \frac{PRd}{2\eta l},\tag{1.11}$$

neglecting d^2 relative to Rd. The current I_s is then at once given by

$$I_s = 2\pi R \sigma v_d = \frac{\pi R^2 P \sigma d}{\eta l} = \frac{PR^2 \zeta D}{4\eta l}.\tag{1.12}$$

The potential V_s builds up until $I_s = I_v$ with $I_v = V_s/R_c$, where R_c is the electrical resistance of the capillary given by $R_c = \frac{l}{\lambda} \frac{1}{\pi R^2}$. Thus

$$I_v = V_s \pi R^2 \lambda/l, \quad \text{whence}\tag{1.13}$$

$$I_s = \zeta D R^2 \frac{P}{4\eta l} = V_s \pi R^2 \frac{\lambda}{l} = I_v.$$

Thus

$$V_s = \frac{\zeta D P}{4\pi \eta \lambda} = \frac{\sigma d P}{\eta \lambda}.\tag{1.14}$$

The conditions of flow indicate that this applies to non-tubulent flow only. It has been pointed out by J. J. THOMSON and by HARDY that the double layer increases the resistance to the flow. For the case of a cataphoretic particle of radius a SMOLUCHOWSKI has indicated that η is increased by $\Delta\eta$ given by $\frac{\Delta\eta}{\eta} = \frac{1}{16\lambda\eta_0} \left(\frac{\zeta D}{\pi a}\right)^2$. A corresponding relation must apply to the flow through a capillary though it has not been derived.

Again since V_s, D, P, η, and λ are known, or measurable, ζ or its equivalent σd may be evaluated.

3. Cataphoresis

For the cataphoresis of solid particles, the following simple relation may be derived. STOKES' law states that the resisting viscous force on a particle of radius a moving through an infinite uniform viscous medium of coefficient of viscosity η with a velocity v is given by

$$F = 6\pi\eta\, a\, v. \qquad (1.15)$$

In an electrical field X in a liquid the particle acquires an electrical surface charge q as a result of the mobile section of the elctrical double layer. The value of $q = 4\pi a^2 \sigma$, where σ is the surface charge density equivalent to the mobile portion of the double layer. In a uniform electrical field X within the liquid, and assuming that the field will not alter the charging of the particle by the double layer, the liquid portion of the double layer d cm thick with its mobile charge will be displaced towards one electrode leaving the inner charged layer of opposite sign on the particle uncovered. Thus the particle will be urged toward the opposite electrode with a force $F_x = qX$. The particle will them migrate in the field X with a velocity v_x given by

$$F_x = X q = 4\pi a^2 \sigma X = 6\pi\eta\, a\, v_x \qquad (1.16)$$

such that

$$v_x = \frac{2}{3}\frac{a\sigma X}{\eta} \quad \text{and} \quad k = \frac{v_x}{X} = \frac{2}{3}\frac{a\sigma}{\eta}. \qquad (1.17)$$

Since for a layer of thickness d the potential is $\zeta = \dfrac{4\pi\sigma d}{D}$, then $\sigma = \dfrac{\zeta D}{4\pi d}$ and

$$v_x = \frac{1}{6\pi}\frac{\zeta X D}{\eta}\frac{a}{d}. \qquad (1.18)$$

This obviously assumes that the mobile double layer is completely displaced unmasking the charge q on the particle. Experimentally, for particles of radius a denser than water, the vertical fall velocity yields a from the relation

$$\tfrac{4}{3}\pi a^3 (\varrho - \varrho_l)\, g = 6\pi\eta\, a\, v_g. \qquad (1.19)$$

Here v_g is the vertical fall under gravity and ϱ and ϱ_l represent the densities of particle and liquid, with g the acceleration of gravity. The horizontal velocity v_x then allows of solution for σ directly, this then also evaluates ζ/d. If the surface is that of a minute gas bubble its radius can be estimated from its rate of ascent. Otherwise, it can be centered along the axis of a cylindrical tube of liquid rotating about its axis.

If the equations apply and capillaries can be made of the same material in a given solution as the small spheres it should be possible to evaluate ζ from endosmosis and σ from cataphoresis and thus to estimate d.

However, the quantity d is a fictitious quantity which assumes that the potential gradient is linear. From endosmotic and streaming potential studies alone only ζ and/or σd can be estimated, while if $q = 4\pi\sigma a^2$ is fully disclosed in cataphoresis, the quantity σ can be evaluated.

The calculations made on the basis of this theory require the use of cgs and electrostatic units. This gives P in dynes/cm^2, ζ, V_s and X in esu, λ in esu, I_s and I_v in esu, σ in esu/cm^2 or the number of electrons/cm^2 by multiplying by 4.8×10^{10}.

As an example, QUINCKE observed for soda glass against distilled water that $k = 3 \times 10^{-4}$ cm²/volt sec compared with a mobility of common ions at infinite dilution of about 5×10^{-4} the mobilities of H⁺ and OH⁻ ions being about an order of magnitude larger. The value of $\zeta = 4 \times 10^{-2}$ volts and the moment $\sigma d = 1.6 \times 10^{6}$ electrons × cm/cm². If d had been $\sim 10^{-5}$ cm then there were essentially 1.6×10^{11} mobile univalent ions of charge/cm² in the double layer. Since there are on a cm² of geometrical glass surface some 10^{15} water molecules it indicates that about one in 10^4 of the possible bound surface ions at the glass interface were in the mobile layer.

It may next be noted that the resulting potential $\zeta = 4 \times 10^{-2}$ volt is very small compared to potentials which develop in galvanic cell action which are of the order of 25 times as large. This leads to the consideration of the significance of the *electro* kinetic potential defined by ζ and measured in such studies. In fact, the theory given is relatively primitive and oversimplified as deduced, a matter which as might be inferred from the discussion at the beginning of this chapter.

F. The more accurate calculation of the double layer characteristics including ions of both signs

In the previous considerations, the double layer has been analyzed assuming *ions of one sign only* and largely that they were emitted by the metal or other surface. In a much larger majority of cases, the ions of the solution of one sign or the other are adsorbed. The double layer is then composed of ions of one sign on the surface and of both signs in the solution. This introduces a more complex charge distribution. Again in discussing the thickness of the layer d and the ζ potential relative to the true potential of the ions at the surface, it was considered that the superficial ions were bound and that for some reason the binding forces on ions further from the surface were such that they could not readily be wiped off or moved off by flow. This implies somewhat more than immobilization of the ions by forces, it implies that the water itself in virtue of the surface fields and the induction along chains of polar water molecules is sufficient to hold the layer rigid out to a distance t. Beyond this the ions are mobile and there is an equivalent to the old distance d which will be given by a new constant $1/K$ yielding the "thickness" of the mobile double layer. It is now necessary to derive these relations, essentially as developed by DEBYE and HÜCKEL[22].

Assume the solid surface to acquire an electro-chemical potential of the order of V_E in consequence of loss or gain of ions. The charge is on a basically solid but a molecularly uneven surface. The opposite charge is in the liquid, but is not uniformly distributed in space nor is it all free to be displaced as the liquid moves. The ions of the outer layer form a diffuse cloud according first to GOUY[21] and later to DEBYE and HÜCKEL[22], distributed in the force field of the fixed charge on the solid layer in consequence of their energy of thermal agitation. Thus the diffuse ionic layer has a potential ψ which varies with the distance x from the surface in consequence of electrostatic forces and the Boltzmann distribution law as indicated in Fig. 1. If the ions have a valence z and the electronic charge be designated by e then the potential *energy* of the ion at x cm from the surface, where the potential is ψ is $z\psi e$. The number density of ions, N_i of valence Z_i is then by BOLTZMANN's law given by

$$N_i = N_{oi}\, e^{-Z_i e\psi/kT}, \tag{1.20}$$

where ψ has the same sign as the potential to which it is due; Z_i is $+$ for $+$ ions, $-$ for $-$ ions, e has numerical significance only. The charge density of all ions is

$$\varrho = \sum_i N_i Z_i\, e = e \sum_i N_{oi} Z_i\, e^{-Z_i e\psi/kT}. \tag{1.21}$$

ϱ will have a sign opposite to that of the potential ψ. One must then apply POISSON's equation

$$\nabla^2 \psi = \frac{-4\pi\varrho}{D}. \tag{1.22}$$

For large plane surfaces applying in general where $R \gg d$ the distribution along the x direction only need be considered. If the valence of the ions are all the same, $+$ and $-$ ions are of the same valence and only ions of the same valence types, e.g. all 1, 2, or 3 are considered the equation, considering positive and negative ions may be written.

$$\left.\begin{aligned}
\nabla^2 \psi &= \frac{4\pi e\, N_0 Z}{D}\left[e^{-Z e\psi/kT} - e^{Z e\psi/kT}\right] \\[2mm]
&= \frac{8\pi e\, N_0 Z}{D}\sin\mathrm{h}\,\frac{Z e\psi}{RT} \sim \frac{8\pi e^2 N_0 Z^2 \psi}{DRT}.
\end{aligned}\right\} \tag{1.23}$$

The last approximation for sin h comes under the usual conditions when the ratio of electrical to thermal energy $e\psi/kT \ll 1$ for even with $Z e\psi/kT < 0.25$ the error is less than 1%. The differential equation may be written

$$\nabla^2 \psi = K^2 \psi \quad\text{with}\quad K = \sqrt{\frac{8\pi e^2 N_0 Z^2}{D\,k\,T}}. \tag{1.24}$$

The coefficient $1/K$ in the approximation has the dimensions of L, a length. Here N_0 is the concentration of dissociable molecule giving the $+$ and $-$ions molecules per cm³*. For the case of the plane surface, or for a sphere very large compared to d, the equation simplifies to the solution of which with boundary conditions

$$\frac{d^2\psi}{dx^2} = K^2 \psi \quad \psi = A\, e^{-kx}.$$

The value of A is determined by the level chosen to represent the extreme inner face of the double layer beyond which ions of the double layer cannot penetrate. Since the total charge of the diffuse outer layer just neutralizes the total charge on the inner layer, the net surface charge σ on the inner layer is given by

$$\sigma = \int_i^\infty \varrho\, dx = \int_i^\infty \frac{D\,K^2}{4\pi}\, A\, e^{-kx}\, dx = \frac{D\,K}{4\pi}\, A\, e^{-kt}. \tag{1.25}$$

Thus the potential ψ varies as

$$\psi = \frac{4\pi\sigma}{D\,K}\, e^{k(t-x)}. \tag{1.26}$$

---

* Sometimes the chemical term ionic strength $\Gamma/2$ is used in place of N_0. The concentration is expressed in moles/liter, designated by $C = 1000\, N_0/N_A$ with N_A the Avogadro number. Then $\dfrac{\Gamma}{2} = \dfrac{1}{2}\, C\,Z^2 = \dfrac{1}{2}\,\dfrac{1000\, N_0}{N_A}\, Z^2$, $N_0 = \dfrac{N_A}{1000}\,\dfrac{\Gamma}{2}\,\dfrac{1}{Z^2}$.

Here t is *the thickness of the firmly adsorbed solvent on the surface in question;* i.e., that *layer of liquid and ions bound so firmly to the charge of opposite sign on the surface, that it cannot move.* It might also represent the radius of the ions of the diffuse layer. If now $\psi_{x=t} = \zeta$ then

$$\zeta = \frac{4\pi\sigma}{DK},\qquad(1.28)$$

and $1/K$ *is equivalent to d* the *thickness of the mobile double layer previously used.* The distance $1/K$ is the distance at which the potential falls to $1/e$ of the value it has at $x = t$. In water at 25° C, the value of $K = 0.33 \times 10^8 \sqrt{\Gamma/2}$ cm^{-1}.

For $\Gamma/2$ of value 1.0, 10^{-1}, 10^{-2}, and 10^{-3} respectively, $1/K = x$ is 3.0, 9.6, 30.3 and 95.8×10^{-8} cm respectively. Thus the thickness of the double layer and the ζ electrokinetic potential vary with electrolytic concentration even though σ does not do so. It is noted that for dilute solutions the layer can become of the order of 10^{-6} to 10^{-5} cm thick. Strong solutions reduce $1/K$ since the oppositely charged ious play a more prominent role owing to Coulomb forces.

The deduction above may be altered to fit the case for spherical particles. Thus for a spherical particle of radius r

$$\psi = \frac{q}{D}\frac{e^{-kx}}{x} = \frac{4\pi r^2 \sigma}{D}\frac{e^{-kx}}{x}.\qquad(1.28)$$

with q the total charge on the sphere, σ is the charge density on the sphere and x is measured from the center of the sphere, the charge being homogeneously distributed over the volume, or over the surface.

The potential at the *surface of the sphere*, coincides with the inner layer of the double layer and represents the potential of the double layer

$$\psi_{x=r} = \zeta = \frac{q}{e\,r}e^{-kr} = \frac{4\pi\sigma r}{D}e^{-kr},\qquad(1.29)$$

or for dilute solutions and small particle size

$$\zeta = \frac{q}{Dr}\left(\frac{1}{1+Kr}\right) = \frac{4\pi\sigma r}{D(1+Kr)} = \frac{4\pi\sigma r}{D}\frac{1/K}{(1/K+r)}.\qquad(1.30)$$

If the small sphere obeys STOKES' law, the electrophoretic mobility is

$$\frac{v}{X} = k = \frac{q}{6\pi\eta r}\frac{1}{1+Kr} = \frac{4\pi\sigma r^2}{6\pi\eta r}\left(\frac{1}{1+Kr}\right) = \frac{2}{3}\frac{r\sigma}{\eta}\left(\frac{1}{1+Kr}\right).\qquad(1.31)$$

as derived by HÜCKEL and for a very large sphere. The Smoluchowski theory yields

$$\frac{v}{X} = k = \frac{q}{4\pi\eta r}\frac{1}{1+Kr} = \frac{r\sigma}{\eta}\left(\frac{1}{1+Kr}\right).\qquad(1.32)$$

The elementary theory of Sect. E yielded $k = \frac{2}{3}\frac{r\sigma}{\eta}$ (1.17)

The Hückel theory implies a particle conductance the same as that of the medium, or that the particle is so small that no distortion of the external field occurs in the region of the double layer. This means that with zero conductance the value of λ must be small compared to $1/K$. For intermediate spheres, HENRY[23] obtained

$$\frac{v}{X} = \frac{\zeta D}{6\pi\eta}f(Kr),\ f(K,r) = 1.5.\qquad(1.33)$$

When the sphere is large compared to the thickness of its atmosphere $1/K \equiv d$ and is 1 when it is small by comparison. The simple theory makes the mobility k too large as it neglects the influence of r/d or Kr. The two agree when $r \gg d$ as they should.

The nature of the double layer in regard to the electrokinetic and electro-chemical potentials is nicely illustrated by Fig. 1. The scheme of the figure was taken from Treatise on Physical Chemistry by TAYLOR and GLASSTONE[24] in which the excellent condensation of the Debye theory serving as a model for the presentation above is also be found. There the curves A, B, and C represent respectively the potential ψ the field X and the charge density ϱ as a function of x. The peak of the positive charge is seen to be at the solid boundary and the peak of the negative charge is at $x = t$, the distance of the bound layer. The *electro-chemical* potential is V_E as seen in 1 (a) and the value of the *electro-kinetic* potential ζ is that at $x = t$ and represents but a small fraction of V_E, e.g., of the order of say some 5 to 10 % of V_E. It is thus to be expected as indicated earlier that charging by electro-chemical effects will be somewhat less than can be estimated with all ions of the double layer removable.

W. G. EVERSOLE and P. H. LAHR[25] extended the calculations above to the end that known measurements of ζ as a function of concentration could lead to evaluations of σ_0 at t, of the poten-

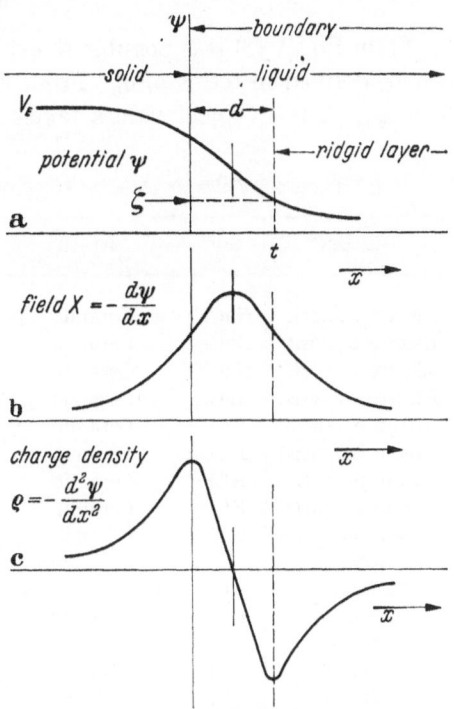

Fig. 1. Electrokinetic potential, fields, and charge density as a function of distance from surface as indicated from the Debye-Hückel theory

tial ψ at t and of $1/K$ for solutions of KCl, NaCl and K I moving through capillaries in ceramics and borosilicate glass. Methods used were endosmosis and streaming potentials. They start with the equation initially derived:

$$\frac{d^2 \psi}{dx^2} = \frac{8 \pi N Z e}{D} \sin \mathrm{h} \frac{Z e \psi}{kT} \tag{1.34}$$

Letting $(8 \pi N_0 Z^2 e^2/D kT)^{\frac{1}{2}} = K$ (1.24) and calling $ez/kT = a$ they perform the integration calling $\psi = \psi_0$ at $x = 0$ and $\psi = \zeta$ when $x = t$. This leads to the relation:

$$2 K t = \log \frac{\cos \mathrm{h}\, a\, \psi_0/2 - 1}{\cos \mathrm{h}\, a\, \psi_0/2 + 1} - \log \frac{\cos \mathrm{h}\, a\, \zeta/2 - 1}{\cos \mathrm{h}\, a\, \zeta/2 + 1} \tag{1.35}$$

since

$$\frac{d \psi}{d x} = - \frac{4 \pi \sigma}{D} = - \left[\frac{16 \pi N kT}{D} \left(\cos \mathrm{h}\, \frac{e Z \psi}{kT} - 1 \right) \right]^{\frac{1}{2}}$$

$$\sigma = \frac{4 \pi N Z e}{K} \cdot \sin \mathrm{h} \frac{a\, \psi_0}{2} \tag{1.36}$$

or

$$\sigma = 35\,300\ C^{\frac{1}{2}} \sin h\ 19.5\ \psi_0 \tag{1.37}$$

for water with $T = 289.2$ and $D = 78.8$.

This applies strictly only if the charge giving rise to ψ_0 lies in the ψ_0 plane.

If further ψ_0 is not a function of concentration, Eq. (1.35) in terms of ζ for two concentrations C_1 and C_2 becomes

$$2t\,(K_2 - K_1) = \log \frac{(\cos h\ a\zeta_1/2 - 1)\ (\cos h\ a\zeta_2/2 + 1)}{(\cos h\ a\zeta_1/2 + 1)\ (\cos h\ a\zeta_2/2 - 1)}. \tag{1.38}$$

From Eq. (1.38) it is possible to evaluate t from the data. From Eq. (1.35) ψ_0 can be derived in terms of σ_0. From Eq. (1.36) the value of σ_0 can be evaluated from ψ_0. A few typical values taken from their tables are shown in Table 2.

Table 2. *Table of observed ψ_0, ζ, σ_0 and t from* EVERSOLE *and* LAHR [33]

Method	Salt	Material	Conc. C Molar	ζ Volts	ψ_0 Volts	σ_0 esu/cm²	t in cm
Electro osmosis	NaCl	Ceramic	0.316	0.010	0.042	24 300	7.5×10^{-8}
Electro osmosis	NaCl	Ceramic	0.10	0.024	0.056	13 700	7.5×10^{-8}
Electro osmosis	NaCl	Ceramic	0.0316	0.036	0.060	7 680	7.5×10^{-8}
Electro osmosis	NaCl	Ceramic	0.010	0.04	0.053	4 320	7.5×10^{-8}
Stream potential	KCl	Ceramic	0.010	0.005	0.028	1 980	4.69×10^{-7}
Stream potential	KCl	Ceramic	0.005	0.0104	0.032	1 390	4.69×10^{-7}
Stream potential	KCl	Ceramic	0.001	0.0156	0.026	624	4.69×10^{-7}
Stream potential	KCl	Ceramic	0.0001	0.0228	0.027	198	4.69×10^{-7}
Stream potential	KCl	Ceramic	0.00005	0.0262	0.029	139	4.69×10^{-7}
Osmosis	KCl	Pyrex spheres	0.01	0.04	0.135	3 040	2.57×10^{-7}
Osmosis	KCl	Pyrex spheres	0.0001	0.095	0.111	1 100	2.57×10^{-7}
Osmosis	KCl	Pyrex capillaries	0.010	0.053	0.107	4 330	1.5×10^{-7}
Osmosis	KCl	Pyrex capillaries	0.0001	0.098	0.107	1 160	1.5×10^{-7}
Osmosis	KCl	Glass	0.001	0.049	0.130	1 230	6.26×10^{-7}
Osmosis	KCl	Glass	0.0001	0.079	0.113	780	6.28×10^{-7}

The assumption of a ψ_0 that is independent of concentration may be the cause for some discrepancy. Comparing Pyrex spheres and capillaries ψ_0 is the same but ζ and t differ. This could be caused by a greater variation of t with the streaming in a capillary. This is possible since the strength of flow current will determine how much of a film of decreasing rigidity will be torn off. As seen, the assumption of a rigid multi-layer of solvent which encloses part of the ions does make it possible to account for the variation of ζ with concentration.

It is seen that on the average ζ potential is of the order of 0.20 to 1 times ψ_0. ζ in general increases relative to ψ_0 with decreasing ion concentration which might be expected. The value of ψ_0 behaves in an irregular manner relative to concentration and may not really change. The value of σ_0 decreases as concentration decreases. The value of t range from 7.5×10^{-8} to 6×10^{-7}. This range of distances for a rigid aqueous double layer at an interface is of interest in that it indicates the thickness to which water films may be drawn out before they break, $\sim 7 \times 10^{-7}$ cm. The value of t is also commensurate with the critical droplet sizes for negative and positive charges in spray electrification as will be seen. The values of the maximum charge densities σ_0 of around 2.4×10^4 esu/cm² correspond to 8×10^{-6} coulomb or about 5×10^{13} electrons, or monovalent ions, per cm². These figures are reasonable and in keeping with what might be expected. Generally speaking, the calculated ζ potentials are a little large compared to the values of ψ_0 in terms of other observations. Otherwise these data are reasonable and satisfactory.

G. The Workman-Reynolds freezing potential

There is now a new phenomenon to add to the electrolytic phenomena responsible for static electrification. This phenomenon was discovered in the search for possible mechanisms of producing electrification in thunderstorms. It had been observed that the very effective turbulence leading to strong electrification in thunderstorms occurs in the violent convective turbulence of the active cell. It begins when precipitation begins to form and involves passages of ice and water up and down through the freezing isotherm.

In 1946, GUNN and DINGER[26] looked for electrification produced in a dish during the freezing of liquid water, with one electrode in the water and the other in the ice phase. Unfortunately, the water they used did not have the correct amount of impurity so that only trivial potentials were observed. They did observe, however, than when the ice melted, the *air above the melting ice* was *positively* electrified. This they correctly attributed to the production of spray from minute bubbles of air released from the melting ice. This phenomenon will be treated in extenso under spray electrification. This bubbling mechanism has newly been observed in the breaking of spray bubbles at sea. Quite independently of GUNN and DINGER, in New Mexico, WORKMAN and REYNOLDS[27] looked for a charging produced when super-cooled water drops splashed to produce glaze ice by impinging on an insulated cold metal plate. Although results were erratic, occasional very heavy charging occurred. Most noteworthy was the observation that a potential of 50 volts or more was measured on the cold plate when unfrozen water at the lower tip of the ice was connected to ground. It was ultimately found that with a pan refrigerated at one end that was grounded, an electrometer probe in the insulated liquid registered up to 230 volts relative to ground *as long as the water was freezing*. The freezing was stimulated by a cold copper block at from -5 to $-30°$ C.

The large potentials were obtained when the concentration of impurities in the water ranged from $10^{-3} - 10^{-6}$ normal. It varied with the electrolyte used. 5×10^{-3} normal NH_4OH solution gave 230 volts with water negative to ice. Relatively pure NH_3 free water gave only a few volts. On the other hand, the fluorides yielded up to $+34$ volts at around some 10^{-5} normal solutions, the water being positive. Alkaline chlorides and bromides also gave positive charging with, however, decreasing potentials. All ammonium salts gave more or less strong negative charge to the water. NaCl counteracts the action of ammonium ions at about 10^{-5} normal concentrations. At 10^{-4} normal NaCl gives $+30$ volts.

The current continues during freezing and 90 000 esu of charge were separated for each cm³ of solution frozen. Above 5×10^{-4} normal, the potential went to zero.

In general, it is the negative ion that appears to be absorbed into the freezing ice leading to the potential. The NH_4^+ ion being isomorphous with the hydronium ion appears to be also incorporated into the ice structure. The F⁻ ions have a position of preference since they are highly electro-negative and are similar to the OH⁻ ion in structure. On the other hand, the NO_3^- and CH_3COO^- ions are rejected by the ice on freezing.

The phenomenon belongs to a class of phenomena which are encountered in solid state physics for which only speculative interpretations exist. Charge separation measurements indicate that the number of centers per cm^3 in the ice of dilute solutions approximates the number of impurity centers in semi-conducting crystals used as rectifiers. In the largest transfer one foreign ion per 3×10^7 water molecules was involved, approximately 10^{15} ion centers per cm^3 of ice. The conductivity of the ice samples prepared in this way was extremely sensitive to the concentrations of contaminants. For example, using NH_4OH conductivity changed by a factor of 200 when concentration varied from 6.5×10^{-5} to 9.5×10^{-5}. Exploratory tests showed that the ice samples prepared from dilute solutions possessed significant rectifying properties.

Potential versus time curves were taken during freezing. Some ten seconds were taken between freezing at the bottom of the dish and the time that measurable potentials occurred. This was the time taken for an insulating layer of ice to form to insulate the water from the dish. Slight, but sharp inflections were noted some 15 seconds after measurable potentials were observed and at about the time the freezing was completed in distilled water. The first remains unexplained, but the latter was caused by accumulated charge dissipation following freezing. Resistivity of ice from ordinary distilled water may be a thousand-fold that for 10^{-4} normal NaF. Very dilute NH_4OH and distilled water yielded a relatively wavering rise compared to the rise of the fluorides. The characteristic inflections appear to be the results of changes in kind and number of ions in solution in the liquid phase during freezing.

The following tests leave little doubt as to the *ionic* character of the effect. Samples of NaCl 7×10^{-5} normal were frozen in a He atmosphere. An electrometer shunt of 20 megohms was used during freezing to permit flow of current produced at a potential of several volts. The unfrozen portion was poured into a clean vessel and the ice dropped into another when freezing was half complete. When both portions were at room temperature p_H measurements were made on each portion. Before freezing the p_H was 6.3. In the frozen fraction it was 6.2 and in the unfrozen fraction it was 7.0. Thus Cl ions were trapped selectively in the ice while Na+ ions discharged into the liquid phase creating NaOH and thus decreasing the hydrogen ion concentration.

This change in p_H represents an increase of 2.4×10^{14} OH⁻ ions, or discharge of this number of Na^+ ions per cm^3 in the unfrozen solution. Discharge of these ions would liberate 102000 esu of charge. This is in good agreement, considering leakage, with the value of 90000 esu of charge per cm^3 in freezing NaCl solutions of 10^{-4} normal.

The energy involved in such selection with 200 volts indicates an energy of more than 3×10^{-10} ergs. This is about three thousand times the latent heat of fusion per molecule. Such energies suggest collective action on the part of the crystal elements of the water at the interface.

WORKMAN and REYNOLDS further speculate as follows:

"ERRERA has found justification for the hypothesis that liquid substances having permanent dipole moments, (highly associated), pass through a 'colloidal' transition stage in freezing. He reports 'Thus associated structures would be formed in such a manner that their dimensions would cause them to come under the category of *media in the state of colloidal division*, that is to say of which the particles have a diameter between a few and 500×10^{-7} cm. Further we must suppose that the arrangement of molecular dipoles inside the colloidal structures is such that these have a high electric moment, higher indeed than that of molecular dipoles*. It seems reasonable to suppose that these associated colloidal particles in process of formation might avoid an increment in potential energy by incorporating ions of compatible form into their structure in such a way as to form a non-polar electrical domain. Questions of why ice does not exhibit polar properties have been the subject of much speculation by investigators. It is not likely that one foreign ion in ten or a thousand million crystal units will by its position alone neutralize the polarization of a crystal viewed as a static system. Collective action at a growing crystal face would produce adequate energy of attraction to maintain the observed potential barrier and once there, the ion may facilitate the arrangement or modify the resonance pattern of crystal units."

"Assuming that the phenomena were related in a primary way to structural changes accompanying freezing, (PAULING estimates that 15% of the hydrogen bonds must be broken on change of state of water), similar effects were predicted for other compounds undergoing marked change in bond structure in liquid to solid transition. Salol, (phenyl salicylate), is such a compound. An electrical effect, believed to be similar to that for water was observed when salol solidified. The flow of current during freezing was detectable but small (the electrical conductivity of solid salol is low). Potential differences of 45 volts with solid positive with respect to liquid were measured. The nature of the impurities present was not determined and the effect of known impurities in this substance has not been investigated. Confirmation of this prediction for salol gives at least circumstantial evidence as to the true nature of the effect."

The remainder of this very excellent study is devoted to the role played by this process in the static electrification in thunderstorms. In principle, if the falling hail and graupel pellets strike super-cooled droplets of water, then the freezing and shattering or blowing away of excess water in the up drafts can carry the appropriate charge upward leaving the residual charge below. The predominant impurities in rain over the New Mexico area are NaCl and $CaCO_3$ at about 10^{-4} normal which make the water positive to the ice. The $CaCO_3$ owing to CO_2 in the atmosphere goes to calcium bicarbonate, an active charging agent. Thus positive charge is carried aloft to the upper regions of cloud leaving the negative ice and larger negatively charged particles in the region of the thunderstorm where they are observed to accumulate. As the hailstones and ice melt in falling to the lower regions, the water drops encountered will add to the larger charged water droplets and will share the charge. In this region of melting falling hail and

---

* It may be added that evidence for oriented polar molecular groupings in small water droplets, pseudo-crystalline structure was obtained by the author from cloud condensation studies. See L. B. LOEB, A. F. KIP, and A. W. EINARSSON[28]. Jour. Chem. Phys. 6, 265 (1937).

water droplets any spray particles and water carried upward from below the freezing isotherm will carry up the shared negative charge thus segregating the large bulk of the negative charge cell in the region of the freezing isotherm and above. Laboratory tests with a simulated hailstone in a specially constructed cold chamber showed that potentials consistent in sign and magnitude with those achieved in the bulk freezing of water in cups could be attained by grazing collision of water drops and ice pellets. After the pellet became warmed to 0° C by collision with water drops subsequent collision resulted in carrying away the accumulated charge. The thunderstorm requires above 1 esu per cm³ of water and perhaps more nearly 5 esu per cm³. The electrical effects in the laboratory were adequate to produce such effects by a one thousandfold margin, if repeated collisions could be realized.

In attempting to relate the contact electrification of two pieces of ice at different temperatures, in which process freezing and regelation occur, to the freezing potentials discussed above further studies of the mechanism were carried out by WORKMAN and REYNOLDS[29]. These studies added one more fact. This is that the sign of the potential difference between liquid and solid state and the amount of charge separated during freezing of dilute solution are *not controlled completely by the kind and amount of ions in solution in the water*. Experiments show that the nature of the surface upon which the solution is frozen may control electrification occurring during freezing. That is, freezing on substrates of ice or on electrolytically treated Pt or Ni affect the sign and amount of charge separated.

In a much later report resulting from an intensive study of *contact charging of ice particles*, the authors conclude[29]*, "the sign and the amount of charge separated during the freezing of dilute aqueous solutions are *not* controlled *completely* by the kind and amount of material in solution, but also by the nature, or condition, of the surface upon which the solution is frozen. Furthermore, the studies of the effects of the *substrate* on the freezing potentials support the hypothesis that the electrification caused by frictional contact between two ice formations results from the resolidification of a liquid film created at the ice-ice contact by pressure and frictional heating. The hypothesis is based primarily on the fact that the *polarity observed during the freezing of a substrate ice is transmitted unchanged to the freezing of a second solution on the substrate, regardless of the kind of ions involved in the second solution*."

Further conclusions amplify the current views as follows: "It has been established clearly that charge separation from the freezing of dilute aqueous solutions

---

* It is to be noted that this represents a study of great importance in a new field. Only one paper has appeared in print, in 1950. The later work on contact ice-ice electrification and its primal cause the freezing potential mechanism was carried out over four year and reported on in the form of many involved and disconnected studies. The results and conclusions were presented in a number of progress reports, including the final report cited. Space limitations preclude detailed discussion of the lengthly and groping experimental investigation. The only recourse is to quote in the authors' own words their latest summary and conclusions which reveal enough of the underlying data in a systematic fashion to be convincing. It is at this stage of information impossible for an outsider to do more than those familiar with the intricate background in presenting this material so that despite some repetition, the conclusions of the report will be quoted in extenso.

depends upon the selective incorporation of one or the other of the solution ions. If the charge separation during frictional contact depends upon the re-freezing of a thin liquid layer, there must be an adequate number of ions present in the layer. (It had earlier been shown that *carefully prepared distilled water shows no measurable electrification on freezing*.) If such water is frozen after ordinary pouring or handling in the laboratory, the liquid becomes negative with respect to ice. This behaviour is attributed to incidental contamination with traces of ammonia absorbed from the air. Certain assumptions about the liquid layer are required to consider this point. When two clean ice formations are rubbed together, approximately 1 esu of charge is separated in a single rubbing contact along a 10 cm path. Assuming that the liquid layer formed at the point of contact is 10^{-2} cm wide and 10^{-3} cm thick, 10^{-7} liters of water are melted and refrozen in a single stroke. The concentration of ions in this water must be *3×10^{-6} molar* to acount for the charge separated, if the process is 1% efficient. This concentration is excessive for conductivity water but reasonably could be expected in the surface layer of the ice-formation. Some incidental contamination is expected because water inevitably is exposed to air in the handling and in forming the ice on the rods. Since it is well known that contaminants are generally rejected by a growing ice crystal it is reasonable to expect the contaminants to be somewhat concentrated in the surface layer of the ice formation. Further evidence of the dependence of frictional electrification upon solution ions is gained from the observation that the amount of charge separated during frictional contact is enhanced greatly (2—3 esu or more to each rubbing contact), if one of the ice formations is prepared from NaCl or NH_4OH solutions of concentrations ranging from 10^{-4} molar."

In conclusion WORKMAN and REYNOLDS[29] make the following suggestions concerning the mechanism of freezing potentials:

"The electrical effects accompanying the freezing of dilute aqueous solutions suggests that as the crystal grows it presents at its growing surface a charge requirement which is met by an ion from the solution. The ions thus attracted become entrapped (substitutionally or interstitially), in the growing surface and the crystal becomes electrically charged. This process may be viewed as one of selective adsorbtion at a surface which refreshes itself continuously.

The experiments give ample evidence that the electrical polarity which occurs during growth is determined in a large measure by conditions at the substrate-solution interface prior to freezing and that the influence of these conditions is propagated for distances of at least a millimeter into the crystal which is grown on the substrate. The solid substrate apparently exerts an orienting effect on the liquid layer adjacent to it. Electrical fields at the substrate-solution interface might easily affect the orientation of the water molecules which have large dipole moments.

There is ample evidence from electro-kinetic studies for the existence of electrical double layers at a solid-electrolyte interface. The origin and details of the structure of this double layer remain partially obscure. Modern theory indicates that the double layer is composed of a layer of ions of one sign firmly attached to the solid and a second layer of an equivalent number of ions of the opposite sign in the solution. Some of these ions are held fixed by their opposite numbers in the fixed part of the double layer and the rest form the different portion of the double layer." This leads to the ζ potential the theory of which appears preceding this section. "The magnitude and sign of this potential difference depend on the nature of the solid and the kind and amount of ions in solution. In the diffuse portion of the double layer the ions move freely under thermal agitation but the distribution is not uniform because of the electrostatic field of the surface ions bound in the solid.

J. C. HENNIKER[30], has pointed out that potential gradients in the double layer may be 243 kv/cm *. He tentatively has explained the retardation of flow of water in capillaries on the basis of orientation of multi-molecular layers of the water molecules in this field. HENNIKER[31] in a review of the field, presents other evidence for the formation of oriented layers of water molecules extending for hundreds of Å into the liquid at interface boundaries. W. G. EVERSOLE and P. H. LAHR[25], have derived an equation for the ζ potential which leads to the assumption of oriented layers of H_2O molecules between 8 and 63 Å in thickness as shown above.

If a substantial portion of the water molecules in the initial layer of ice are similarly oriented, the dipole field of these molecules will account for the selective incorporation of ions of one sign into the growing ice. The dipole fields of these oriented water molecules in the ice will also result in the orientation of liquid moleules adjacent to the advancing ice solution interface and thus account for the continuing selective incorporation of ions throughout the growth of the ice structure. This is indicated by the fact that the dipole field of an H_2O molecule is 1.5×10^6 volts/cm at a distance of 10 Å from the molecule."

WORKMAN and REYNOLDS pointed out that the energy required to cause a single ion to cross the barrier represented by the measured potential difference between ice and water is equivalent to the latent heat of freezing of 3000 water molecules. The mechanism by which a single ion could achieve this energy is difficult to imagine. It is consistent with the foregoing discussion to take the view that the electrical forces are propagated to considerable distances into the liquid in much the same manner that the force of a relatively weak magnet may be extended a considerable distance by an intermediate chain of pieces of iron. As the chain of water molecules extends from the ice into the liquid, a liquid ion may become bound in the dipole field of the water molecule at the terminus of the chain and held until it is incorporated by the advancing ice surface. Thus the energy required for the ion to cross the potential barrier is provided by the solidification process which causes the charged ice surface to advance the electric field of the bound ion.

It is unfortunate that at present insufficient information concerning the electrical structure at the substrate-surface interface under the conditions of the studies on freezing potentials exists to permit the prediction of the orientation of the water molecules in the initial layers of ice. Studies are underway, but the problem is complex. Thus the orientation of the water molecules may not be determined uniquely by the sign of the ζ potential, the molecule might conceivably be directly adsorbed onto the substrate, or it might be oppositely oriented by

---

* The data for maximum σ_0 given above for solutions indicate fields owing to the surface charge of the order of 120 kv/cm.

adsorbtion on the solution side of the double layer when ions of a different kind are involved. The relative number of molecules adsorbed at these two locales may be determined by the kind and amount of ions in solution and the nature of the substrate.

Preliminary attempts made to measure the ζ potentials between Pt and dilute NH$_4$OH and NaCl with the techniques of KLEEMAN and FREDERICKSON indicated as might be expected that the sign of the ζ potential was the same with either solution in the range from 0° C to 22° C. This leads to the conclusion that either the water dipoles are adsorbed in different portions of the double layer in the two solutions or that the degree of orientation of the water dipoles at the metal-liquid interface *is insufficient* to be transmitted to the initial layers of ice. The latter assumption appears to be the more reasonable and it follows that the interphase potential at the ice-electrolyte boundary arises almost instantaneously after freezing begins and affects the orientation of liquid water molecules adjacent to the advancing ice surface in accordance with the nature of the ions in the solution. If the metallic refrigerating surface of the cup has been polarized electrically before freezing, the orientation of the water molecules at the metal electrolyte interface apparently is transmitted to some degree for distances of several mm into the growing ice. Such an effect appears to the author to be logical since electrical polarization of the surface may be on Pt in dilute solutions resulting in the ζ potential.

The spectacular nature of the effects of the NH$_4^+$ and F$^-$ ions in the freezing potential studies require some comment. If certain ions can readily be built into the structure of the solid, they may have profound effects on the structure of the double layer. There is a notable similitude between the NH$_4^+$ and F$^-$ ion and the hydronium and hydroxyl ions respectively. Thus these ions might selectively be built into the ice structure. These ions are also ions that are most strongly adsorbed into the air-water interface and are active in lowering surface tension as will later be seen. Thus such behaviour in the freezing potential phenomenon is not surprising.

The phenomenon of freezing potentials has been confirmed by observations of V. J. SCHAEFER [32], and by GILL and ALFREY [33], although no new data of any significance were added to the very thought provoking studies of WORKMAN and REYNOLDS.

H. Summary and conclusions

In what has preceded the foundation has been laid for understanding the role of electrolytic effects in static charging phenomena. They will occur whenever the electrolytic liquid phase can remove ionic charges either *emitted by the surface*, or else the residual ionic charges left in solution *of sign opposite to those retained by the surface* in consequence of adsorbtion. The surface can be metal, a neutral semi-permeable membrane, or any non-metallic surface capable of giving up or receiving ions. The phenomenon can be occurring quite generally, but notable charge collection may be precluded because of neutralization by the conductivity of the liquid through some return path. Conditions where charging by such processes has been noted are in the electro-endosmosis and streaming potential phenomena through capillaries, over solid surfaces in the flow or else in the separation from solid surfaces of sufficiently electrically resistant liquids to prevent loss by return currents, in the separation of electrolytic films of sufficiently small thickness from surfaces, and in the freezing potentials of ice exhibited in the Workman-Reynolds potentials. The appearance of electrolytic charging as an important mechanism in ordinary contact charging experiments, in the "frictional" electrification of some dusts, and in the observed electrification by shattering of conducting liquids, notably water, against surfaces while very probably active in many of the older ill-controlled experiments has not surely been verified as a contributing or confusing factor. Studies of BOWDEN and others confirm the possible presence of liquid films of appropriate thickness under conditions of many of the earlier studies and with the lack of cleanliness and humidity control existing these films were undoubtedly present. With oxidized and slightly soiled surfaces adequate aqueous films can exist above 60% relative humidity at room temperatures. Otherwise aqueous films are often less than 2 molecules thick . The older studies using salts, acids, and bases many of them hygroscopic are highly suggestive. Obviously static electrification studies are not too successful at great humidities. This is not necessarily because the essential ionic films are not present, but because these films also form on all the insulating surfaces needed in the experiments thus reducing static electrification by leakage currents over the insulators. However, it is clear that in many instances along with these electrolytic charging actions, the normal contact charging processes are occurring simultaneously if sufficient pressure between solid surfaces breaks down the liquid film in areas. Likewise, the spray electrification observations are simultaneously affected by electrolytic and by true spray electrification mechanisms which are very difficult to separate experimentally, but are intrisically different though perhaps related processes.

There can be little doubt but that certain aspects of precipitation static generation are definitely electrolytic in nature especially since now WORKMAN and REYNOLDS associate the newly observed ice-ice electrification on contact with temperature differences, frictional melting, and regelation which introduces freezing potentials, a definite electrolytic mechanism. However, even neglecting all the doubtful manifestations of electrolytic processes in static charging, there is ample evidence at present that very high and dangerous electro-static charge accumulations are produced by electrolytic processes such as in the charging of inflammables like gasoline, ether, etc., etc., and in precipitation static. Further evidence associates all of these phenomena with the formation of electrical double layers and electro kinetic potentials.

II. The contact potential difference or Volta potential

A. Introduction

Sometime in 1801, VOLTA, who, together with GALVANI, was responsible for the development of the galvanic cell, Voltaic pile, or, in short, the electrolytic cell, believed that there was an intrinsic potential difference between dry metals in no way related to the electrolytic phenomena occurring between metals in solutions. He even went so far as to arrange the dry metals in a potential series based on some simple electrostatic experiments. VOLTA's belief also supported by HELMHOLTZ[34] went largely unrecognized for many years through confusion of the phenomenon with the electrolytic actions also earlier discovered by VOLTA as well as with the Peltier effect potentials. Controversy was rife for many years until finally in 1898, Lord KELVIN[35] presented to the Royal Institution and published in the Philosophical Magazine his studies carried forward from 1859 to 1861 onward that definitely established the existence of an intrinsic contact difference of potential between different metals. The potentials bore no relation to electrochemical effects and were independent of the medium in which the metals were immersed as long as the metals were not altered by the medium. More clarification relating to the contact or Volta difference of potential came from the basic studies of R. A. MILLIKAN[36] in 1915 on the Einstein photoelectric law in which he identified this quantity with the difference between the work functions of the two metals in question. Using light on various metal surfaces shaved in vacuum, he established the energy required to remove electrons from different metals and showed that this energy difference represented the differences in work function, for *free electrons in the metal*, without any additional work being required to remove the electrons from atoms in the metal. Actually, at an earlier date, the electron theory of metals set up by H. A. LORENTZ after J. J. THOMSON's identification of the electron led those active in the field to accept a characteristic work function needed to remove electrons from metals. Thus as early as 1901, H. A. WILSON[37] and later, WILSON and O. W. RICHARDSON, set up thermodynamic analyses of the thermionic current of electrons liberated by hot bodies following J. J. THOMSON's proof that these particles were electrons in 1899. Experimentally, O. W. RICHARDSON[38] in 1901 began his measurements of thermionic emission as a function of temperature to test the theory of WILSON. In consequence, by 1916, there was enough data to indicate that in principle the contact potential differences between clean metals were closely related to the thermionically and photoelectrically measured differences in work functions.

Properly manipulated, the contact potential differences can be used as a means of causing static electrification as was early indicated by Lord KELVIN. Since the principles underlying contact electrification with metals are possible involved in contact electrification in general, perhaps with complications, it is essential that the character of these processes be presented in detail to facilitate discussion of other processes and to establish this mechanism in its own right as a mechanism producing such electrification which is not to be confused with others.

B. The metallic state energy bands, the Fermi law and basic principles

When isolated metallic atoms are brought together to form a solid metal, interatomic forces are of such magnitude that in general the centers of the positive atomic ions, including their completed inner shells, are in closer proximity than the average radii of their equivalent outer valence electron orbits as isolated atoms permit. Thus in consequence of the crowding and the Pauli exclusion principle, the narrow energy levels in the free atomic state broaden and become diffuse bands with each electron in different energy states in a fine structured near continuum which fills up until all available free valence electrons are accommodated. In Fig. 2 there are shown schematically the situation for Na and Mg atoms. In Na the sharp isolated atomic levels are shown to the left to a somewhat compressed vertical scale at 100 Å separation of atoms. The broadening of the levels is shown as the centers approach to within $\sim$3 Å. Note that with only 1 valence electron, the 3 S band is only half occupied. For Mg, the levels on the right show that while the 3 S level would normally be filled with the two 3 S electrons, the 3 P level is so broadened and is of such energy, that it overlaps the 3 S level. Thus the 3 S electrons spill over and fill up the 3 S and 3 P to a common level leaving considerable unoccupied space above. Thus overlapping can make available space for more than the two electrons per state normally following for single levels and the strict application of PAULI's principle.

One must now discuss the way in which the electrons are distributed in these broadened states. In this case, the electron momenta and position coordinate p_1, p_2, and p_3, q_1, q_2, and q_3 can be expressed each in terms of 3 dimensional Cartesian coordinates. Thus the state of each electron is given by values of p_x, p_y, p_z and q_x, q_y, q_z, or six coordinates of position and momentum which constitute a 6 dimensional phase space. Since the electrons are free in the lattice, (with minor restrictions), the position coordinates are not essential to their energy. That is the effect of potential of the individual ions in the lattice is neglected. On HEISENBERG's uncertainty principle, values of position and momentum are specified only to a precision given by $\Delta p\, \Delta q \geq h$ where h is the natural unit of moment of momentum first designated by PLANCK. Thus the ends of the moment of momentum vectors of allowed momentum and position states in phase space can only be specified as lying in cells of the size $\Delta p\, \Delta q = h$ or larger. Thus the product of the uncertainties in all 6 coordinates of phase space is

$$\Delta p_x\, \Delta p_y\, \Delta p_z\, \Delta q_x\, \Delta q_y\, \Delta q_z \geq h^3 \tag{2.1}$$

or the uncertainty in the location of the end of the momentum vector defined by $p^2 = p_x^2 + p_y^2 + p_z^2$ will be confined to a cell

$$\Delta p^3\, \Delta v \geq h^3 \tag{2.2}$$

in phase space. Thus the crowding of momentum vectors for electrons of low m with $\sim$10^{22} electrons per cm^3 will necessitate a very large number of cells of h^3 in phase space for the vectors p in dv cannot end in cells less than h^3.

In order to understand the filling of the level, one must revert to the Fermi distribution law which governs the assignment to levels. The classical law for distribution of energy among molecules in a gas, that of MAXWELL, is derived from the elastic impacts of free particles with no restrictions on their motion since the densities of molecules is 10^{-3} that of electrons and since the momentum vectors of molecules are large—more than 1000 times those of electrons, so that quantum restrictions are not imposed on the vectors of molecules in phase space. In contrast to this, the Fermi law derives from the fact that there are far more moment of momentum vectors p than there are cells of different values of h^3 to accommodate them. Thus the electrons must have vectors in phase space that end with no more than two electrons of opposite spins per cell of h^3. Since there are not enough cells of h^3 of low energy available, the *electrons must assume momenta and energies of much hicher values than they would in a free distribution* in order to have their tips end in the elements h^3. This is an alternative way of describing the broadening of the band indicated in Fig. 2. The successive cells h^3 being filled out to a limiting energy μ_0 at the top of the band at 0 °K, electrons below this level are not *free to move or exchange* energy, i.e., they are degenerate unless there is a vacant cell h^3 available for them to move to and they receive the energy to move into it. The Maxwell-Boltzmann law derived from the fact that particles could move from one state to another only provided they received the energy. The Fermi law has the added restriction that there must be an empty cell available. As electrons continually exchange energy, equilibrium exists when as many electrons pass into one state in unit time as in the reverse sense. Assume two quantum states 1 and 2. Through impacts, they go to states 1' and 2'. On the other hand, electrons in states 1' and 2' go back to 1 and 2. Call the number of transitions from 1 and 2 by impact to states 1' and 2', $W_{1,2}^{1',2'}$ and the reverse change $W_{1',2'}^{1,2}$. Call E_i the energy in the i-th state. Then $f(E_i)$ is the symbolical form of the energy distribution law giving the probability of the number of electrons in this state. Let $P_{1,2}^{1',2'}$ represent the quantum mechanical transition probability from states 1 and 2 to 1' and 2' and $P_{1',2'}^{1,2}$ that for the reverse change. Then $W_{1,2}^{1',2'}$ depends on the product of $P_{1,2}^{1',2'}$ the number of particles $f(E_1)$ and $f(E_2)$ *and* on the number of vacancies

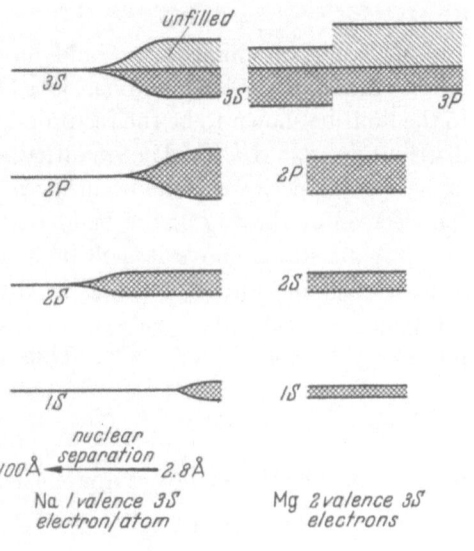

Fig. 2. Energy level diagrams of atoms as a function of distance of separation, showing filled and vacant levels adjoining

$1 - f(E_{1'})$ and $1 - f(E_{2'})$ in the state to which they are going. Similarly the quantity $W_{1',2'}^{1,2}$ depends on $P_{1',2'}^{1,2}$ on $f(E_{1'})$, $f(E_{2'})$ and $1 - f(E_1)$ and $1 - f(E_2)$. Thus equilibrium by the principle of detailed balancing requires that

$$W_{1,2}^{1',2'} = P_{1,2}^{1',2'} f(E_1) \left[1 - f(E_{1'})\right] f(E_2) \left[1 - f(E_{2'})\right] \left.\vphantom{\begin{matrix}1\\1\end{matrix}}\right\}$$
$$= W_{1',2'}^{1,2} = P_{1',2'}^{1,2} f(E_1^1) \left[1 - f(E_1)\right] f(E_{2'}) \left[1 - f(E_2)\right]. \quad (2.3)$$

Since $P_{1,2}^{1',2'} = P_{1',2'}^{1,2}$ by means of a general theorem of quantum mechanics it follows that:

$$\frac{f(E_1)}{1 - f(E_1)} \quad \frac{f(E_2)}{1 - f(E)_2} = \frac{f(E_{1'})}{1 - f(E_{1'})} \quad \frac{f(E_{2'})}{1 - f(E_{2'})}. \quad (2.4)$$

For the Maxwellian case, the relation would have read

$$f(E_1) f(E_2) = f(E_{1'}) f(E_{2'}).$$

Conservation of energy requires that $E_1 + E_2 = E_{1'} + E_{2'}$ in both cases. Thus if the sums of the energies are to be equal while the products of functions above must also be equal the terms $\dfrac{f(E_1)}{1-f(E_1)}$ must be of the form $\dfrac{f(E_1)}{1-f(E_1)} = A\,e^{-\beta E}$

where A and β are constants, and the negative exponent appears since $\dfrac{f(E_1)}{1-f(E_1)}$

must decrease to zero as E_1 increases indefinitely. Thus the Fermi distribution law probability has the form

$$f(E_1) = \frac{1}{\dfrac{1}{A}\,e^{\beta E} + 1}$$

or more generally

$$f(E) = \frac{1}{\dfrac{1}{A}\,e^{\beta E} + 1}. \tag{2.5}$$

The Maxwell-Boltzmann law would have been $f(E) = B\,e^{-\beta E}$. The constant β in both laws must have the dimensions of a reciprocal energy and it can by going to the limit be shown to be the reciprocal of the most probable velocity in classical distribution, $\beta = 1/k\,T$. The quantity A is determined by μ the energy of the highest occupied level. The quantity μ the potential plus the kinetic energy of the electron at the top of the band that can exchange energy because it is free with vacant states above it, will be a function of T. At absolute zero, $T = 0$, it has a constant value μ_0 which is the energy of the top of the filled levels. Since the potential energy is negative, we write it $-\mu$ the constant A is then given by $A = e^{\mu/kT}$ and $1/A = e^{-\mu/kT}$. Thus

$$f(E) = \frac{1}{e^{(E-\mu)/kT} + 1} \tag{2.6}$$

with E the kinetic energy of the electrons,

$$E = \tfrac{1}{2}m\,(u^2 + v^2 + w^2) = \tfrac{1}{2}m\,(p_x^2 + p_y^2 + p_z^2).$$

When $T = 0°\,K$ $f(E)$ can easily be derived. At this temperature $\mu = \mu_0$. If $E > \mu_0$ with $T = 0$, $f(E) = 0$ since $e^{E/kT}$ is indefinitely great. If $E < \mu_0$ we must write $e^{-\frac{\mu+E}{kT}}$ and at $T = 0$ the exponential becomes $e^{-\infty}$ or zero. Then $f(E) = 1$ and remains constant from $E = 0$ to $E = \mu_0$. That is beyond $E = \mu_0$ the probability is 0. All levels are occupied to μ_0 and with no heat present there are no electrons of $E > \mu_0$.

To evaluate the number of electrons in the interval between E and $E + dE$. Viz. $N(E)\,dE$, $f(E)$ must be multiplied by the number of cells $n(E)$ existing in the interval and by the occupancy factor G usually (2).

To evaluate μ_0 and get $n(E)$ note that all electrons below

$$E = \frac{p^2}{2m} = \frac{p_x^2 + p_y^2 + p_z^2}{2m}$$

lie within a sphere of radius p. The volume of this sphere is $\tfrac{4}{3}\,\pi\,p^3$. In this sphere, the number of cells of volume h^3 is $\tfrac{4}{3}\pi p^3/h^3$. Now two or better G, electrons of opposite spin go to each cell h^3. The chance of an electron of energy E in an elementary volume of metal

$$dx\,dy\,dz = dv = \Delta q_x\,\Delta q_y\,\Delta q_z,$$

is proportional to $f(E)\, dx\, dy\, dz$. In a volume of metal V there is a chance

$$\int_0^V f(E)\, dx\, dy\, dz$$

of an electron of energy E. Since potential is constant and since above $E = 0$ and below $E = \mu_0$, $f(E) = 1$.

$$\int_0^V f(E)\, dx\, dy\, dz = V.$$

Thus in V there are $GV(1)\,\tfrac{4}{3}\pi\, p^3/h^3$ electrons in the filled levels of momenta from 0 to p. Thus at absolute zero since the levels are filled out to an energy. μ_0 wide which is all kinetic $\mu_0 = p_0^2/2m$ where p_0 is the radius of the sphere in momentum space. Thus $p_0 = (2m\,\mu_0)^{\frac{1}{2}}$. Now the number n of electrons in a volume V of this level must be given by the number of occupied phase cells out to μ_0. Thus $n = G\,\tfrac{4}{3}\pi\, V p_0^3/h^3$. Since $N = n/V$ is the volume density of electrons

$$N = \tfrac{4}{3}\pi\, G\, \frac{p_0^3}{h^3} = \tfrac{4}{3}\pi\, G\left(\frac{2m\,\mu_0}{h^3}\right)^{\frac{3}{2}}$$

so that

$$\mu_0 = \frac{h^2}{2m}\left(\frac{3N}{4\pi\, G}\right)^{\frac{2}{3}}. \tag{2.7}$$

The population of the levels, i.e., the form of the energy distribution at $0°$ K, can be obtained. Since the probability at any energy below μ_0 is the same, that is, since $f(E) = 1$ from 0 to μ_0 the number of electrons of a given energy

$$N(E_e) = \frac{p_e^2}{2m}$$

will be that of the number of cells h^3 in unit volume between p_e and $p_e + dp$ in the spherical shell dp thick and $4\pi\, p_e^2$ of surface, multiplied by G, or 2, whichever is appropriate. This will be

$$N(E)\, dE = G\,(4\pi\, p^2\, dp)/h^3 \quad \text{with} \quad p = (2m\,E)^{\frac{1}{2}} \quad \text{and} \quad dp = \left(\frac{m}{2E}\right)^{\frac{1}{2}} dE.$$

This yields $N(E)\, dE$ as

$$N(E)\, dE = 4\pi\, G\, \sqrt{2}\,\frac{m^{\frac{3}{2}}}{h^3}\, E^{\frac{1}{2}}\, dE. \tag{2.8}$$

Had T been greater than $0°$ K, this would require that $f(E)$ be included, yielding

$$N(E)\, dE = 4\pi\, \sqrt{2}\, G\left(\frac{m^{\frac{3}{2}}}{h^3}\right)\frac{E^{\frac{1}{2}}\, dE}{[e^{(E-\mu)/kT} + 1]}. \tag{2.9}$$

This is the Fermi distribution law. The quantaty $N(E)\, dE/N$ to be used in comparison with MAXWELL's law is then

$$\frac{N(E)\, dE}{N} = \frac{3}{2\mu_0^{\frac{3}{2}}}\, \frac{E^{\frac{1}{2}}\, dE}{e^{(E-\mu)/kT} + 1} \tag{2.10}$$

for FERMI's distribution, and

$$\frac{N(E)\, dE}{N} = \frac{2}{\sqrt{\pi}}\, (kT)^{-\frac{3}{2}}\, E^{\frac{1}{2}}\, e^{-E/kT}\, dE. \tag{2.11}$$

for MAXWELL's distribution.

At $0°$ K
$$\frac{N(E)\,dE}{N} = \frac{3}{2}\left(\frac{E}{\mu_0}\right)^{\frac{1}{2}}\frac{dE}{\mu_0}. \tag{2.12}$$

This law is plotted as the solid curve of Fig. 3a for

$$\frac{dE}{\mu_0} = 0.01, \; E = \mu_0 = 1 \text{ ev at } T = 0.$$

The dashed line indicates T modification when T is $1500°$ K. Fig. 3b shows the corresponding Maxwellian distribution at a temperature of $1500°$ K. As noted in the Fermi distribution, as temperature rises, the electrons near to the left step into vacant cells of the unoccupied level and form the asymptotic foot to the right. The average energy in the Fermi distribution at $0°$ K is $\frac{3}{5}\mu_0$.

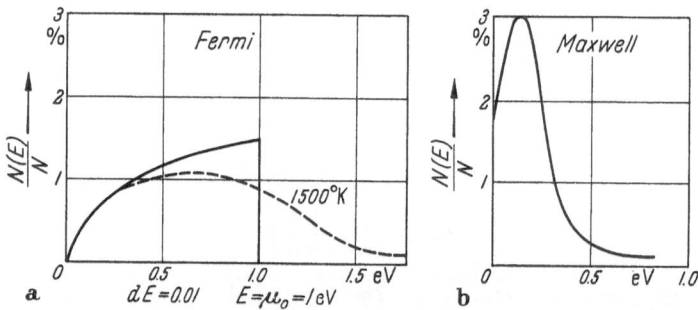

Fig. 3a, b. Fermi law for energy distribution

It should be remarked that if the temperature is high enough, the Fermi distribution goes over to the Maxwellian. There is a critical temperature separating the two extremes T_c which is given by

$$T_c = \frac{h^2}{2\,m\,k}\left(\frac{N}{2\pi\,G}\right)^{\frac{2}{3}}.$$

For electrons introducing the values of h, m and k, this becomes

$$T_c = 5\times10^{-11}\left(\frac{N}{G}\right)^{\frac{2}{3}}, \tag{2.13}$$

with N the number of electrons/cm³. At $N = 2\times10^{21}$ and $G = 2$

$$T = 5\times10^3 \text{ °K.}$$

As indicated actually μ is a slowly varying $f(T)$ which has been evaluated as

$$\mu = \mu_0\left[1 - \frac{\pi^2}{12}\frac{(k\,T)^2}{\mu_0} + \cdots\right]. \tag{2.14}$$

This correction is so small that below $1300°$ K μ is within 1% of μ_0.

The number of electrons n in an element of volume in phase space

$$d\,p_x\,d\,p_y\,d\,p_z\,d\,V \quad \text{at } 0° \text{ K} \quad \text{is given by}$$

$$n = (G\,d\,p_x\,d\,p_y\,d\,p_z\,d\,V)/h^3.$$

But

$$d\,p_x\,d\,p_y\,d\,p_z\,d\,V = m^3\,d\,n\,d\,v\,d\,w\,d\,V$$

where m is the electron mass and u, v, and w are the velocity components along x, y and z. Thus $n(u, v, w) = G m^3 \, du \, dv \, dw \, dV/h^3$. Above $0°$ K this must be multiplied by $f(E)$ where, $E = \dfrac{m}{2} c^2 = \dfrac{m}{2} (u^2 + v^2 + w^2)$. Thus FERMI's law for the case of a *directed velocity vector* c, where $c^2 = u^2 + v^2 + w^2$ ending in an element of velocity space $du \, dv \, dw$ per unit volume is

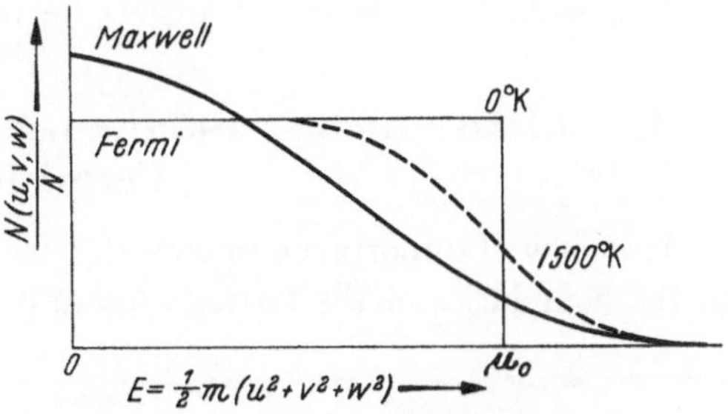

$$\frac{n(u, v, w)}{dV} = N(u, v, w)$$

$$= \frac{G m^3}{h^3} f(u, v, w) \, du \, dv \, dw$$

and

$$N(u, v, w) = G \left(\frac{m}{h}\right)^2 \times$$

$$\times \frac{du \, dv \, dw}{\left[e^{\left\{ \frac{m}{2} (u^2 + v^2 + w^2) - \mu \right\}/kT} + 1 \right]}.$$

Fig. 4. Fermi distribution and Maxwell distribution laws for directed velocity vectors

$$\frac{N(u, v, w)}{N} = \frac{3}{4\pi} (2 m h \mu_0)^{\frac{3}{2}} \frac{du \, dv \, dw}{\left[e^{\left\{ \frac{m}{2} (u^2 + v^2 + w^2) - \mu \right\}/kT} + 1 \right]} \qquad (2.15)$$

to be compared with

$$\frac{N(u, v, w)}{N} = \left(\frac{m}{2\pi kT}\right)^{\frac{3}{2}} e^{-\frac{m(u^2 + v^2 + w^2)}{2kT}} \, du \, dv \, dw, \qquad (2.16)$$

in the Maxwellian form. These are shown plotted to different scales in Fig. 4.

It is of importance to be able to estimate the number of electrons per unit volume. If A_w is the atomic weight of the metal, ϱ its density in g/cm³ and θ the number of free electrons per atom with N_A the avagadro number, viz the number of atoms for A_w grams then

$$N = \theta \cdot \frac{\varrho}{A_w} N_A \text{ per cm}^3.$$

For alkali metals θ is 1, for alkaline earths perhaps 2, etc. $\varrho \sim 10 \, \text{gm/cm}^3$. Let $A_w \sim 100$ with $N_A = 6 \times 10^{23}$ this gives 6×10^{22}

Table 3. *Table of computed values of the number of free electrons in different metals after* HERMANN *and* WAGENER [39]

	$\mu_0/\theta^{2/3}$		$\mu_0/\theta^{2/3}$		$\mu_0/\theta^{2/3}$
Li . . .	4.7	Be . . .	9	Th . . .	3.5
Na . . .	3.2	Mg . .	4.5	Ta . . .	5.3
K . . .	2.1	Ca . . .	3.0	Mo . . .	5.8
Rb . . .	1.8	Sr . . .	2.5	W . . .	5.8
Cs . . .	1.5	Ba . . .	2.4	Fe . . .	7.1
Cu . . .	7.0	Al . . .	5.6	Ni . . .	7.4
Ag . .	5.5			Pd . . .	6.1
Au . . .	5.5			Pt . . .	6.0

electrons/cm³. In this event, $T_c \sim 5 \times 10^4$ °K. If $G = 2$, as for an s band the limiting energy μ_0 may be calculated from N in terms of A_w, ϱ and θ. Thus

$$\mu_0 = 26.1 \left(\frac{\varrho \theta}{A_w}\right)^{\frac{2}{3}} \qquad (2.17)$$

in electron volts.

Exact values for θ are not available. However, a table for θ due to HERRMANN and WAGENER [39] is given in terms of $\mu_0/\theta^{\frac{2}{3}}$. Some of the values in electron volts are shown in Table 3.

Temperature dependence of μ is not alone limited to the small change above noted, for since volume increases with T, N will also decrease. If the linear coefficient of the thermal expansion is α,

$$\mu_T = \mu_0 (1 - 2\alpha\,T). \quad e + \alpha \text{ is } 1 \times 10^{-5}, \quad \mu_{1000} = 0.98\,\mu_0.$$

The temperature dependence differs from this if the highest bands are overlapped and if the Fermi level is close to the top or bottom of the band. The difference there is caused by electron transfers between bands.

C. Surface structure and the work function in relation to the Fermi level

It is now of importance to consider the structure of the energy levels relative to the boundaries of the lattice. Again as earlier, the potential differences pro-

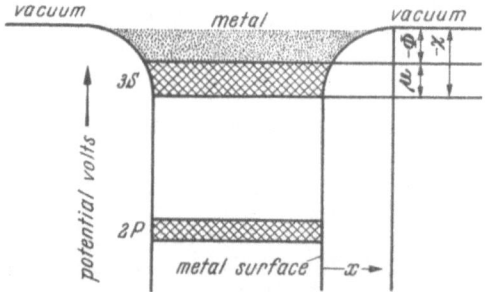

Fig. 5. Conventional representation of the energy level diagram of a metal showing the zero potential at the surface, the top of the Fermi level at $-\varphi$, the bottom of the band at $-\chi$ and μ the width of the Fermi band

duced by individual metal ions of the lattice will be ignored and all electrons will be considered to move at a constant potential inside the metal. However, as electrons approach the surface of the metal, their potential must rise; otherwise, they would leave the metal. There is, therefore, a potential barrier created by the forces of attraction of the metallic ion lattice on the electrons which keeps them within the metal, even in vacuum. This potential barrier must obviously be higher than the energy of the top of the Fermi level for electrons generally do not leave metals at room temperatures but require materially higher temperatures before measurable thermionic emission occurs. The situation is nicely represented in Fig. 5. There are depicted a completely filled lower level $2p$ and above it the S band of a univalent element with $G = 1$. The cross hatched area to the dashed line represents the top of the filled Fermi level of height μ above the bottom of the band. Above the dashed line is the region of unfilled levels occupied by a distribution of electrons forming the asymptotic foot of the Fermi distribution above $0°$ K shown as dots.

The potential is measured along the vertical and distance from the surface is indicated by x to the right. The *potential at the surface* is 0 and the potential energy of the electrons is negative downward. The top of the filled Fermi band lies at a potential φ below the top of the potential barrier; or 0 potential. The total potential χ represents the distance of the bottom of the S band from below the top of the potential barrier. Thus one can set $\varphi = \chi - \mu$ multiplied by the electronic charge as the *work to remove* an electron from the top of the filled band. That is, the work to remove a free non degenerate electron from the metal, is then given by $W = \varphi\,e$. It is the work to remove an electron from the metal and is restored to the lattice whenever an electron enters the lattice from outside. The quantity φ is the potential against which the electron must work. $\varphi\,e$ is called the *work function*.

The origin of the potential barrier may be described in two ways. One is to assume that an electrical double layer at the boundary is produced by the polarization of the metal ions near the boundary due to the inward direction of the electric field. This can create a potential barrier such as observed. A more convenient picture is that of the image force of the electron in the conducting metal surface as the electron leaves the surface since this force is open to calculation. If an electron is emitted from the surface and is at a distance x, it induces a charge of equal magnitude and opposite sign, i.e., a positively charged mirror image inside the metal. The induced charge and electron attract each other by Coulomb forces. The work per electron may then be associated with the depth of φ and thus helps to evaluate χ.

The image force using C.G.S. units is

$$F(x) = \frac{e^2}{(2x)^2}$$

and the corresponding field strength

$$X(x) = \frac{e}{4x^2},$$

which leads to a potential

$$V(x) = \int X(x)\, dx = -\frac{e}{x}.$$

Here x ist the distance from the surface.

In the expression for $V(x)$ it is noted that $V(x)$ would become infinite at $x = 0$. This indeed would be so for an ideal image force due to an infinitely fine structured charge distribution at the surface. However, as the electron approaches the surface and x becomes small, the effect of the discrete character of the surface ionic charges cannot be neglected. SCHOTTKY solved this problem in an approximate fashion by drawing a tangent to the curve for V applying to higher values of X at a point x_0 which then intersects the surface at a value of $V = \varphi$ corresponding to the top of the Fermi level. This then permits a rough calculation of φ and accounts for its value thus yielding a value for the total barrier height χ. The variation of V with x is linear from the intersection at $-\varphi$ up to the point of tangency at x_0, thereafter it follows the image force relation. The relation between $\varphi\,e$ the work function and the image force field is then obtained by adding the constant field strength $\frac{e}{4x_0^2}$ multiplied by $e\,x_0$ to the potential between x_0 and infinity multiplied by the charge e. Thus we can set $\varphi\,e$ as

$$\varphi\,e = -\frac{e^2}{4x_0} -- \frac{e^2}{4x_0} = -\frac{e^2}{2x_0} \text{ ergs.} \tag{2.18}$$

This makes

$$W = \varphi\,e = 7.2 \times 10^{-8}/x_0$$

electron volts and the potentials $\varphi = 7.2 \times 10^{-8}/x_0$ volts.

This calculation introduces a critical distance x_0 which is somewhat arbitrary unless a value can be chosen with some justification. Since the interatomic distance, or the lattice spacing d, is a characteristic measurable representative of the inhomogeneity of the surface, the distance x_0 may be set as of the order of d, since at this distance, the electron effectively senses the discrete ionic charges.

Thus to a first approximation one may set $x_0 = d$. This approximation is in very rough agreement with observation since φ for the alkalies is low and the grating space d for these is large.

Observationally φ is rather successfully given in better than order of magnitude by setting $x_0 = d$ for some metals. Thus for Cs $d = 6.05$ Å which makes $\varphi = 1.9$ev, its correct value. However, for $W, d = 3.16$ Å making $\varphi = 2.3$ ev according to the relation above, while actually the observed vallue is for W 4.5 ev.

Actually the work functions are more properly calculated quantum mechanically from the energy of the metal lattice with all electrons included and again with one electron excluded. These calculations confirm the general relation between φ and $1/d$ and yield good values for the alkali metals.

In consequence of the foregoing, it is clear that the nature and origin of the potential φ and $\varphi e = W$ the work function is relatively well established and in the case of simple lattices can quantitatively be accounted for. In this case, there is basically no mystery.

A few further comments should be made. The value of χ can be measured, but not calculated from theory by electron diffraction studies. Electrons in motion regarded as waves are diffracted by crystal lattices. The diffraction patterns of such waves were first observed by DAVISSON and GERMER and G. P. THOMSON and have extensively been studied since. The positions of the diffraction maxima are influenced by the coefficient of diffraction n_e of the waves given by

$$n_e = \sqrt{\frac{E + \chi e}{E}}$$ with E the electron energy outside. From the observed position

of the maxima, it is possible to calculate n_e and thus χ. The value of for Ni is 14.8 volts, the measured value of φ for Ni is 4.9 volts leading to a value of μ of 9.9 volts. If the value of θ for Ni is 1 then Table 3 yields $\mu = 7.4$. Since θ for Ni is surely more than unity, the agreement is satisfactory.

D. Influences modifying the work function

The major question as to the nature of the work function being settled it must be remarked that the quantity is subject to minor but not negligible variations and modifications. In all that has gone before the discussion considered a pure clean metal-vacuum interface. Obviously contamination of the surface, distortion of the lattice by foreign atoms, etc., etc., will alter the properties and thus affect φ. Aside from such disturbances φ will depend on the temperature of the metal, on the magnitude of existing electrical fields outside the surface and on the orientation of the surface to the crystal lattice of the metal. The latter would be obvious since as the surface orientation alters, the parameter $1/d$ will alter.

The temperature dependence follows from the temperature dependence of the Fermi barrier μ. If the potential χ were independent of T then the sign of the temperature variation of φ would be opposite to that of μ, that is φ would increase as μ decreases with T. The potential χ however depends on the lattice dimensions which also change with T. Various calculations of this variation show that the temperature coefficient of χ is negative and of the same order as μ. It is seen that in any case, the temperature dependence of φ is not very great.

The effect of an external field on the value of φ must next be discussed. This best done in terms of Fig. 6.

The surface of the metal is shown at $x = 0$, with metal to the left including the top of the Fermi level μ and the surface at a potential φ_{x_0}.

The image force barrier indicating x_0 is shown in curve b. Line a represents the potential due to a uniform field of strength X imposed from without, with the surface negative to space, the negative potential declining with x along a. By subtracting the potentials represented by a along x from the normal potential barrier b, the dashed curve c

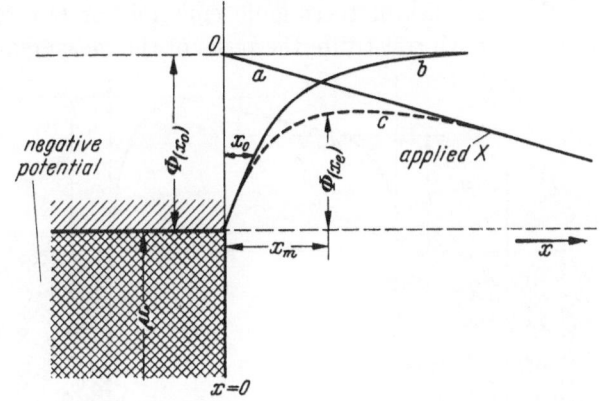

results. This represents the new potential barrier. It is at once noted that the height of the barrier above the top of the Fermi band is reduced from φ_{x_0} to φ_{xe} and that there is now a decline in potential beyond the point x_m of the maximum which will aid electrons in escaping.

It is at once possible to compute x_m from the condition that at x_m the potential

Fig. 6. Effect of an external field on the potential barrier at the surface of the metal. Note the reduction in height of barrier and decrease in width

due to the imposed field $X(e)$ and that of the natural potential barrier caused by the image force are equal and opposite, and that at the maximum the electrical field is zero. Thus

$$X(x) + X(e) = -\frac{e}{4\,x_m^2} + X(e) = 0$$

whence $x_m = \sqrt{\dfrac{e}{4\,X(e)}}$ when $X(e)$ is in esu/cm and e is in esu. Putting in numerical values and expressing $X(e) = X_V$ in volts/cm, then

$$X_m = 1.9 \times 10^4 \sqrt{\frac{1}{X_v}} \tag{2.19}$$

Thus if X_V is 100 volts/cm, $x_m = 1.9 \times 10^{-5}$ cm and if $X_V = 10000$ volts $x_m = 1.9 \times 10^{-4}$ cm. The potential at the maximum is given by getting the potential of the external field at x_m. It the external field is X_V in volts then V_V the reduction in potential height in volts is $2V_V = 2x_m X_V$. The value of the reduction in barrier height is $\Delta V_V = 3.78 \times 10^{-4} \sqrt{X_V}$, with an external field X_V expressed in volts/cm, the height of the potential barrier is then merely

$$\varphi_V = \varphi_0 - 3.78 \times 10^{-4} \sqrt{X_V}. \tag{2.20}$$

The work function as indicated above should vary with the lattice spacing and consequently with the crystal planes at the surface. The combination of electron microscopy and the use of the Mueller techniques for studying the emission patterns from minute single crystals surrounded by evacuated spherical bulbs of large radius covered with phosphors has revealed the influence of crystal orientation.

The values for different crystal faces were also directly measured using single crystals with different planes emitting. NICHOLS[40], with W, found φ to be 4.35 volts for the 111 plane and 4.66 for the 112 plane. Theoretical calculation has substantiated such variations in magnitude. In general most surfaces are made of aggregates of microcrystals with all sorts of orientations. Thus most measurements will yield an average or composite value.

E. The contact potential difference

Since studies of thermionic emission, or electron emission from surfaces in general, which constitute the basis of the measurement of work functions always involve a *collecting* as well as an *emitting* electrode it is clear that *differences in work functions* between the two could lead to errors in the currents measured as a function of applied potential, if differences in work function introduce differences in the true potentials between the surfaces for which the applied potentials are not corrected. Thus, for example, if two metals are connected electrically but face each other in vacuum, the *Fermi levels in each metal adjust to the same level producing* the situation shown in Fig. 7. Here A and B are two metals in contact

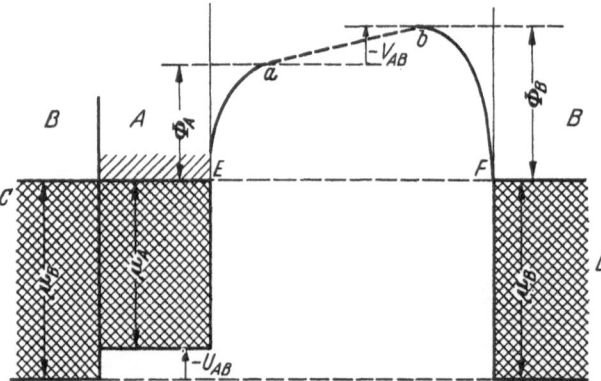

Fig. 7. Conventional description of two metals with different work functions facing each other with Fermi levels equalized by electrical contact between back sides showing the galvanic potential and contact potential difference

between C and D while they face each other at EF. The Fermi levels adjust to a common level along EF. If the two bands μ_A and μ_B do not have the same height, a potential difference $-U_{AB}$ exists at the junction which has the value $U_{AB} = \mu_A - \mu_B$. This potential difference is called the *galvanic potential*. It has little practical significance. On the contrary, it is noted that the potentials at a and b denoting the region of the image force field at the faces of the metal surfaces reach the work function levels of potential φ_A and φ_B so that between these faces the *contact difference of potential*, $V_{AB} = - (\varphi_B - \varphi_A) = \varphi_A - \varphi_B$ exists. This *contact potential* is a direct result of the *difference in work functions and is equal to that difference divided by the electronic charge*. In studying the effects produced between two surfaces resulting from an applied potential difference the *true* potential difference active must be derived by correcting the applied potential differnce by the value of V_{AB} or as it will be designated hereafter as V_C. The difference in work function

$$\Delta W = (\varphi_A - \varphi_B)\, e = e\, V_C. \qquad (2.21)$$

The definition of contact difference of potential assumes that the two metals are at the same temperature. If temperatures differ, the well known Seebeck potential differences are set up which must be added with proper sign to V_C.

These Seebeck potentials are, however, very small $\sim 10^{-3}$ of the values of V_C under most circumstances.

This contact potential often *correctly* called the *Volta potential* is as seen of intrinsic metallic origin and *must never be confused with electrolytic potentials* discussed in Chap. I.

A few comments on the influence of adsorbed foreign materials on the work function and contact potential difference of metal surfaces is now necessary. Adsorbed layers may be of several sorts. These are (a) adsorbed layers of ions, (b) adsorbed layers of dipoles and (c) adsorbtion by the VAN DER WAAL's forces. All of these in one fashion or another create dipole layers at the surface which may lower or raise the work function depending on sign.

a. Let a positively charged metal ion approach the surface. It will be attracted by its image force with increasing magnitude until the very rapidly rising repulsive forces between the electron cloud of the ion and that of the surface, in consequence of PAULI's principle, just balance the image forces. The ion is then trapped at a certain distance x_i from the surface. If ion potential in terms of the image force and repulsive interaction be plotted the ion sits in a potential well at x_i cm from the surface. In being adsorbed, the adsorbtion energy is set free and if the absorbtion energy per ions is divided by the ionic charge, this gives the image force potential of the ion. The energy is then

$$\varphi_i e = \frac{e^2}{4\,x_i} = \frac{3.6 \times 10^{-7}}{x_i} \qquad (2.22)$$

in electron volts. If $\varphi_i e$ is known, then x_i can be determined. It will again be near some 10^{-8} to 10^{-7} cm if $\varphi_i e = 0.5$ volts per ion.

b. Either molecules with permanent, or molecules with induced, dipole moments can be adsorbed on surfaces. Thus molecules like HCl, H_2O, NaCl can be adsorbed by metal surfaces when they approach the surface closely enough to sense the normal surface force field of the ions. All molecules and atoms including the inert gas atoms are polarized by an electrical field which displaces the electron cloud and thus the negative center of charge relative to the positive nuclear charge. In the case of the induced polarization, the magnitude of the molecular dipole is proportional to the electrical field. For atoms like He and molecules like H_2, the value of γ, the polarizability is small, but it becomes very large for larger configurations such as Cs or Xe. Thus the normal dipole moment for HCl is 1×50^{-18} esu $\times$ cm while that say, for the Ne atom will be of the order of 5×10^{-18} esu $\times$ cm at 30 Kv/cm and 5×10^{-12} at 3×10^8 volts/cm such as might be found at the surface of metals. The forces with which dipoles are attracted in such fields depend on dX/dx, the rate of change of the surface field and on the dipole moment. Thus they vary as a higher power of the distance than the inverse square law for image forces.

c. The VAN DER WAAL's forces result from the continual variation of the surface field of the atom owing to surface charge density fluctuations in space and in time. Thus one atom in a transient polar phase will induce polarization in the other atom and the mutual polarizations will cause attraction. These forces vary as a still higher power of the inverse distance than the dipole forces and the forces are rather weak. These adsorbed configurations also lead to potential curves of the same sort as for the ions, except that in general, binding energies are less, so that the potential trough at the surface distance x_i is less and the forces become small very rapidly as one recedes from the surface.

The adsorbtion of ions or dipoles always produces an electrical double layer at the surface. For the ions, it consists of the charged positive ions outside and the charges induced in the metal on the inside. If m is the dipole moment and σ the surface density of dipoles, the double layer gives a jump in potential the magnitude of which is $\Delta V = m \, \sigma$.

In the case of the ions the charge of the dipoles cannot be below the surface of the metal so that only the parts of length x_i lying *outside* the metal will be effective. The work functions is changed by ΔV. It is decreased if the positive pole of the double layer lies outside while it is increased if the negative pole lies outside. If the potential at the surface of the pure metal is φ the potential with double layer φ_D will be, $\varphi_D = \varphi - \sigma \, m$.

To indicate how such dipoles work concrete examples may be chosen. The atoms of the metal Cs have an ionization potential $E_i = 3.88$ v which is 0.66 ev lower than the work function for W. Since the vapor pressure of Cs at room temperature is 5×10^{-6} mm Hg then Cs in a tube will quickly diffuse to a W surface. There it forms ions by giving up its electron and the ions adhere to the surface, heat being gained by the W in the amount of 0.66 ev per Cs atom together with the heat of adsorbtion of the ions.

If the temperature of the cathode becomes too high, the kinetic energy of the lattice becomes greater than the heat of adsorbtion of the ions on the surface. At that point, the ions evaporate and the pure W surface remains. In the case of Cs on W. LANGMUIR and KINGDON [41] found an emission current of 8.8×10^{-5} amp per cm² at 690° K with Cs present while the value was 3.6×10^{-26} amp/cm² for pure W at that temperature. Using the thermionic emission these workers found that the value of $\varphi_D e$ was 1.38 ev while φ_e for pure W was 4.54 ev. This decrease in φ_e of 3.2 ev produced a very large increase of emission at low temperatures. BECKER [42], investigated the relation between the emission current and the fraction θ of the surface covered by Cs. The W cathode was flashed and the increase of emission current as a function of time was observed for a given vapor pressure, p, of Cs. If the Cs does not re-evaporate, a maximum of emission is reached at a certain time t_m. The quantity $p \, t_m$ was observed to be constant. Thus the maximum of emission is reached when a certain fraction of the surface θ_m is covered. It then appears for smaller times t that the fraction θ is governed by $\theta/\theta_m = t/t_m$. Further studies were carried out by I. LANGMUIR and J. P. TAYLOR [43], by DE BOER [44] and VEENEMANS, who, among other things, studied the potential variation in front of the cathode during adsorbtion. It appears that the cause of the maximum is that *only the Cs first deposited goes on as ions.* The later deposit of Cs is atomic and does not lower the work function. The value of θ_m for Cs on W occurs at 0.67. Similar results occur for Ba on W. Another interesting type of double layer formation comes from Th on W. LANGMUIR [45] found on examining W wires with 1 % ThO₂ that the ThO₂ is reduced to Th by the W at 2600° K. It permeates the metal and diffuses to the surface to yield a dipole layer of enhanced emission. Unlike those of Cs and Ba, if the W is flashed to remove the double layer more Th diffuses from within and reforms the layer. This has very obvious advantages for thermionically emitting filaments. At complete coverage at one temperature $1274°$ K, the emission increased from 10^{-10} to 10^{-4} amp/cm².

Another interesting example is that of O_2 on W. Here the atoms of O form *negative ions* that adsorb. This gives a dipole layer with *negative charge outside*. This increase φ_D and decreases emission. The chemistry of W indicates that adsorbtion in this case is not of the electrostatic type. In this case, an oxide of W forms as a monolayer with the O^- ion outward and the W^+ ion in the surface.

By making a W—O dipole surface and then admitting Cs, an extremely high emission can be obtained from a layer showing much more stability than Cs on W. Here according to DE BOER, the Cs^+ ions are bound to the negative O^- ions of the W^+O^- dipoles. They are bound more strongly than by image forces, there is larger surface coverage and far less ready evaporation. The maximum of thermionic emission of W—O—Cs occurs at $1000°$ K the current density being 0.35 amp per cm², which is 4×10^3 times that of W—Cs alone.

Observations of the emission of surfaces with the electron microscope show distinct differences in the activation of different areas by adsorbtion of atoms, atoms going preferentially to certain areas. These have been observed for Th on W and for Cs and Ba on an Ni cathode. These differences of work functions in different adjacent areas of a surface are more significant than just indicating composite surfaces and average work functions. Because of different densities on different faces, the potential jumpf $\Delta V_D = \sigma_m$ will have different values in different surface areas. Thus potential differences between adjacent areas on the surfaces will appear producing additional, or *surface* fields in front of the cathode. These fields superpose on the image force field and on the external field. They will influence the work function in the same way as the external field influences the work function in the Schottky effect. Such surface fields exist to a lesser degree even in pure metals as φ varies over the crystalline forces.

If the work function can be measured as a function of the anode voltage V_A applied the distance at which internal and external fields are equal can be derived from $\dfrac{d\Phi}{dV_A} = -\dfrac{x_m}{d}$ with d the anode distance. Then if the external field can be calculated from the geometry of the system, if roughness is not too great and if fields are below field emission values the internal fields can be calculated. L. B. LINFORD [46] carried out some studies on W—Th cathodes. He observed that between 10^{-4} and 10^{-5} cm the internal fields are very much larger than the image force fields. Thus at 10^{-4} cm the internal fields are about 1000 volts/cm while the image force field is only 3.5 volts/cm at that point. Thus the external fields affect the surface fields much more than they do the image force field. This causes external fields to influence emission more than for pure metals.

Existence of emission patches influences the velocity distribution of emitted electrons since the slow electrons will be preferentially prevented from overcoming surface fields. Calculation shows that differences between the lowest value of the work function and the mean value are not very great. Thus the areas of smallest work function contribute most to the emission of a cathode of non-uniform work function. The mean value of the work function varies with temperature, for as the latter increases the areas of larger work function contribute relatively more to the total emission at high temperature.

F. The measurement of contact potential and/or work functions

As seen the work function of a metal is affected by external fields, slightly by temperature change, varies with crystal face exposed and is altered by adsorbed atoms and the surface force fields. Thus measurement of contact potential differences and work functions by different methods may lead to average values having different significance as a result of the differing influence of the varied factors on the phenomena used to evaluate them. All data should be corrected for external field where possible. Otherwise in the main, for equally pure samples with no contamination agreement between data from various methods is satisfactory.

1. Thermionic emission

The first estimates were evaluations of φe using the thermionic emission equation of RICHARDSON, or better, the equation as later modified quantum mechanically and thermodynamically for G and for the transmission coefficient D_e of electrons at the surface. This equation reads, J_s, the current density is

$$J_s = 2\pi G D_e \frac{m e}{h^3} k^2 T^2 e^{-\frac{\varphi e}{kT}} = 60 G D_e T^2 e^{-\frac{\varphi e}{kT}}. \tag{2.23}$$

amperes per cm² per deg.². Since in practice the constant $60 GD_e$ varies considerably, especially since G and D_e are not known solution of the full equation from J_s is not a profitable procedure. Incidentally, the symbol J_s represents the saturated space charge limited current in fields so high that the space charge current limitation is removed. Since the value of φ appears in the exponent, if we take the logarithms of both sides

$$\log \frac{J_s}{T^2} = -\frac{\varphi e}{kT} + \log 60 G D_e. \tag{2.24}$$

Accordingly $\log I/T^2$ plotted against $1/T$ is a straight line with slope $-\frac{\varphi e}{k}$, or more conveniently $-5040 \varphi/T$, (φ in volts), the value of $60 G D_i$ being given by the intersection with the axis of ordinates. This yields φ at the given saturation value of the field strength. In consequence, one must correct back to φ_0. This is done by measuring φ for various values of applied field above saturation plotting and extrapolating back to φ_0. Correction for T may also be made if the temperature variation of μ with T is known.

This extrapolation method for the field is termed that of the Schottky line. Owing to surface roughness and differences in work function, the extrapolation ist not too accurate. The value obtained is that of the mean work function Φ, or better *apparent* work function. Values of Φ for W range from 4.63 to 4.52 ev while $60 G D_e$ ranges from 22 to 212, where measured. Probably $\varphi = 4.54$ is about as good as can be chosen. For Mo values range from 4.15 to 4.38 ev with the constant ranging from 24.6 to 175. The high constants appear to go with the high work function measurements, i.e., high currents and higher work function values appear associated in measurement. The best value for Mo is probably around 4.24 ev. A proper evaluation of $60 D_e G$ is largely limited by the accuracy of the measurement of T, since a small error in T in the exponent causes large errors in the evaluation of the constant. Thus owing to this cause alone, the constant is uncertain by $\pm 50\%$.

2. Calorimetric measurement of work function

Work function can be measured by calorimetric means. The cathode is cooled by the emission of electrons, the energy coming from the heating current in thermionic emission. Thus if the heating current required to maintain a certain temperature with no thermionic current shown has a potential applied between emitting wire and collector in vacuum yielding saturated electron current, the cathode cools. The reduction in emission of radiant energy produced is that needed for electron emission. The emitted electrons have a Maxwellian energy distribution. The normal component has an energy of kT while the two tangential components have energies of $kT/2$ each. The energy associated with the velocity of emission of electron emission is then $2kT$ and the total energy required is

$$E = \varphi e + 2kT.$$

The cooling power which is equal to the decrease of energy radiated from the cathode per unit time is then

$$W_{\text{cool}} = I_s \varphi + 2 I_s \frac{kT}{e} \quad \text{watt}$$

whence

$$\varphi = \frac{W_{\text{cool}}}{I_s} - 1.72 \times 10^{-4} T \quad \text{ev}.$$

The measurement of W_{cool} is not very accurate or satisfactory. A more instrumentally elaborate method determines W_{cool} by the amount of heat, i.e., the current increase needed to restore the temperature. The restoration of the temperature is determined by the resistance of the cathode as divulged by a bridge measurement. Satisfactory data have been obtained in this fashion.

3. The photoelectric studies

Quite early in the study of contact potentials and the concept of the work function was MILLIKAN's[34] accurate verification of the Einstein photoelectric law. Earlier attempts in this direction had been made by numerous workers, but it was MILLIKAN's very accurate study for which technically he received the Nobel Prize that established the facts. This law was enunciated by EINSTEIN in 1905 after P. LENARD reported that by means of retarding potential studies, he had found the energy of the photoelectrically emitted electrons depended on the frequency of the *incident light* and *not* on the square of the amplitude of the incident light waves as called for by classical mechanics and electromagnetic theory. EINSTEIN accepted the Planck quantum theory and wrote the law as:

$$Ve = E_e = \tfrac{1}{2} m v^2 = h \nu - h \nu_0. \tag{2.25}$$

Here E_e is the kinetic energy and is equal to Ve, the potential energy to be applied via a retarding potential V, just to reduce the emission current to zero. The quantity $h\nu$ is the energy of the incident photons, ν being the frequency and h the Planck constant. The quantity W is the energy required to remove the electron from the metal, i.e., its work function. By using a monochromator and Increasing wave length, or decreasing frequency, until $h\nu$ equals some lower iimiting energy at which the emission ceases, W is observed. If the electrons are free in the metal then $h\nu_0 = W = \varphi e$ the work function. Various workers attempted to verify the law experimentally, using both X rays and light waves with intermediate success. MILLIKAN[36] resorted to freshly shaved surfaces of alkali metals in vacuum and evaluated Ve, and $h\nu_0$ for the different metals using a quartz monochromator. Photoelectric yields from these surfaces were adequate for the feeble light. Together with W. H. SOUDER, quantum efficiency was also measured by measuring the incident energy with a bolometer together with the photo current. Current densities were sufficiently low such that space charge limitations were of no consequence. At that time, controversy as to the character of the work function was still raging. It was not even certain that the electrons were free, i.e., that the light might not first have to remove electrons from metal atoms and then give them energy to escape from the surface. Furthermore, since Millikan collected currents on anodes which were not freshly shaved surfaces of the same alkalies, but some metal such as Ni, the *contact* potentials between alkali metal and anode came into the picture. These questions were resolved by MILLIKAN's study. Referring to the Eq. (2.25)

$$\tfrac{1}{2}m v^2 = h\nu - h\nu_0 = h\nu - W,$$

the $\tfrac{1}{2}m v^2$ requires that a potential $+V$ be applied to plate A, e.g., the shaved Na surface, so that electrons do not escape. Call the collecting plate B and let B have an escape work $W_B = \varphi_B e$ while A has one. $W_A = \varphi_A e$. Now the $+V$ active to prevent escape of electrons from A is not the value of the whole applied potential since the contact potential difference $\varphi_A - \varphi_B$ is acting to drive electrons from B to A, i.e. is also acting to retard electrons assuming $W_A > W_B$. Thus

$$\tfrac{1}{2}m v^2 = [V - (\varphi_A - \varphi_B)]\, e = h\nu - h\nu_0.$$

If the measurement be repeated with metal A' instead of A leaving B unchanged, and if $W_{A'} < W_B$ one would observe for A'

$$\tfrac{1}{2}m v'^2 = [V' - (\varphi_{A'} - \varphi_B)]\, e = h\nu - h\nu_0'$$

subtracting the two expressions for A and A' there results

$$e\, [V - V' - (\varphi_A - \varphi_{A'})] = (h\nu_0 - h\nu_0'). \tag{2.26}$$

MILLIKAN's measurement always showed that the applied retard potentials V and V' were the same, i.e. $V - V' = 0$ and $V = V'$. Thus

$$(\varphi_A - \varphi_{A'})\, e = h(\nu_0 - \nu_0') \quad \text{and} \quad (\varphi_A - \varphi_{A'}) = \frac{h}{e}\,(\nu_0 - \nu_0'). \tag{2.27}$$

Hence EINSTEIN's energy of escape is the work function or contact potential of VOLTA. Direct measurements of the Volta potential difference showed that

$$e\,\varphi_A - e\,\varphi_{A'} = h(\nu_0 - \nu_0')$$

as measured by the red wave length limits. The equivalence of the quantities above also proved that the *electrons were free* in metals as $h\nu_0$ is the total energy to liberate and $e\,\varphi_A$ is the work done against the image force field. For 1 ev the wave length λ_0 is about 12340 Å lying well in the infra red and for 4.9 ev it is 2537 Å in the quartz ultra violet.

When $h\nu$ exceeds $h\nu_0$ then the electrons that are emitted do not all have the same energy $\frac{1}{2}mv^2$. For some of the electrons below the top of the Fermi band can also be lifted up if $h\nu - h\nu_0$ is large enough. The velocities emitted thus will be distributed in such a fashion as to have many electrons below the limit $\frac{1}{2}mv^2 = h\nu - h\nu_0$. The curve for the distribution of velocities emitted may be obtained by the shape of the retarding potential curve. Differentiation of this curve gives the form of the velocity distribution.

The curve obtained begins at about $\frac{1}{2}mv^2$ and generally has a peak at about 0.6 ($\frac{1}{2}mv^2$ instead of at $\frac{1}{2}mv^2$ showing the extent to which electrons below the top of the level are drawn out. Actually electrons with an energy below $h\nu_0$ can be emitted by those few conduction electrons having thermal energies above the top of the Fermi band. These, however, are so few in number at $300°$ K that the rise of current at a fixed potential above $h\nu_0$ observed as $h\nu$ increases while not very sharp is such as to give reasonable but not accurate values of ν_0. R. H. FOWLER[47] has derived an exact equation for photo electric emission for temperatures above zero° K.

This equation reads

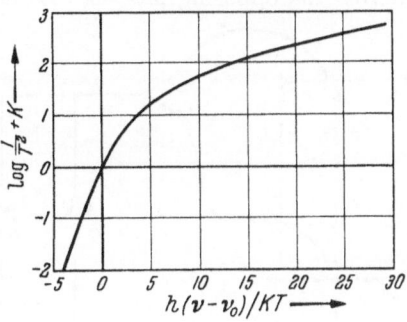

Fig. 8. FOWLER's function $f\dfrac{h\nu - h\nu_0}{kT}$ relating photo current I as a function ν and plotted as a function of $h\nu/kT$

$$\log \frac{I}{T^2} = K + f\left(\frac{h\nu - h\nu_0}{kT}\right) \qquad (2.28)$$

with K a constant and $f\left(\dfrac{h\nu - h\nu_0}{kT}\right)$ a complicated function of the argument and I the photo current. The shape of this function is shown in Fig. 8. Measurement of I for a given T as a function of ν and plotted a function of $h\nu/kT$ as indicated yields a curve of the shape of that shown but displaced from it by an amount $h\nu_0/kT$ in the $h\nu$ direction. The displacement gives ν_0 and thus φ.

As noted in MILLIKAN's treatment

$$\tfrac{1}{2}mv^2 = [V_s - (\varphi_A - \varphi_B)]\,e = h\nu - h\nu_0$$

and since $h\nu_0 = e\,\varphi_A$ it follows that $h\nu - eV_s = e\,\varphi_B$ where V_s is the retarding potential and ν is the impressed frequency. This at once evaluates the work function of the anode.

4. Contact potential difference measurements

The direct observation and measurement of contact potentials was most clearly establihed as to what was being observed by Lord KELVIN[35] in his famous paper before the Royal Society which was printed in 1898. While the phenomenon had been observed and even measured on several occasions before this date but subsequent to VOLTA's original observations, there had been confusion with electrochemical effects and other matters. In principle, the observation of VOLTA and KELVIN's first demonstration are performed as follows: An insulated condenser of plates AB say of Zn and Cu either with their opposing falls coated with varnish or clean and *dry* were arranged so that B could be raised and moved to some distance from A as shown in Fig. 9. The plates CD represent the capacity of an electrometer or electroscope system with its associated capacity. The electrometer, or electroscope, of capacity represented by CD registers potential across it. Start with AB in contact, except for the varnish film, or separated a very small fraction of a mm, but not touching if bare. Now close the switch K which shorts the electrometer condenser assumed to have no contact potentials, and connects the metals A and B. If A and B are electrically connected, their Fermi levels are made to coincide. Electron current flows from Zn at B to Cu at A, or positive charge from A to B, building up a contact potential difference across the opposing faces of AB as shown in Fig. 7. The quantity q_1 is then on

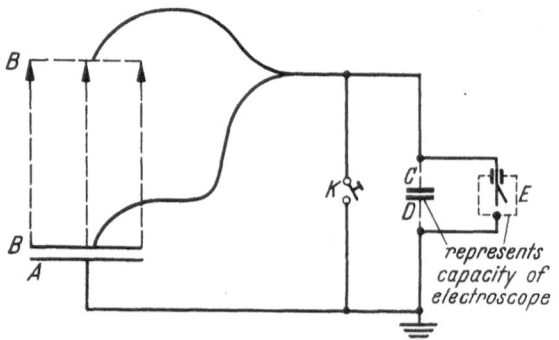

capacity C_1 at the contact difference of potential V_c. Then the key K is opened. Now the two condensers AB and CD are in series. Condenser CD has a capacity C_E assumed constant. The quantity q_1 is now on the combined system. Now let B be raised from A reducing the capacity of AB from C_1 to a small value C_F. The potential across CD will rise to a value V_E. The value of V_E can be computed

Fig. 9. Diagram of KELVIN's original condenser system for demonstration of volta or contact potentials

as follows: $q_1 = V_T C_T$. Where V_T is the potential across the two series condensers and C_T is the series capacity. Now $V_T = V_{CF} + V_E$, where V_{CF} is the potential across AB after separation. Thus as q_1 is constant

$$V_T = \frac{q_1}{C_T} = \frac{q_1}{C_F} + \frac{q_1}{C_E}.$$

Whence

$$C_T = \frac{C_E C_F}{C_E + C_F}.$$

Again

$$q_1 = V_{CF} C_F = V_E C_E \quad \text{whence} \quad \frac{V_{CF}}{V_E} = \frac{C_E}{C_F}.$$

Thus

$$V_{CF} + V_E = V_E \frac{C_E}{C_F} + V_E = V_E \left(\frac{C_E + C_F}{C_F}\right) = \frac{q_1}{C_T} = q_1 \frac{C_E + C_F}{C_E C_F}.$$

Yielding

$$V_E = q_1/C_E, \quad \text{but} \quad q_1 = C_1 V_C.$$

Thus

$$V_E = V_C \frac{C_1}{C_E}.$$

As carried out both by VOLTA, KELVIN and others not much attention was paid to details in improving this experiment. Usually C_E was not neglegbile and $V_C \sim 1-4$ volts. While C_1 could have been large, it was rarely so. In fact, the Zn and Cu plates used by KELVIN were small. When used without varnish the *shorting by K*, was not used. Instead the plates were brought into contact at the several high spots and while distance was small at the low points not in contact, since the plates were separated by hand they were often tilted so that at the *last contact*, equivalent to closing K and opening it, C_1 was generally commensurate with C_E. Hence the value V_E observed was of the order of V_C. In the use of shellac, condensers AB of small plate area were used and often the distance of separation of AB were not great. Where electrometers were used the plate system was in an enclosed chamber the initial plate separation was large and final separation was not too much greater. Thus in all the early electrostatic tests C_1 was commensurate with C_E and the resulting Volta potential observed was $\sim V_C$ or a few volts. In fact, in order to demonstrate the phenomenon to the audience, KELVIN repeatedly touched the Zn and Cu plates together and separated them, the upper movable Cu plate being touched each time to the lower plate of a *large condenser*, connected to an electroscope, whose upper plate was connected to the Zn plate. After some hundred repeated separations he had transferred enough charge q_1 at a potential V_C to the upper condenser of large capacity, such that on re-

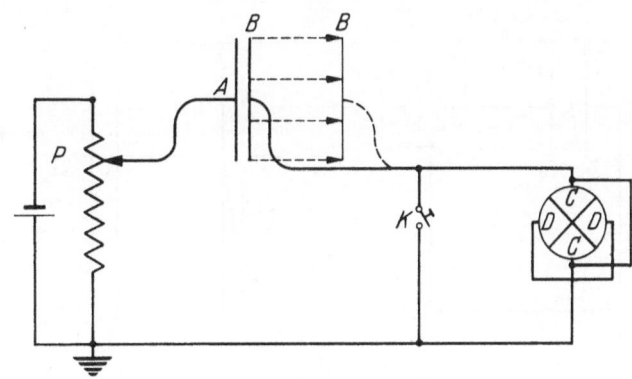

Fig. 10. The Kelvin null method for contact potential measurement using a potentrometer for compensation

moving the upper plate of the large condenser some 300 volts or so caused the gold leaves to move. The sign of the contact potential can be established by bringing a charged body of known sign near the isolated plate attached to the electrometer and noting whether it increased or decreased the deflection. With varnish between A and B as operated in Fig. 9 the charge observed was generally greater since C_1 was larger owing to the proximity of A and B when K was closed and opened.

Repetition by KELVIN of the same experiments using an electrometer with a drop of water on the plates revealed the *electrolytic charging* which followed an entirely different behaviour pattern so that once and for all the electrolytic charging was differentiated from the Volta potential.

For quantitative measurement of the Volta, or contact potential difference the scheme used was that shown in Fig. 10.

The arrangements were the same as in Fig. 9 except that now the fixed plates A was connected to the ground and the electrometer plate D through *a potentiometer* P and plates A and B were in a shielded case with B motile. Then with B near A K was closed and opened and the contact on P was moved in a fashion such that as B was moved away, the electrometer showed no deflection. That is, that charge was placed on plates D of the electrometer in such a way as to neutralize the quantity q_1 that flowed from B to C as the capacity of AB decreased. The potential needed to yield a null deflection at CD was then just the potential V_C placed across AB when the Fermi levels as shown in

Fig. 7 adjusted on closing K. Here the quantity q_1 is not kept constant but is adjusted to maintain the potential across CD or $V_E = 0$. This, in principle, but with various modifications, is KELVIN's original quantitative method and is used today. One modification due to W. A. ZISMAN vibrates the plate B of the condenser relative to A at a fairly high frequency. This generates an AC which can be amplified and detected. P is adjusted until the signal due to the AC vanishes. It is claimed to be good to 10^{-3} volt and takes but a few seconds to measure.

That the potential exists between A and B can be shown in several ways. If A and B are connected by K so that the Fermi levels coincide the field $X_c = V_c/d$ produced by V_c between A and B separated a distance d can be measured by the deflection of an electron probing beam in vacuum.

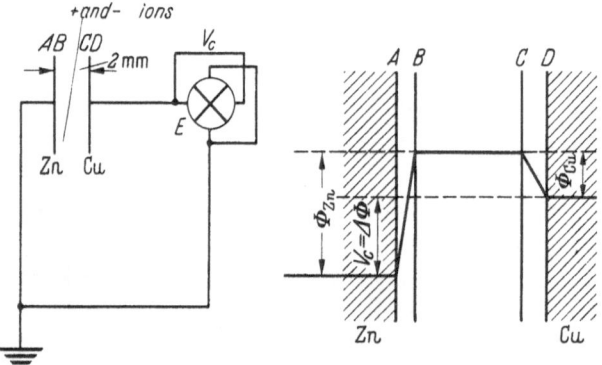

Fig. 11. The description of two metals with different work functions when facing each other with air rendered conducting by ions from radioactive radiations showing that Fermi levels on the faces adjust to zero leading to electrometer potentials with sign reversed

Probably one of the more direct measurements is one first mentioned by Lord KELVIN in his report made in 1898. Very shortly after the discovery of the radioactive properties of Uranium and the conductivity of air produced by its radiations in 1896, Lord KELVIN used the ionization produced by U to measure the contact potential difference. Reagard Fig. 11 in which the two metals Cu and Zn face each other in a non-reactive, gas e.g., an inert gas. The electrical double layers are indicated at the surfaces of Zn, AB, and Cu, CD. The potentials φ_{Zn} and φ_{Cu} are indicated by the linear gradients in the layers. The plate separation is some 2 mm. The circuit is ideally simple—merely the Zn and Cu opposite each other and connected across the electrometer. Nothing happens until the radioactive source is introduced between the plates. At this point, conduction is established and the potential difference between B and C equalizes. This throws the potential difference $\varphi_{Cu} - \varphi_{Zn} = \Delta\varphi$ across the electrometer E the potential $V_c = \Delta\varphi$ being registered. In this case *Zn is negative* to Cu. In the original Kelvin method by which the Fermi levels equalize inside the metal, the Zn appears positive to Cu, i.e., the polarity of V_c is reversed. This method was investigated by various workers, perhaps best by COMPTON[49]. It is not very accurate. Radioactivity makes the surrounding atmosphere conducting so that leakage currents appear in the wrong places. The measurement is not made in vacuum and the surfaces are likely to adsorb ionized gases and charge the surface potentials. Its accuracy does not compare with that of other methods.

In conclusion a table of the most satisfactory average values of φ for pure metals as reported in 1951 by G. HERRMAN and P. S. WAGENER in "The Oxide Coated Cathode" Vol. II, are given as Table 4. These values are corrected to zero field and are in ev:

Table 4. *Table of work functions of some pure metals as reported by* HERMANN *and* WAGENER *in 1951[3]*

Li . . .	2.48	Be . . .	3.32—3.92	Th . . .	3.38	Fe . . .	4.49
Na . . .	2.28	Mg . . .	3.67	C . . .	4.35—4.60	Ni . . .	4.96
K . . .	2.22	Ca . . .	3.20—3.71	Si . . .	3.54	Pd . . .	4.98
Cs . . .	1.93	Ba . . .	2.51	Ta . . .	4.13	Pt . . .	5.36
Cu . . .	4.45	Zn . . .	4.29	Cr . . .	4.60		
Ag . . .	4.46	Al . . .	4.20	Mo . .	4.24		
An . . .	4.89	Zr . . .	3.73	W . . .	4.54		

In the preceding discussions, the development of the theory of contact potential difference and its measurement have been presented. In addition, its use in charging of condensers has been indicated in the presentation of its measurement. It should, however, be indicated that its appearance in electrostatic charge generation other than as indicated has *not* been *frequently* recorded, or established without question owing to electrolytic, or other, effects. However, VOLLRATH[50] showed that iron and antimony powders blown through a copper tube gave particularly strong contact charging. Similarly KUNKEL[51] showed that Ni dust impinging on Pt also gave strong charging with the correct sign of potential to be expected from their relative work functions. In both studies, the metals could have had oxide layers present. However consideration of BOWDEN and TABOR's[52] studies on friction indicate that under the conditions of these observations the oxide films played no great role. This fact was borne out by KUNKEL's studies in general and in particular the Pt surface of his duster was actually etched by the impact of the powders used. In an attempt to make a careful studyof the phenomenon HARPER[53] carried out a series of investigations using contact between steel spheres electroplated with various polished metal films. No great care was taken in cleaning the spheres since again oxide layers were thin and broken down. Elaborate precautions were taken to avoid lateral motion at contact. The elaborate metal apparatus prevented any thorough outgassing and cleaning techniques by contact potential study standards were far from satisfactory. The data obtained yielded charging that paralleled the differences in work function. The theory involved was very elaborate and inadequate publication space was allotted to permit sufficiently detailed calculations to be given so that the reasoning could be followed. The *amount of charge separation* observed in view of the surfaces in contact *was far less* than would naturally follow from the known values of the contact potentials. An elaborate calculation which again is not given in sufficient detail accounts for this deficiency on the basis that fields at separation are of such magnitude that there is considerable back discharge between separating surfaces due to field emission. The theory indicated justifies charging being ascribed to contact potential even though charge separated is only a small fraction of that corresponding to the true contact potential difference. On this basis HARPER concludes that the static charging observed was nearly quantitatively in conformity with theory and the true contact potentials of the metal pairs concerned provided rubbing between surface was eliminated on contact. Any errors then obtained were caused by the presence of surface roughness which could not be controlled and varied with the different films. This elaborate and careful study leaves little doubt about the reality of contact electrification of metals and static charging thereby, if there was any before. It also indicates that metal dust blown against dissimilar metals can prove a serious hazard for static electrification in industrial processes.

G. Discussion and conclusions

It is seen that inherent in the physics of the metallic state and on the basis of quantum mechanical considerations, there is a segregation of electrons from the outer valence states of free atoms to energy bands of electrons capable of moving relatively freely in the spaces between the ions of the metal lattice in-regions of nearly constant potential. These electrons are assigned momenta and energies which fill the band to a given level. Either because the band is partly empty, or because of overlapping bands of different states, there lie at lower negative energies below the zero energy at the metal surface vacant states which the electrons at the top of the Fermi band may occupy. These states are more and more occupied as the temperature rises above absolute zero. These electrons account for electrical and heat conductivity and may escape from the surface at high temperatures, or by action of high imposed external fields if they get energy enough to escape against the image field, or in high fields they can tunnel through the potential barrier. Between the free electrons at the top of the Fermi band and the outside, there exists the image force field, or potential field at the surface of the metal which prevents their escape. The effective potential from the top of the Fermi band to the surface of the metal to be overcome by escaping electrons is called the intrinsic potential φ of the metal and when the potential is multiplied by the electronic charge φe represents the *work function W* of the metal. Increase in temperature alters the value of φ by changing the Fermi band, (increases φ), and by expanding the metal lattic (decreases φ). External fields lower the potential barrier and decrease φ below φ_0. Adsorbed positive ions place a new double layer and reduce φ strongly. Adsorbed negative ions increase φ. These adsorbed ions also produce surface fields that effect variation of φ with external fields. In addition the different grating spaces of the different crystal planes influence the image forces near the surface and change φ. Most of the changes noted except those by adsorbed ions, are relatively small.

Two different metals A and B each having a characteristic value of φ, φ_A or φ_B opposed to each other, will disclose these differences as follows, assuming $\varphi_A > \varphi_B$.

(1) If facing each other as a condenser and if connected electrically so that charges can flow the Fermi levels will adjust by current flow to a common level. Thus metal A with $\varphi_A > \varphi_B$ will have its face opposite B at a positive potential relative to B. The potential difference $\Delta\varphi_{AB} = \varphi_A - \varphi_B = V_c$ represents the *contact potential difference* of the two metals. If the metals A and B face each other with their back sides connected to an electrometer ionization or conductivity *between the two faces* will place the surfaces A and B facing each other at the same value and the metal A with the higher φ_A will appear negative to metal B on the electrometer, the value of V_c again appearing. Countless studies have confirmed the reality of this potential difference, the *Volta* potential and the model on which it is predicated. They have also established it as a characteristic property of the metallic and quasimetallic states as in semi-conductors, depending on electronic carriers. It is *not* to be confused with the *electrolytic*, or *galvanic*, potential difference also discovered by VOLTA as *was clearly demonstrated by* KELVIN.

By bringing two metals A and B with a contact potential difference close together to create a condenser with a capacity C of high value, connecting the two to equalize the Fermi levels and breaking contact, reduction in C by separating AB will reveal a considerable potential difference. In fact, as KELVIN showed in 1898, *it can be used to generate static electrification.* In this case, the phenomenon is well understood and owing to the importance of work functions for many industrial applications data are at hand so that the characteristics can or could be used for static generators in a predictable fashion.

It is noted that this process in metals involves the transfer of electrons and is associated with the electrical image force field characteristic of the lattice and the existence of a Fermi band and its associated properties. This situation does not necessarily or properly apply to insulators or to substances having ionic lattices with no large numbers of free electrons. Thus contact electrification even between a metal and an insulator may not necessarily involve the loss or gain of electrons by the metal. Again it was noted that owing to many influences ions positive and negative, neutral atoms and molecules with induced dipoles, molecules with permanent dipoles or atoms and molecules in consequence of VAN DER WAAL's forces may be adsorbed on the surfaces of metals. Further many more reactive atoms may be chemi-adsorbed as uni or poly atomic layers at the surfaces of some metals, e.g., O to form oxide layers. Such layers may be polarized at the surface. The atoms or molecules of the layers in many instances will choose certain preferential sites on the lattice structures. These produce profound changes on the values of φ. Since metals have the proclivity for adsorbing atoms and molecules all studies of contact potentials of pure substances require excellent vacuum techniques as well as intensive surface treatment to remove the layers that have formed. Under these conditions it is clear that contact electrification between metal and a non-metal surface can occur not only through the exchange of electrons but through the adsorbtion of ions of the non-metallic substance by the metal. Thus electrons can wander into vancancies in the non-metallic lattice, *possibly* electrons from the non-metallic lattice can wander into the metal. Certainly ions of one sign or another without the intervention of a dielectric ionizing liquid can leave the non-metallic surface and adsorb on the metal surface. Thus contact electrification between non metal and metal—the next most simple case to that between two metals can be most likely caused by electron transfer from the metal and ion transfer of either sign from the non metal. In each of these instances the forces and potentials between electrons and non metal surface or ions and metal surfaces must be of such magnitude together with the action of kinetic heat energy that charge transfer can occur against the surface forces, image forces, or forces of binding existing in the free metal and free insulator surfaces. It is clear that in these instances the matter is not dependent on any equalization of Fermi levels on contact for the non metal has no such level, though an electron on a non-metal surface may sit in some state in a shallow potential well. Thus although the parallelism cannot be closely drawn from the more common form of contact electrification, a great deal has been learned in the study of surface behaviour through the understanding gained on the nature of adsorbtion, the effect of double layers on surfaces and the action of such complicated systems as with the W—O—Cs cathode.

One added remark may be made in this connection namely that since in contact electrification of the metal—non-metal system one system is not a conductor static electrification processes are much more effective owing to the inability of the separated charges from one part of the system to leak away during separation. Thus the amount of charge separated is more likely to be proportional to the effective areas of the surfaces in contact and the charges per unit surface contact area transferred unless separation fields produce back discharge of some sort. The potentials achieved can be greater since the capacity of the condenser created on contact is larger and transfer of charge may be large without the back flow loss.

One aspect of contact electrification and separation has not been touched on either in the metal—metal, or the metal—non-metal case where surfaces are not just opposed but brought into actual contact. With the contact between clean metal surfaces in the few asperities where contact is made, it is possible that on separation some of the surface of the softer metal will come off on the harder one as the contact points break as observed by BOWDEN ahd TABOR[57]. In KELVIN'S first experiment, this will have no effect, the important factor is the opposing surfaces *not* in contact, but near *at the instant that the last contact is* broken, since the Fermi levels equalize. With the case of metal—non-metal the contact electrification requires actual surface contact for transfer of electrons or ions and the charge transfer occurs only at the points of contact. If contact of such dissimilar surfaces leads only to transfer of ionic layers and electrons, then the charging will depend on the true surface areas in contact and the number of transferable charges per unit area. If actual adhesion of one material on the surface of the other occurs, the situation can become involved since possible homogeneous charging superposes on the other process especially with shearing of the non conductor. In addition, the question might be raised as to how much Cu or Zn will adhere to glass or quartz for example, on contact and separation. If thick layers, e.g., 100 Å thick are torn off, the metal lattice one situation arises. If the non-metal can pick up a *single layer* of the ions composing the metallic lattice then a definite positive charge transfer could occur, especially if the surface making contact had a negatively charged strongly bound ionic layer that attracted the positive lattice ions and repelled the electrons. Forces between glass and quartz spheres rolling on Ni vacuum are very great owing to electrostatic charging and such actions could occur. It is, however, probably unlikely that negative layers in say, quartz, could be so strongly bound as to remove ions from the metal lattice. This point requires direct investigation but the probability is high that such contact charging depends on electron transfer from metals or transfer of loosely bound ions from the non-metals.

III. Static charging by spray electrification
A. Introduction

The production of large amounts of static electrification and high potentials through the spraying or bubbling of water and possibly other liquids as an independent mechanism in its own right and quite distinct from the electrolytic processes discussed in Chap. I, must be accepted as an experimental fact. It was discovered in consequence of observations of ELSTER and GEITEL[54] in 1890, on the intense electrification and static sparks occurring in association with certain waterfalls in the Austrian Alps. The phenomenon was investigated by LENARD[12], reported in 1892, and named by him *Wasserfall-Elektrizität, Waterfall Electrification*, or better in English *Spray Electrification*. He observed that the finer mist carried aloft by updrafts was predominantly negatively charged, while nearer to the water surface, there appeard to be some positively charged spray. The water appeared to be the source of the phenomenon, negative charge being carried away on the finer mist produced by the shattering of water and by bubbling of the water surface. Studies later carried on in the laboratory led him to conclude that pure water was endowed with an electrical double layer at a water gas interface. The double layer was inherent in the water and independent of the nature of the gas. It consisted of an outer negative layer with, on the average, one extra electron to 10^4 or 10^5 water molecules and of a thickness of about 5×10^{-7} cm. Deeper in the liquid resided a positive layer. Introduction of electrolytic ions into pure water appeared to cause positive ions to be drawn up into the double layer in different amounts and to different depths depending on the valence and type of ion. Numerous and varied experiments confirmed this action such as the cataphoresis of small gas bubbles in an electrical field, the bubble being suspended by spinning the water in a horizontal tube at high speed. Collecting spray produced by bubbling of very dilute solutions of NaCl under some conditions, yielded Na_2O deposited on the cathode and nothing on the anode, the electrons or OH^- ions in the finer negative spray leaving no residue but the positive Na^+ ions at the cathode on neutralization yielded NaOH. This initial picture has been confirmed in its broad outline, but has been modified in detail by later work.

Despite all the work which has been done on water and aqueous solutions there is still much confusion and relative to other liquids much remains to be done. Interpretation has been hampered by occasional falsification of data by the electrochemical effects discussed in Chap. I. It has been further complicated by the very sensitive nature of the phenomenon to the influence of ions and impurities and their seemingly erratic behaviour. Interpretation of the data are complicated by the critical role of droplet size in influencing the charges exhibited, and the very different number-size distributions and nature of droplet formation caused by the various mechanical production processes. The condition and humidity of the atmosphere in which the drops find themselves during study, alter the size and thus drift velocities of the carriers, either by evaporation leading to groups of small ionic carriers of unique velocities or by some condensation lowering mobilities and smearing them out. Great care must be used in studying the carriers produced, techniques being such that groups of carriers are not lost by diffusion en route and that adequate resolution be had to cover ranges of mobilities extending from 2 to 3×10^{-3} cm²/volt sec, the areas under the curves truly representing the numbers of carriers involved. Again since in the separation of liquids containing ions into droplets, statistical factors cause equal numbers of droplets of a given size to be charged with an excess of q ions of either sign, *symmetrical charging*, it would be expected that the statistics of the true spray process depending on the double layers would in some measure be falsified by the symmetrical process.

Another factor that has often been overlooked in such studies is the influence of electrical fields on the charging of droplets by electrostatic induction during the time of their separation. The principle of this charging mechanism was strikingly illustrated by the electrostatic charging device known as the Kelvin Water Dropper. The complete neutralization of inducing fields in the study of such processes is often difficult since fields caused by contact potentials are difficult to eliminate but quite effective. That such fields cause charging can be seen in the use of charge produced in the water dropper technique for measuring the potentials at water, or contaminated water, surfaces both by FRUMKIN[55] and by CHALMERS and PASQUILL[56].

Fortunately in the final analysis it will be seen as this chapter develops that much more is really known about the nature of spray electrification than previously suspected, since the cumulative but hitherto uncoordinated studies of a number of independent workers, including the names of COEHN[57], McTAGGART[58], CHALMERS, and PASQUILL, especially ALTY[59], FRUMKIN, CHAPMAN[60] and DODD[6] yield a unique and consistent picture of the processes involved. This picture permits of some extrapolation and prediction as to results to be expected under various conditions, certainly for aqueous solutions.

The important question as to whether this process is unique to that peculiar substance water with its high dielectric constant, its high surface tension, its dipolar character with hydrogen bond linkages, may now tentatively be answered in the negative. The theory of ALTY and the studies of FRUMKIN and perhaps of COEHN, indicate that certain classes of organic substances possessing dipole moments and contaminated by adequate quantities of the right kinds of ions should produce similar charging. COEHN and his associates indicate that the surface electrical double layer is present at all liquid gas interfaces, the outside layer being negative. The extent to which these potentials can lead to separation of charges as the surfaces are disrupted into small droplets will depend on the possible rupture of bond linkages, if present, or to the separation of the "uncovered" portions of ALTY's double layer produced by the presence of ions. The double layers produced in such liquids have not to the present been revealed in cataphoresis studies of gas bubbles as with water, chiefly because in the high fields required other disturbing and masking actions caused by ionic currents intervened. COEHN's own studies on substances fairly uniformly and well contaminated with ionic impurities indicated such uncovered ionic double layers to exist. HARPER[61] in a very careful study of the charging produced by bubbling on a number of non-polar, non-conducting hydrocarbons observed that when they were purified such that their specific conductivity was less than 10^{-13} mho no electrification was observed. If impure, electrification was observed on bubbling, ions of both sign being present sometimes in unequal amount, at other times in equal amounts. The ion collecting potentials were about 1500 volts/cm and carriers of mobilities as low as 3×10^{-4} cm^2/volt sec were collected. It is unfortunate that a mobility analysis of higher resolving power was not used by HARPER to yield a spectrum of carriers. Symmetrical charging of the larger droplets must have occurred, but unless ion concentrations were above 10^{+12} per cm^3 droplets under 10^{-5} to 10^{-4} cm diameter would have shown little symmetrical charging. The commercial grades of hexane, octane, benzene, xylene, and even carbon tetrachloride had sufficiently low conductivities that no charging was observed with them. However, pinene required removal of acid and di iso amyl decane required removal of alkali before it could be distilled to yield charge free droplets.

The dangerous charging of commercial grades of ethyl ether and of gasoline has never been properly resolved. Ether would, on the basis of ALTY and FRUM-KIN's work, show a tendency to charging. Gasoline should not. In the cases of electrification through flow and spraying of these liquids, resulting in heavy charging and explosions, conditions have always been such that contact of the liquids with metals and flow thereover could have created the charges by electrolytic processes. The generation by such action is the more likely since these substances are good insulators relative to water and the return flow of accumulated charge by conduction is small as shown by COEHN's study. The final answer can be had only by using a mobility analyzer, bubbling ether of various grades of purity with N_2 gas and studying the carriers with an Erickson tube analyzer.

With this introduction, it is now of interest to present the accumulated result on water and aqueous solutions, to interpret them in the light of this information and to indicate their role in meteorological and other processes.

B. The existence of an electrical double layer at gas-liquid interfaces

The existence of an electrical double layer at gas liquid interfaces, is unquestionably established. Again there is much confusion in the literature and various properly conducted measurements appear to give such confusing results that interpretation is to say the least, difficult. The various studies must now be presented, interpreted, evaluated and correlated.

1. The studies of COEHN on double layers as related to dielectric constant

Probably one of the most frequently quoted investigations and the law there revealed in 1898, is one due to COEHN [62]. It is most convinvingly demonstrated experimentally in a later paper of COEHN and RAYDT. In this study, various "pure" liquids, largely organic, but including conductivity water were subjected to an endosmotic analysis. They were placed in a cell having two parts separated by a short capillary of some tenths, (0.274—0.065), mm diameter and 0.5 mm long and for various applied potentials of the order of 440 volts. The height to which the endosmotic pressure pushed the liquid surfaces was measured by an inclined manometer read with a micrometer eye piece. The height difference with reversed potential was measured and all heights were relative to that for acetone. Either being unfamiliar with the Helmholtz theory of endosmosis or because this study preceded it, no use was made of the theory developed in Chap. I. Empirical observation, however, led COEHN to conclude that the endosmotic pressure as indicated by difference in the height of the liquid produced by the same potential, assumedly corrected for density, was proportional to the difference between the dielectric constant of the liquid and the glass, quartz, or diamond capillaries used. Their measurements of 1909 showed that for these liquids

$$\frac{h_x}{h_A} = \frac{D_x - D_g}{D_A - D_g}.$$

(3.1)

Here the subscript A represents a standard substance, actually acetone, and x represents any one of some 30 substances, including pure H_2O, while g stands for glass. The symbol h represents the measured height of endosmotic rise which, when corrected for liquid density is proportional to the pressure P, while D is the dielectric constant. In general, the method *appeared* satisfactory so that D_x could be evaluated to some 10% for a fairly varied group of substances, (with $D = 2$ to $D = 81$). For some of the nitrobodies such as nitromethane, agreement was good to less than 50%. These were with some justification presumed impure. Impurities altered the observed values and even reversed the sign. The action of impurities was specific and probably associated with substances lowering surface tension and thus altering the surface. Purification by distillation and refrigeration was kept up until the "*right* value" of h_x was observed. When this appeared, further purification produced no change. Some nitrobodies never were adequately purified on this basis as they seemed to alter on standing and heating. Electrolytic conduction affects the phenomena in a different fashion. He observed the currents accompanying flow and noted that the electrolysis causing flow was not proportional to the specific conductivity of the liquid, but showed characteristic action. He observed that the height decreased directly with the current through the liquid. He observed that HCl markedly reduced the height in H_2O in proportion to concentration increase while the same conductivity produced by KOH gave no change. HCl is known to alter the negative charge of the double layer as see infra. The currents observed varied from $4—50 \times 10^{-7}$ amp.

These measurements formed the basis of the oft quoted *Coehn's law* which states "*Substances with a high dielectric constant in contact with substances of a lower dielectric constant charge positively relatively to the substance of lower dielectric constant*". That signifies that there is an electrical double layer at the boundary. This layer COEHN assumed moves a liquid in a field according to the sign of its charge relatively to the fixed boundaries. It could not be ascribed to dissolved electrolytes dissociated in the liquid. Quoted was the fact that $C_6H_5NO_2$ of high D showed the effect and is known not to cause dissociation of dissolved electrolytes. He asserted that the phenomenon is due to some "electrokinetic potential", but did not know which.

If one regard Eq. (1.9) for the endosmotic pressure P, it reads: $P = \dfrac{\zeta D I W}{4\pi \eta \lambda}$ $= \dfrac{\zeta V W D g}{4\pi \eta l}$ (1.9), where $\zeta = \dfrac{4\pi \sigma d}{D}$ (1.4) is the electrokinetic potential, D the dielectric constant of the liquid, $W = 8\eta \, l/\pi R^4$ is the POISEUILLE's law resistance to the flow through the capillary, g is the area of cross section of the tube, λ the specific conductivity of the liquid, and η its coefficient of viscosity. Inserting the value for W

$$P = \frac{2\zeta VD}{\pi R^2} = \frac{8V}{R^2}\, \sigma d = A_1 \sigma d \qquad (3.2)$$

with $A_1 = \dfrac{8\pi V}{R^2}$ a constant of COEHN's measurement. The measurements on a relative basis showed that observationally

$$\left(\frac{h_x}{h_A}\right)_\varrho = \frac{P_x}{P_A} = \frac{(\sigma d)_x}{(\sigma d)_A} = \frac{D_x - D_g}{D_A - D_g} \quad (3.1),$$

and assumedly for air, or space

$$\frac{(\sigma d)_x}{(\sigma d)_A} = \frac{D_x - 1}{D_A - 1}. \tag{3.3}$$

This relation implies that the dipole moments per unit volume of the surface double layer are proportional to the dielectric constants diminished by unity.

If an electrical field X be applied to a block of substance of dielectric constant D, of surface area A and depth d, then

$$DX = X + \frac{4\pi\sigma}{A} = X + \frac{4\pi\sigma d}{Ad} = X + \frac{4\pi\sigma d}{V} \tag{3.4}$$

$$D = 1 + \frac{4\pi\sigma d}{XV}, \quad D - 1 = \frac{4\pi\sigma d}{XV}. \tag{3.5}$$

Neglecting the inner field correction at the surface, which is, perhaps not completely justified, it is seen that

$$D - 1 = 4\pi \frac{\sigma d}{VX} = 4\pi \delta N \tag{3.6}$$

where δ is the molecular polarizability and N is the number density of molecules. Thus if

$$\frac{D_x - 1}{D_A - 1} = \frac{(\sigma d)_x}{(\sigma d)_A} = \frac{\delta_x N_x}{\delta_A N_A}. \tag{3.7}$$

This says that the moments per unit volume of the electrical double layers leading to the ζ potentials causing endosmosis must be in proportion to the product of the polarizabilities δ of the molecules and the number densities N in the surface.

This is a rather naive or oversimplified inference though formally correct. For, as will be seen later in the case of water and doubtless also, in the case of some other liquids, in consequence of the electrochemical forces between molecules which are responsible for the surface tension the dipolar molecules are oriented in the surface layer to a rather high degree. In water, these dipoles have the O^- end of the dipole outward and the protons inward. The dipoles so oriented produce a field X' on the inside of the liquid which bind largely OH^- ions but also anions of impurities to the inner face of the dipole layer. The remaining protons and *most* of the impurity cations are also fairly strongly held by the anions and OH^-ions. However, in the decreasing inward fields, some of the cations are sufficiently loosely bound so that the heat motions detach them and thus "uncover" their opposite but firmly bound anions. It is the "uncovered" anions and their partner cations deeper in the solution that constitute the endosmotic ζ potential double layer of HELMHOLTZ and the mobile dipole layer $(\sigma d)_x$ observed by COEHN. The potential at the surface resulting from dipole orientation is the much larger immobile layer representing the potential ψ of Fig. 1 and as will later be seen, was measured for water by CHALMERS and PASQUILL. However, the ζ potential does not depend alone on the field X' created at the surface by the double layer which is, in turn, a function of the $(\sigma d)_w$ of water molecules and the surface tension forces γ. It depends as well on the effect of the anion impurities on surface tension and on the concentrations of such impurities.

Under these circumstances, it is hard to understand how COEHN with a large range of different substances, not only varing in D, but in surface tension and in concentrations and character of ionizable impurity, should have found the $(\sigma d)_x$ or ζ potentials to depend solely on $\delta_x N_x$ and not varying with unknown impurity solute concentration, etc. It is to be noted that impurity concentrations could not have been the same in each case, though it is true that the conductivities of the substances indicated a relatively small range of variation by a factor of not much more than 10. In favor of COEHN's observations is the fact that experimentally, the effects of known added ions were not general, but highly specialized. Thus while addition of a certain concentration of HCl and NaOH produced the same change in conductivity, the HCl produced a marked lowering of P while NaOH did not affect the result. This could well be expected in aqueous solution on ALTY's theory. As H^+ ions "cover" the anions while OH^- may replace some anions, but this is not likely unless large concentrations are used. He found that the most active agencies in disturbing the potentials were those that lowered surface tension and thus collect at the surface a fact also noted by FRUMKIN. It is, however, difficult to imagine that the polarizability should be so great an influence on the complex conditions to create proportionality in $(\sigma d)_x$ or ζ, despite the possible wide differences in surface tension forces, the nature of impurity ions, degrees of dissociation and ion concentrations to exhibit such an accurate correlation, if it indeed be a correct observation.

One must then consider that while COEHN's law as stated might, on the basis of modern knowledge concerning the endosmotic potentials, have some theoretical justification the accuracy of his experimental findings must, for the present, be considered fortuitous for liquid-air interfaces. There is further no justification in assuming that the polarity of the dipoles at the surface will always be negative, especially in view of FRUMKIN's [55] observations for different organic compounds in water and formation of surface layers for such compounds. There thus remains only the empirical experimental observation on thirty liquids, the purity of which was *largely established by the liquid conforming to the law,* although the constancy of behaviour on purification beyond this point cannot be ignored. It is indicated that further careful work with substances purified with modern techniques be carried out in the light of the theories of HELMHOLTZ and ALTY [59] to establish the true significance of the law and the prediction as to the sign of the double layer. Certainly careful studies by other methods to verify the nature of the potentials at the surface must be carried out.

In a later paper, A. COEHN and M. MOZER [57] investigated the charges created by the bubbling of various liquids to note whether charging paralleled the intensity of the surface fields as indicated by COEHN's law. They used substances of fair purity and clean gases. They avoided contacts between liquid and solid surfaces by using a bubbler of large dimensions. Charging appeared the greater the higher the dielectric constant except for benzaldehyde which gave a higher charge than anticipated. The fine spray was negative. Their method of detection was insensitive so that they did not observe positive as well as negative droplets, the negative predominating. It should be noted that W. R. HARPER, with far more sensitive detecting equipment and somewhat purified organic non-polar liquids, e.g., benzene, carbon tetrachloride, toluene, etc., observed no asymmetrical charging on bubbling though substances had values of $D \sim 2-3$. They corrected some earlier conclusions of DE BROGLIE [63]. They verified the existence of a double layer for gas-water by letting fine gas bubbles rise vertically in water applying a horizontal field and noting that the gas bubbles moved towards the positive plate as if the water surface were negatively charged. They succeeded in making a very important observation. It had been noted that the spray in air produced by electrolysis of acid and alkaline solutions differed in sign. They noted that the difference came from the size of the bubbles. In alkaline solution, the alkali ions collect at the cathode and increase the concentration locally. The Na^+ ions lose their charge at the cathode and Na atoms react with the water at once to give NaOH and H_2 which escapes from the body of the solution as fine bubbles from the liquid in front of the cathode. As will later be noted, the coarser spray from the finer bubbles, especially with alkali ions present, appears to yield largely positively charged droplets in the airs as indicated by the high speed photographs of bursting bubbles by BLANCHARD [64]. If the solution electrolysed is acid, the H^+ ions discharge at the cathode and adsorb temporarily on the surface building a layer of H_2, which rises as large gas bubbles. These bubbles as they break, (see infra) generally yield an excess of very fine droplets with predominant negative charge. In fact, this study was the first of a series of many studies to reveal the nature of the charge in the air over liquids and solid surfaces undergoing chemical reaction. They are caused by fine bubbles created by gases from chemical reaction. Likewise GUNN and DINGER [26] observed the air over freezing water with dissolved air to be charged by the action of microscopic bubbles.

COEHN and MOZER investigated various gases and found no change in charge except in so far as the mechanical action of the gas altered particle size distribution in spraying. They observed with KCl that as the concentration decreased, they got more negative charge than positive, though at around 0.1 normal, the net charge changed sign.

2. The electrical double layer in cataphoresis

Probably the next important contribution concerning the double layer was the study of McTAGGART [58], working in J. J. THOMSON's laboratory. He suspended various small gas bubbles of different gases in a tube of water spun around its horizontal axis with an electrical field applied along the axis.

It will be recalled that the expression for the cataphoretic drift velocity v_x as given by VON SMOLUCHOWSKI, Eq. (1.32) for spheres of radius r large compared to $d \equiv 1/K$ for the mobile layer is

$$v_x = \frac{X r \sigma}{\eta (1 + K r)} \quad (1.32) \quad \text{which for} \quad K r \gg 1$$

approaches a value

$$v_x = \frac{X \varrho}{K \eta} \quad \text{with} \quad \frac{1}{K} \equiv d.$$

Thus McTAGGART observed that v_x was sensibly independent of r over a relatively narrow range of values. Thus he was able to write

$$k = \frac{v_x}{X} = \frac{\sigma d}{\eta}.$$

The value of
$$k \sim 4.1 \times 10^{-4} \text{ cm}^2/\text{volt sec}.$$
This gave

$$\sigma d = 1.2 \times 10^{-3} \text{ esu/cm}^2 \times \text{cm}$$

$$= 4 \times 10^{-13} \text{ coulomb/cm}^2 \times \text{cm} \sim 2.4 \times 10^6 \text{ electrons/cm}^2 \times \text{cm}.$$

He set $d = 10^{-8}$ cm and got $\sigma = 1.2 \times 10^5$ esu/cm^2 = 4×10^{-5} coloumb/cm^2 = 2.4×10^{14} electrons/cm^2. Actually $d \sim 10^{-6}$ cm so that these values should be 1.2×10^3 esu/cm^2 = 4×10^{-7} coulomb/cm^2 = 2.4×10^{12} electrons/cm^2. From the expression $\zeta = 4\pi\sigma d/D$ McTaggart evaluated ζ the electrokinetic potential of double layer as 0.055 volts, for the pure water-air interface. The water used was good conductivity water. Other than this McTaggart spent much time on ionic influences and thus missed the data of significance observed by Alty.

3. Cataphoresis and true nature of the double layer

We must now turn to the work of Alty[59] which likewise started in J. J. Thomson's laboratory and which has done more to clarify the whole issue than any other study. Alty again studied cataphoresis of gas bubbles in a tube similar to that of McTaggart, but with improved techniques, eventually studying really pure water in a quartz tube and varying bubble size. Alty found that the *charge* of the bubble was *constant* above a certain diameter range 0.033 to 0.2 cm, but that it decreased with bubble diameter below that value. He observed changes in charge with the reduction in bubble size by solution of the gas in the water. He found that *considerable time was required for the bubble to attain its equilibrium charge*. In fact, the bubble charge at first increased in pure water and then decreased. If water was really pure, the bubbles appeared to exhibit nearly zero charge. In consequence of his careful study extending over a much larger range of variables in pure water than had that of McTaggart, he was forced to abandon in its *simple form*, the Helmholtz theory of a double layer as for small colloidal particles in solutions, since velocity is always proportional to the field. He began by calculating the velocity of his bubbles from the applied field X, the bubble charge E and Stokes law, i.e., from

$$XE = 6\pi\eta r v \tag{3.8}$$

where η is the coefficient of viscosity, r is the bubble radius and v its observed velocity.

He pictured the situation as follows: At the water surface the water dipoles are oriented with the O$^-$ ions directed outward and the H$^+$ ions inward. This produces a potential at the water-gas interface, the absolute magnitude of which was measured by Chalmers and Pasquill and which indicated that about one in twenty-five of the water molecules was so oriented. These *dipoles* are *fixed* and *charges cannot be separated* in *this layer*. Possibly dipoles are disoriented and others take their place so that the layer remains fairly constant. However later work attributes remarkable rigidity to this layer. In consequence, these dipoles produce a strong surface field, possibly related to the dielectric constant, which creates the mobile double layer responsible for Coehn's law and the observed cataphoresis and spray electrification. Now in water and if COEHN'S observations are correct, the dipoles orient with negative charge outward and positive inward. In consequence of the influence of the fields of the positive charges of the *fixed surface dipoles*, any *free anions* in the liquid will migrate to the fixed positive charges and be more or less tightly bound. This fixes the *negatively charged section of the double layer* of the interface at the surface, and *the binding is fairly rigid*. This layer of free negative charge which may *not occupy each oriented dipole end*, (see infra), but will occupy a certain number, then *leaves the oppositely charged cation in the body of the solution* free to move around, but still attracted by the field of its dipole bound partner. These positive ions then are somewhat more loosely bound than their dipole bound negative counterparts, but never-

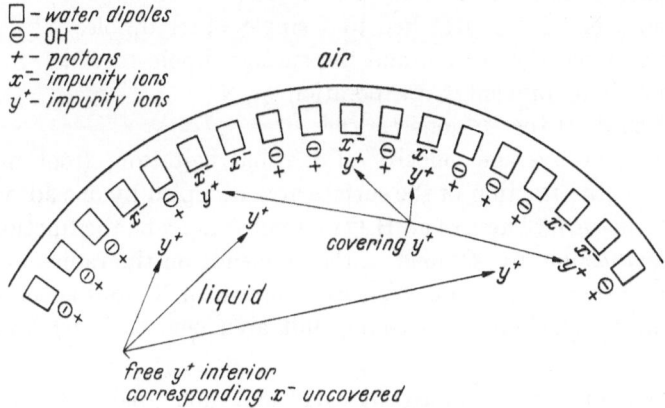

Fig. 12. Schematic illustration of dipole orientation, anion binding and cation covering yielding the mobile surface layer of ALTY

theless, "shield" or "cover" the large proportion of the negative counterparts. In the measure *that the negative counterparts are not "shielded" or "covered" by positive ions* the surface shows a net negative surface charge. If the surface is fixed as in an experiment such as COEHN'S, then *it is the loose ions of the "covering" layer that by heat motions are alternately "covering" or escaping that move*. The *electric kinetic potential* in this case would then be the consequence of *uncovered dipole bound negative charge*. Schematically, the situation is represented in Fig 12 which is self-explanatory.

In his second paper, ALTY develops the theory of the process and through observation of charge E with time he evaluates the constants. He then applies this to the bubble dissolving in time and shows all data to agree with them. It is of importance to follow ALTY in his reasoning. To do this assume that X^- and Y^+ are impurities present in such small quantities that the exact number do not change conductivity of the pure H_2O for which 10^{13} OH$^-$ and H$^+$ ions are present per cm^3. OH$^-$ and H$^+$ attach tightly to inner ends of dipoles through H bond linkages. They act to limit the number of vacant places to which X^- may go. Rate of acquisition of charge varies widely even for "pure" water. The time varies from 1000 to 100 seconds and only in the purest water.

Experiments show that in equilibrium the number of uncovered ions/cm^2 is inversely proportional to the square of the bubble radius r. This holds true only if the numbers N_x and N_y per cm² of surface, each varies as $1/r^2$. N_x and N_y depend on age of the bubble surface.

$$N_x = \frac{f(t)}{r^2} \quad \text{and} \quad N_y = \frac{g(t)}{r^2} \tag{3.9}$$

where $f(t)$ and $g(t)$ are functions of t.

The total charge $M\varepsilon = E$ on the bubble measured by Stokes law and the field is $4\pi r^2 (N_x - N_y) = 4\pi [f(t) - g(t)]$, when multiplied by the electron, which is independent of r in accord with experiment. The total bubble charge $E = M\varepsilon$ can be calculated at any time t after formation.

Assume: 1. That the fraction α of surface is alone capable of absorbing X^- ions from the solution. The fraction $(1-\alpha)$ is occupied by OH^- and H^+ in loose combination, which exclude X^- ions at the surface.

2. Each adsorbed X^- is attached to a single water dipole.

3. Forces between the X^- ion and the surface dipole to which it is attached are great enough to prevent re-evaporation of X^-.

Let there be n_x of the X^- ions *per cm³*.

Let v_x be their average velocity of thermal agitation, (root mean square value). Let θ be the fraction of the surface area occupied by one adsorbed X^- ion an let it equal the surface area of an H_2O molecule. Let α be the fraction of surface *occupied in equilibrium*, by X^- ions. α then depends on the radius of the bubble and by observation $\alpha = \beta/r^2$. Let N_x be the number of X^- ions adsorbed per unit area at a time t. The number striking unit area per sec is by kinetic theory $n_x v_x / \sqrt{6\pi}$.

The fraction of area occupied by OH^- and unavailable to X^- is $1-\alpha$. The fraction of *area per cm² occupied* by X^- ions at t is θN_x.

The fraction of whole surface *unoccupied* at t is $\alpha - \theta N_x$. The number of free X^- ions striking *these unoccupied spaces* per second is

$$\frac{n_x v_x}{\sqrt{6\pi}} (\alpha - \theta N_x).$$

If all collisions between the unoccupied surface molecules and free X^- ions result in absorbtion and if re-evaporation is negligible as assumed, then

$$\frac{d N_x}{d t} = \frac{n_x v_x}{\sqrt{6\pi}} (\alpha - \theta N_x).$$

The solution of this equation is

$$N_x = \frac{\alpha}{\theta} \left(1 + C e^{-\frac{n_x v_x \theta t}{\sqrt{6\pi}}} \right).$$

When $t = 0$, $N_x = 0$ thus

$$N_x = \frac{\alpha}{\theta} \left(1 - e^{-\frac{n_x v_x \theta t}{\sqrt{6\pi}}} \right) \tag{3.10}$$

which represents the adsorbed X^- ions at t. Now let the same symbols but with the suffix y refer to Y^+ ions. Since these ions are bound to the surface only when they strike an adsorbed X^- ion, the chance of any Y^+ ion being bound is proportional to the fraction of the surface occupied by uncovered X^- ions. The fraction of the surface occupied by *uncovered* X^- ions is

$$\theta (N_x - N_y).$$

Thus the increase in the number of Y^+ ions attached to unit area of surface in dt is

$$\frac{n_y v_x}{\sqrt{6\pi}} (N_x - N_y) \theta \, dt.$$

At the same time some of these Y^+ already attached are removed by thermal agitation as these ions are loosely bound.

Let k be the fraction of positive covering ions removed per second in this fashion. Then the expression for N_y is

$$\frac{dN_y}{dt} = \frac{n_y v_y \theta}{\sqrt{6\pi}} = (N_x - N_y) - k N_y \tag{3.11}$$

when

$$t = . \infty \quad \frac{dN_y}{dt} = 0 \quad \text{and} \quad N_x = \frac{\alpha}{\theta}.$$

Thus N_y is given by

$$N_y = \frac{\alpha/\theta}{1 + \dfrac{\sqrt{6\pi}\, k}{n_y v_y \theta}}. \tag{3.12}$$

The number of uncovered negative ions in equilibrium is therefore $(N_x - N_y)$ per cm² which is

$$N_x - N_y = \frac{\alpha}{\theta} \left\{ 1 - \frac{1}{1 + \dfrac{\sqrt{6\pi}\, k}{n_y v_y \theta}} \right\} = \frac{k\,\alpha}{\theta} \left\{ \frac{1}{k + \dfrac{n_y v_y \theta}{\sqrt{6\pi}}} \right\}. \tag{3.13}$$

This is the surface density on the bubble at equilibrium. In consequence Eq. (3.11) becomes

$$\frac{dN_y}{dt} + N_y \left(\frac{n_y v_y \theta}{\sqrt{6\pi}} + k \right) = \frac{n_y v_y \theta}{\sqrt{6\pi}} N_x = \frac{n_y v_y \theta}{\sqrt{6\pi}} \frac{\alpha}{\theta} \left(1 - e^{-\frac{n_x v_y \theta t}{\sqrt{6\pi}}} \right)$$

in virtue of Eq. (3.10). Thus

$$N_y = m_y v_y \alpha \left[\frac{1}{n_y v_y \theta + \sqrt{6\pi}\, k} - \frac{e^{-\frac{n_x v_x t}{\sqrt{6\pi}}}}{\left(n_y v_y \theta - n_x v_x \theta + \sqrt{6\pi}\, k \right)} + \right.$$
$$\left. + \frac{n_x v_x \theta \, e^{-\left(\frac{n_y v_y \theta}{\sqrt{6\pi}} + k \right) t}}{\left(n_y v_y \theta + \sqrt{6\pi}\, k \right) \left(n_y v_y \theta - n_x v_x \theta + \sqrt{6\pi}\, k \right)} \right].$$

Also

$$N_x = \frac{\alpha}{\theta} \left(1 - e^{-\frac{n_x v_x \theta t}{\sqrt{6\pi}}} \right).$$

But the number of uncovered negative ions per cm² is $N_x - N_y$ whence M the total number of uncovered negative ions on the bubble surface is $4\pi r^2 (N_x - N_y)$

$$M = \frac{4\pi\beta}{\theta} \left[\frac{k}{\dfrac{n_y v_y \theta}{\sqrt{6\pi}} + k} + \frac{n_x v_x - \sqrt{6\pi}\,\dfrac{k}{\theta}}{n_y v_y - n_x v_x + \sqrt{6\pi}\,\dfrac{k}{\theta}} e^{-\frac{n_x v_x \theta t}{\sqrt{6\pi}}} - \right.$$
$$\left. - \frac{n_x v_x n_y v_y \, e^{-\left(\frac{n_y v_y \theta}{\sqrt{6\pi}} + k \right) t}}{\left(n_y v_y + \sqrt{6\pi}\,\dfrac{k}{\theta} \right) \left(n_y v_y - n_x v_x + \sqrt{6\pi}\,\dfrac{k}{\theta} \right)} \right]. \left. \right\} \tag{3.14}$$

This then gives the number of free negative ions on the bubble surface at any time t after formation. If the ions are univalent of charge ε then the charge E on the bubble is $M\varepsilon$. The constants β and k are evaluated later from measurement and some approximations. If the values for these indicated below are used Eq. (3.14) becomes

$$M = 1.91 \times 10^6 [0.1096 + 1.19\, e^{-3.23 \times 10^{-3} t} - 1.299\, e^{-5.44 \times 10^{-3} t}]. \qquad (3.15)$$

The agreement between observation and this theory are shown by the points relative to the solid line in Fig. 13.

The constants β and k give information regarding the number of positive ions on the surface at any time. The measurement of the charge gives only the difference between negative ions and positive ions adsorbed. By definition β the total number of adsorbed ions at equilibrium is given by

$$4\pi r^2 \frac{\alpha}{\theta} = \frac{4\pi\beta}{\theta}$$

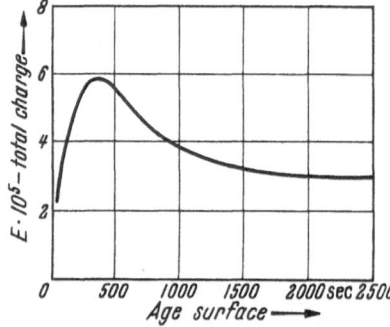

Fig. 13. Agreement between ALTY's theory and experiment for growth of double layer on bubble in time

and may thus be directly obtained if β is evaluated. To obtain β the expression for M is applied to experimental curves. The maximum charge is attained in 300 sec in Fig. 13 and is $5.66 \times 10^{-5}\,\varepsilon$. For monovalent ions, the number, of uncovered negatives at this time was 5.66×10^5. At equilibrium $E = 2.093 \times 10^{-5}\,\varepsilon$. Thus from the observed curve, at

$$t = 300, \quad \frac{dM}{dt} = 0, \quad M = 5.66 \times 10^5 \quad \text{and at} \quad t \cdot =. \infty, \quad M = 2.093 \times 10_2. \qquad (3.16)$$

In Eq. (3.13), there appear v_x and v_y. As the nature of the ions is not known, we cannot accurately fix these. At temperature equilibrium, the kinetic energy of liquid and gaseous molecules are equal. The mean square velocities of gaseous molecules range from 1×10^4 to 5×10^4 cm/sec. ALTY assumed that $v_x = 2 \times 10^4$ cm/sec and $v_y = 3 \times 10^4$ cm/sec. The error in these assumptions can only alter the magnitudes in a minor fashion. Differentiate (3.14) and insert the limiting conditions just indicated in (3.16), making $n_x = n_y = n$. Then the constants have the values

$$n = 7 \times 10^8 \qquad k = 5.96 \times 10^4 \qquad \beta = 1.5 \times 10^{10}.$$

This shows that the impurities are very few in number compared to the H^+ and OH^- ions. It is seen from k that $\sim 6 \times 10^{-4}$ of the Y^+ ions are removed/sec by impact of the surrounding molecules. Since these positive ions are furthest from the surface and thus loosely bound, it is seen that the evaporation of more closely bound X^- ions will be negligible. The *total number of negative ions* on the surface at equilibrium is $4\pi\beta/\theta = 1.85 \times 10^6$. The *total number* of *uncovered ions* at equilibrium is 2.09×10^5. Thus 89% of the adsorbed X^- ions are covered by bound positive ions, while 11% are uncovered and thus charge the surface negatively.

If the number of water molecules is 10^{15} cm^2 the very great majority of them are covered by OH$^-$ and H$^+$ ions so that few "evaporate" to give a resultant charge. This is logical since there are 10^{13} H$^+$ and OH$^-$ ions per cm^3 in the solution while $n = n_y = n_x$ is of the order of 7×10^8 per cm^3. Thus there are only $\sim 2 \times 10^6$ X$^-$ ions/cm^2 bound on the surface of which $\sim 2 \times 10^5$ are uncovered reevaporation of Y$^+$ being at 6×10^{-4} per second. In this way a definite resultant charge is given to the bubble by those negative ions that are momentarily uncovered.

Since the water used is not saturated with air, the bubble slowly dissolves. It gets smaller and the solution of air molecules should aid in removing the covering molecules. Thus the total charge is greater the faster the bubble is absorbed.

If the surface is fully formed and no gas solution occurs, the absorbtion has reached a state of dynamic equilibrium. Thus from Eq. (3.13) and (3.10) when

$$t = \infty, \quad \frac{d N_y}{dt} = 0, \quad N_x = \frac{\alpha}{\theta}, \quad \frac{n_y v_y \theta}{\sqrt{6\pi}} (N_x - N_y) = k N_y.$$

Thus

$$N_y = \frac{\alpha/\theta}{1 + \dfrac{k \sqrt{6\pi}}{n_y v_y \theta}}.$$

The equilibrium charge is then

$$M = 4\pi r^2 (N_x - N_y) = \frac{4\pi \beta}{\theta} \left(1 - \frac{1}{1 + \dfrac{k \sqrt{6\pi}}{n_y v_y \theta}} \right) = \frac{4\pi \beta/\theta}{1 + \dfrac{n_y v_y \theta}{k \sqrt{6\pi}}}. \tag{3.17}$$

This is the charge when air is not being absorbed by the water. The gas molecules during absorbtion pass through the surface in a continuous stream. More positive ions will be removed by collisions with these molecules. The number removed depend on the number of molecules passing through and on the number of ions present. The rate of absorbtion of the bubble is $\dfrac{d (\text{Vol})}{dt} = 4\pi r^2 \dfrac{dr}{dt}$. Therefore the number of air molecules passing through 1 cm^2/sec is dr/dt.

$$\frac{d N_y}{dt} = \frac{n_y v_y \theta}{\sqrt{6\pi}} (N_x - N_y) - k N_y - C N_y \frac{dr}{dt}.$$

with C a constant. In equilibrium

$$\frac{d N_y}{dt} = 0 \quad \text{and} \quad N_x = \frac{\alpha}{\theta}.$$

Therefore

$$N_y = \frac{\alpha/\theta}{1 + \dfrac{\left(k + C \dfrac{dr}{dt}\right) \sqrt{6\pi}}{n_y v_y \theta}}.$$

and the charge per unit area is

$$N_x - N_y = \frac{\alpha}{\theta} \left\{ 1 - \frac{1}{1 + \dfrac{\left(k + C \dfrac{dr}{dt}\right) \sqrt{6\pi}}{n_y v_y \theta}} \right\}$$

$$= \frac{\alpha}{\theta} \left\{ \frac{1}{1 + \dfrac{n_y v_y \theta}{\left(k + C \dfrac{dr}{dt}\right) \sqrt{6\pi}}} \right\}.$$

The total charge is

$$M = \frac{4\pi\beta}{\theta}\left\{\frac{1}{1 + \dfrac{n_y v_y \theta}{\left(k + c\dfrac{dr}{dt}\right)\sqrt{6\pi}}}\right\} \tag{3.18}$$

but

$$M_0 = \frac{4\pi\beta}{\theta}\left\{\frac{1}{1 + \dfrac{n_y v_y \theta}{k\sqrt{6\pi}}}\right\}$$

which equals the total charge when adsorbtion is 0. Then

$$\begin{aligned}\frac{1}{M - M_0} &= \frac{(k\sqrt{6\pi} + n_y v_y \theta)}{4\pi n_y v_y \beta} + \frac{(k\sqrt{6\pi} + n_y v_y \theta)^2}{(96\pi^3)^{\frac{1}{2}} n_y v_y \beta C}\frac{dr}{dt}\\ &= \left(\frac{k\sqrt{6\pi} + n_y v_y \theta}{4\pi n_y v_y \beta}\right)\left(1 + \frac{k\sqrt{6\pi} + n_y v_y \theta}{\sqrt{6\pi}\,C}\frac{dr}{dt}\right)\end{aligned} \tag{3.19}$$

$$\frac{1}{M - M_0} \equiv A + B\frac{dr}{dt}$$

with A and B constant as long as n, v and K are constant. Thus experimentally $\frac{1}{M - M_0}$ plotted against $\frac{dr}{dt}$ should be a straight line. Experimentally a 0.5 cm diameter bubble was used in air saturated water. This gave M_0. The same sized bubble was then studied in water unsaturated by air. With various degrees of unsaturation M could be measured for various values of dt/dr experimentally measured. For a range of values of $\frac{dt}{dr}$ from 25 to 60 with $\frac{1}{M - M_0}$ varying from 5 to 12.5 the points fell on a nice straight line. This appears to strengthen ALTY's theory.

If the bubble gets too small, the coverage is more complete and the *uncovered* charge decreases as r decreases below a critical limit which lies near 0.033 cm. As long as the conductivity of the water is low around 2×10^{-6} mho and preferably around 1×10^{-6} mho the charging time is around 200 seconds as indicated in Fig. 13. If conductivity increases to 6×10^{-6} mho the time becomes very short. The uncovered charge on the bubbles above the critical size is thus $2\times10^{+5}$ electrons or 1×10^{-4} esu, or 3.3×10^{-14} coulomb. At 0.033 cm radius this is a charge density of the order of 1.5×10^7 electrons/cm² and less as the bubble increases in size, far less than estimated by MCTAGGART on the basis of his measurement.

ALTY points out that as the bubble has a resultant charge, the theory of the Helmholtz double layer may not be applied. Nevertheless the electrokinetic potential akin to ζ may be computed in consequence of the concentration of the adsorbed ions on the bubble surface. Let n_1 and n_2 represent the ion concentrations in regions 1 and 2, n_2 at the bubble surface and n_1 in the interior of the liquid. Let the work to move an ion from the covering position to the interior be W. If the potential difference be ζ then $W = \frac{\zeta e}{300}$ ergs. Since the ions are in thermal motion $n_1 = n_2 e^{-W/kT}$. Taking $e = 4.774\times10^{-10}$, $k = 1.372\times10^{-16}$, $T = 300°$ K, $\frac{e}{kT} = 38.65$. Then $\log\frac{n_1}{n_2} = -38.65\,\zeta$ and $\zeta = \frac{38.67}{1}\log\frac{n_2}{n_1}$. Now the ion concentration of impurity causing the potential was 7×10^8 ions/cm³, which fixes n_1. The concentration n_2 represents only the uncovered negative ions. The number of ions on the bubble as derived from its charge E in one instance was 2.09×10^5, the area of that bubble was 5×10^{-2} cm^2. The number of attached uncovered ions per cm$^2 = 4.18 \times 10^6$. The average distance between ions on the surface is $d_{ave} = \left(\dfrac{1}{4.18 \times 10^6}\right)^{\frac{1}{2}} = 4.89 \times 10^{-4}$ cm. The volume ion density n_2 equivalent to this is $\left(\dfrac{1}{4.89 \times 10^{-4}}\right)^3 = 8.55 \times 10^9$.

Applying these data to the Boltzmann equation above $\zeta = 0.064$ volts which is in sensible agreement with MCTAGGART's and other evaluations.

Now the question arises as to why only the impurity ions contribute to ζ and the double layer. There are 10^{13} OH$^-$ ions/cm^3 in the pure water and only 7×10^8 ions/cm^3 of ionic impurity to some 3×10^{22} molecules of water. The charging rate of the droplet dipole layer by the *OH$^-$ ions* would be 1/100 sec. ALTY sets the number of surface H$_2$O molecules as 10^{15} per cm^2, of which most are probably are oriented at the surface as later work shows. ALTY believed that the binding forces, or linkages between the OH$^-$ and H$^+$ in the body of the liquid and the oriented dipoles are such that the greater proportion of the positive ends of these dipoles are not available for the negative impurity ions of which he assumes there are only 2×10^6 per cm^2 linked to the dipoles. Of these, 89% are covered by their positive opposites, there being a rate of "evaporation" of covering ions of 6×10^{-4} per second at equilibrium leaving only about 1×10^5 ions uncovered. Obviously the population of the double layer by negative ions other than OH$^-$, and the degree of coverage fixing ζ and other properties will vary with the valence, nature, and concentration of impurities and for liquids other than water will depend on the dielectric constant, the concentration and nature of impurities. This is no doubt the reason why HARPER with very pure and highly electrically resistant liquids found *no electrification* on bubbling. It is now also clear that COEHN's liquids must have been rather conducting with large numbers of ions. This is clearly indicated by the near equality of the electrical currents observed in all of his liquids. In the case of his organic liquids, it is probable that many of the relatively weaker dipoles must have picked up negative impurity ions and that the degree of coverage must have been correspondingly less to have observed the endosmosis which he did. Here he had no OH$^-$, H$^+$, and hydrogen bond linkages to take up the dipoles. There is now also little trouble in envisaging the character of the double layers involved in spraying and bubbling and in correlating these in a rough measure with LENARD's estimates from drop size.

4. The total potential difference at the water-air interface

Probably one of the most competent investigations of the double layer at the gas-water interface was carried out by CHALMERS and PASQUILL[56]. Considering that MCTAGGART measured only the electrokinetic potential ζ with his bubbles and that FRUMKIN's values were only relative to a standard interface, these workers attempted to get the true total potential difference across the double layer. They considered an inactive hygroscopic neutral substance of rough surface saturated with water as in principle representing the *interior* of a water surface. Then an opposed surface of a water droplet with its water-air boundary should give the same potential difference as between the inside and outside of the interface. They used the Kelvin water dropper technique in which drops fell from a nozzle at regular intervals into a receiving cup in which their charge could be measured. Owing to the field between the wet filter paper and the outside of the forming drop, the drops charge positively as they leave the grounded metal dropper. By applying to the water a potentiometer and dropper and adjusting the potential to zero charge on the drops separated, the potential between drop and filter paper is established. There is, however, a chain of potentials involved since metals are needed for contacts. The zero charge occurred at a potential of 0.26 volts. The chain of potentials is indicated by subscripts to the symbol V, cu for a Cu water container, f for filter paper, w for water, a for air, etc. Observation showed that

$$V_{wcu} + V_{cuf} + V_{fa} + V_{aw} = -0.26 \text{ volts}.$$

Putting a wet filter paper between the dropper and tube gave no change in V,

$$V_{wcu} + V_{cuw} + V_{wf} + V_{fcu} + V_{fa} + V_{aw} = -0.26.$$

Therefore

$$V_{wcu} + V_{cuf} + V_{fw} = 0.$$

Then Zn replaced Cu as the metal

$$V_{wzn} + V_{znf} + V_{fa} + V_{aw} = -0.26.$$

Thus as V_{wcu} and V_{wzn} are not the same

$$V_{wcu} + V_{cuf} = V_{wzn} + V_{znf}$$

whence

$$V_{wcu} = -V_{cuf} \quad \text{and} \quad V_{wzn} = -V_{znf}$$

whence $V_{fw} = 0$ which was anticipated. Thus $V_{fa} = 0$ and $V_{aw} = -0.26$ volts. A very careful check was made of all sources of error.

1. Wet cotton, wet silk, wet glass wool, wet asbestos, wool or paper and wet BaSO$_4$ replaced filter paper with no change in value. Using water running down the tube instead of filter paper, etc., gave *no charge* as was the case for a wet, smooth tube. Such things as drop size, type of material of nozzle, height of fall, time of formation of drop had no influence.

In view of the surface potential difference these workers proceeded to calculate the number of oriented dipoles in order to give this potential difference. Since we consider the surface as a plane parallel condenser of capacity $C = A/4\pi d$ where d is the dipole separation and since the charge density is σ so that $q = \sigma A$ and with $A = 1$ cm^2 $q = \sigma$ then potential $V = q/c = 4\pi \sigma d$ in esu, whence in esu $V = \dfrac{0.26}{300}$. Thus $\dfrac{0.26}{300} = 4\pi\sigma d$, $\sigma d = 6.9 \times 10^{-5}$ esu $\times$ cm/cm^2. The dipole moment of H$_2$O is 1.9×10^{-18} esu $\times$ cm. This gives $n\sigma = 3.6 \times 10^{13}$ H$_2$O dipoles per cm^2 of surface. With each water molecule occupying $\sim 9 \times 10^{-16}$ cm^2 so that $n_w = 1.1 \times 10^{15}$ assuming the surface completely smooth. This indicates that one in 30 of the dipoles is oriented normal to the surface. This is less perhaps than ALTY's picture calls for but is reasonable considering surface roughness.

C. The relative potentials in double layers of water in relation to surface tension for salt, acid and organic solutes

Prior to the later work of CHALMERS and PASQUILL, but after McTAGGART's study, the Physical Chemist FRUMKIN [55] measured the relative potential differences between various aqueous solutions and either a standard dilute solution of H_2SO_4 or KCl. The technique was the water dropper technique initiated by KENDRICK, or the use of a Pt wire coated with ionium. The purpose of the study was to relate the liquid-air interface potentials to the constitution of the surface layer in the light of GIBBS's relation between change of surface tension and the ionic concentrations in the surface layer. The work preceded the later papers of ALTY and the remarkable contribution of this work to that of ALTY and CHALMERS in clarifying and completing the picture of the surface layer has remained unrecognized until today.

FRUMKIN used the surface of a liquid, (one of those to be compared), flowing down the inside of a 2 cm diameter tube of glass. Down the axis of this outer tube fell the stream of water from a glass nozzle so that it broke into drops 4 cm from the glass tip. It was immaterial which liquid was inside and which outside, except for reversal of the sign of charge. Usually for inorganic substances, the outer flowing layer was 0.01 N H_2SO_4 or KCl. The connections of reservoirs of the two liquids were achieved to ground through salt bridges and normal calomel electrodes. The receiving beaker for the drops was connected by a salt bridge to a normal calomel electrode and this to the unearthed quadrants of an electrometer, the other pair of quadrants being earthed. Measured was the potential of the drops broken off in the field between the liquid surfaces. When the dropper liquid as well as the outside stream was 0.01 N H_2SO_4, the potential should have been zero but was often observed to be 1.5 millivolts above or below. This reading was checked in any series of measurements and constituted a zero. With solutions of other salts, etc., the potentials were usually read in the tens of millivolts. They were highly reproducible. Distilled water was used in making up solutions. Solutions used were in the range of 0.01 Normal up to 4 or 5 N. The diffusion potentials from the salt bridges caused some trouble if they were not watched.

It is noted that what was being measured was *not* the *total potential* as with CHALMERS but the electrokinetic potential ζ of McTAGGART. Thus in the light of ALTY's work it is clear that the investigations dealt with the occupation of uncovered water dipole ends by anions and their coverage by the cations. This as we shall see, except for the very concentrated solutions and for the very actively surface adsorbed organic species which lower surface tension is exactly what FRUMKIN's data reveal.

Solutions of 1 or 2 moles/liter for equal concentration of anions were compared.

All salts except ammonium salts, charged the outer surface of the water negatively to that of the standard dilute acid surface. This is just the effect that would be expected if the occupation of positive dipole ends by the negative anions were affected by characteristics of the anions and the concentrations thereof. The effect of the associated cations on coverage cannot always be ignored. The order of effectiveness of various anions follows the sequence below, with the most negative charge being somewhat under 100 millivolts, generally below 70, and running down to 1 or 2 millivolts which is the unit to be used in all further discussions. It will be remembered that McTAGGART's ALTY's potentials ζ were of the order of 55 in those studies with pure but slightly contaminated distilled water. The order is:

$$CNS^- > ClO_4^- > ClO_3^- > I^- > MnO_4^- > NO_3^-, \; CNO^-, > Br^- > BrO_3^- > CN^-.$$

For the divalent anions $Fe(CN)_5NO^= > S_2O_8^= > CrO_4^= > S_2O_3^= > S^= > CO_3^=$.

Thus KCNS ranged between -57 and -87 for 1 and 2 moles/l, while NaCl gave -1 and -4 millivolts respectively.

Using the same anion, K^+, Na^+, Rb^+, Cs^+, Ca^{++} and Ba^{++} act about the same within limits of error. Li^+, Ag^+, and Zn^{++} yield somewhat higher potential. CdI_2 falls outside of the series. The acids charge more strongly negative than the corresponding salts for equal concentrations of the solute in the bulk of the liquid.

The negative charge increases with concentration from 0.1 to 10 N KCNS goes up to -200 millivolts. On the other hand, KF yields a positive charge.

The theoretical explanation is very interesting in connection with ALTY's theory of surface structure. GIBBS's has shown that roughly, the change in concentration of a solute in the surface layer of a liquid W is related to c, the concentration in the body of the liquid by

$$W = -\frac{c}{RT}\frac{d\gamma}{dc}. \tag{3.20}$$

Here γ is the surface tension of the liquid and RT represents the energy of agitation of the molecules. This can be simplified to read

$$RTW = \frac{c}{2}(\gamma_0 - \gamma), \tag{3.21}$$

where γ_0 is the surface tension of the liquid and γ is that of the solution. For salts which *increase surface tension* W is *less* than 0, e.g., salts are forced out of the surface. Thus $\gamma > \gamma_0$. This difference in $\gamma - \gamma_0$ varies very widely in salts. In normal solutions values for $\gamma - \gamma_0$ are

KF	1.83	KNO_3	1.12	KOH	1.8	KNCS	0.9
KCl	1.46	KBr	1.06	KI	0.82		

Thus the more strongly the salt charges the surface negatively, the weaker is its negative adsorbtion. That is the stronger the charge, the more the anion has forced its way to the outer surface and stolen places from the OH^- ions at the dipole end. It is thus clear that it is the surface tension forces that are related to the dipole orientation and to the composition of the surface layer.

In fact, the largest charging effects are from the salts of strong organic acids e.g., such as $KCOOCCl_3$, potassium trichlor acetate where $\gamma_0 > \gamma$. Here charging is strongly positive, of which more later, but the potentials run as high as $+600$ millivolts. CNS^- and I^- salts *raise* γ least and penetrate while CO_3^- and F^- salts raise γ most and therefore do not penetrate to the surface.

The sequence of some of the alkali metals at first appeared to violate this law. Li salts are supposed to increase γ more than Na and these more than K.

However, the charging sequence reads $K > Na > Li$ salts. If one employ the rigorous form of GIBBS's equation reading

$$RTW = \frac{1}{2} \frac{\partial \gamma}{\partial c} \frac{1}{\dfrac{1}{c} \dfrac{\partial \log F_a}{\partial c}} \tag{3.22}$$

where F_a is the activation energy RTW calculates to -1.44, -1.40, -1.30 respectively, for the sequence KCl, NaCl, LiCl respectively and LiCl is the more adsorbed.

If concentrations become very large, they usually give small potentials of the positive sign. It is suspected that this is some sort of a salting out effect in which the dipole orientation is altered by the crowding of ions and undissociated salt molecules into the surface.

The ammonium compounds give very *strong positive* charging. For this substance and definitely for the amines γ_0 is lowered by the substances and NH_4 is forced into the surface distorting the whole orientation of dipoles.

What FRUMKIN calls auto complex formation increases the action of CdI_2 and $ZnCl_2$ relative to the alkali metal salts. These substances may hydrolyze.

The acids all gave greater charging than did salts of those acids, up to 150 milli-volts increasing from 0.1 to 1.0 normal. The order of effectiveness read

$$HClO_4 > HI > HNO_3 > HBr > HCl.$$

In this case, acids decrease γ_0, $W > 0$, and reduction follows. The change in $\gamma - \gamma_0$ for the series is

$$HI, \ HNO_3 > HBr \gg HCl$$
$$1.45 \quad 1.1 \quad\quad 0.75 \quad\quad 0.5$$

Comparing the salt and acid effects in terms of the *same concentration of adsorbed substance* (W) *in the surface layer* the acid effects are the smaller. Thus as with H_2O the presence of the H^+ in the acids renders the double layer for equal numbers of occupied spots less effective, but owing to the effect on γ_0 the concentrations of the acids are greater and thus for the same c the acids are more effective.

FRUMKIN speculates that the reason for this action may lie in the adsorbtion of undissociated molecules with the acids while with salts, ions are adsorbed. This, in principle, states just what ALTY's picture does, the acids in the surface layers have the H^+ tightly bound so that there are less uncovered anions. FRUM-KIN adds that HF and NaF show a great difference and that HF has the lowest dipole moment, e.g., smallest electron shells and less dipole separation. Actually, HF has much more "coverage" than NaF.

H_2SO_4 increases γ_0 up to a certain concentration then γ goes through a maxi-mum and falls. At high concentrations molecules are adsorbed while weak H_2SO_4 at low concentrations acts like the metal sulfates in the potential series.

In the work using organic non-electrolytes, the potential-concentration curve has a form typical of an adsorbtion isotherm, at first increasing proportional to concentration and then tending towards a maximum or saturation value. For the inorganic substances, two circumstances act to modify their behaviour so that saturation is not shown. There must be repulsive forces between adsorbed ions. Thus the adsorbed amount increases more slowly than the concentration and will not act so as to approach a limit. This will occur in greater measure the better the salt is adsorbed, for example the potential increase shown by concentrated KCNS. The second action is the increase in activity coefficient of dissolved electrolytes especially the acids and strongly hydrated salts. The latter causes a rapid increase in the adsorbed quantity, i.e., the anions (or the whole molecules), are "salted out" in concentrated solutions.

The anion sequence of CNS^- to F^- agrees generally with the lyotropic series of FREUNDLICH who arranges certain anions as increasing hydrophiles, i.e. as having increased affinity for water. The increasing affinity towards water of the series $F^- > Cl^- > Br^- > I^-$ has been indicated by the direct calculation of

Table 5. *Data of* FRUMKIN *on the contact potentials of various dissolved substances in aqueous solutions against a standard 0.01 N H_2SO_4 solution*

Subs				
Formic acid HCOOH:				
Conc. (normal)	0.32	1.00	2.5	10.6
Potential mV	− 0.22	− 43	− 61	− 85.5
Acetic acid CH_3COOH:				
Conc.	6.10	15.54		
Potential	+ 2.74	+ 312		
Sodium acetate CH_3COONa:				
Conc.	0.17	2	3	
Potential	− 1	− 1.5	+ 0.2	
Propionic acid C_2H_5COOH:				
Conc.	0.017	4.6		
Potential	+ 5.5	+ 271		
Chloracetic acid $CH_2ClCOOH$:				
Conc.	0.13	0.33	2.6	
Potential	+ 0.50	+ 0.93	+ 126	

hydration energy, by FAJANS. I^- has less affinity for water than F^-, thus I^- goes to the surface and F^- is much less "uncovered" in ALTY's terminology than I^-. The adsorbability of divalent anions is much less than the analogous monovalent ions.

FRUMKIN tried to evaluate the surface potential of the water-air interface. His attempts were unsuccessful as he could not eliminate the glass walls. The difficulties can be understood be reference to CHALMERS and PASQUILL's measurement.

The study of organic substances required some changes to be made in techniques. Sometimes, the solution to be tested was run outside while the standard 0.1 N H_2SO_4 surface was used in the dropper as the potential did not build up in the dropper with sufficient speed. This, in fact, was indicated in ALTY's work to be a problem with weak ionization. However, with the larger concentrations used here, time lags could be of the order of 10^{-2} sec only. They were longer with some organic substances. The non-volatile substances yield definite potentials. At times, 0.001 to 0.01 H_2SO_4, KCl or HCl had to be added to the solutions of organics to render them sufficiently conducting. A few sample data are given in Table 5.

The potential, approximately ζ, follows the rule for a Langmuir adsorbtion isotherm in the form of

$$\zeta = \frac{A\,c}{1 + B\,c} \tag{3.23}$$

very closely for many substances.

The charging is symbiotic with the capillary activity of the substance. Substances that little affect γ_0 such as oxalic acid, glycerine or urea charge only weakly and only in concentrated solution.

As the number of carbon atoms in the aliphatic chain, as represented by the number of CH_2 in R increases, we get rises at always smaller and smaller concentrations. A series of acids increases in activity in forming the saturated layer by a factor of 3.65 for each CH_2 added to the chain in R. This is closely analogous to the doubling of toxicity in biological studies of each alcohol in a series for each CH_2 added no doubt for the same reason. TRAUBE's measurements indicate an increase in γ by a factor of 3.2. per CH_2 added to the R group. This even holds for the series from NH_3 to $N(C_2H_5)_3$. Thus in the series NH_4Cl, $NH_3C_2H_5Cl$, $NH_2(C_2H_5)_2Cl$, $NH(C_2H_5)_3Cl$, and $N(C_2H_5)_4Cl$, the first amines have their R groups forced outward of the surface and are increasingly active. The last one, however, is symmetrical and is forced downward out of the surface.

Table 6. FRUMKIN's *proof that charging and* ζ *potential parallels surface absorbtion*

c	ζ	$10^{10}\,W$	$10^{-10}\,c/W$
0.0069	$+\ 91.5$	0.68	134
0.069	$+390$	2.90	136
0.345	$+565$	4.08	139
0.691	$+573$	4.29	134

In general, the charging can be calculated as an unequivocal measure of adsorbtion. This is shown in Table 6.

It is clear from these studies that for substances as strongly adsorbed in the surface and containing large groupings of molecules the whole double layer normal to water is altered. The reversal of potentials from negative to positive indicates that the normal dipole surface structure of H_2O molecules is altered.

The magnitude of the potentials exceeding the normal potential of the uncontaminated water double layer of negative -260 millivolts goes up to as much as positive 600 millivolts. Quite independently of ALTY, FRUMKIN is led to speculate on the surface structure of the double layers. In this speculation, he was guided by the classical studies of I. LANGMUIR and W. D. HARKINS of 1917 and 1918, on adsorbtion on surfaces. It is above all, clear that the R groups of hydrocarbon chains are hydrophobic and are pressumably forced *outside* the aqueous surface. The surface arrangement is illustrated in Fig. 14.

Chlorinating radicals reduces $+$ charge yielding $-$, acetic acid $+ 285$, monochlor acetic -150, dichlor acetic -280, trichlor acetic -600. If H^+ is replaced by a metal here the effect is still stronger. In amines R can be replaced by H with out much change.

From what has gone before, it is seen that the data here confirm and extend ALTY's picture. They add some more data of a semi-quantitative character in that they indicate that the anion layer attached to the dipoles is influenced by the action of surface tension forces.

According to the extent by which the solute alters γ and thus its anion concentration at the surface the negative potential ζ is altered in the measure that more "uncovered" anions appear. Observations of the action of acids and salts as well as specific ions substantiate ALTY's model. The behaviour of very strong solutions ~ 4 N as well as of strongly adsorbed substances appears to disrupt the normal dipole orientation forming new surfaces of varied polarity and very high potentials. These result from the water repulsive character of the R groups and

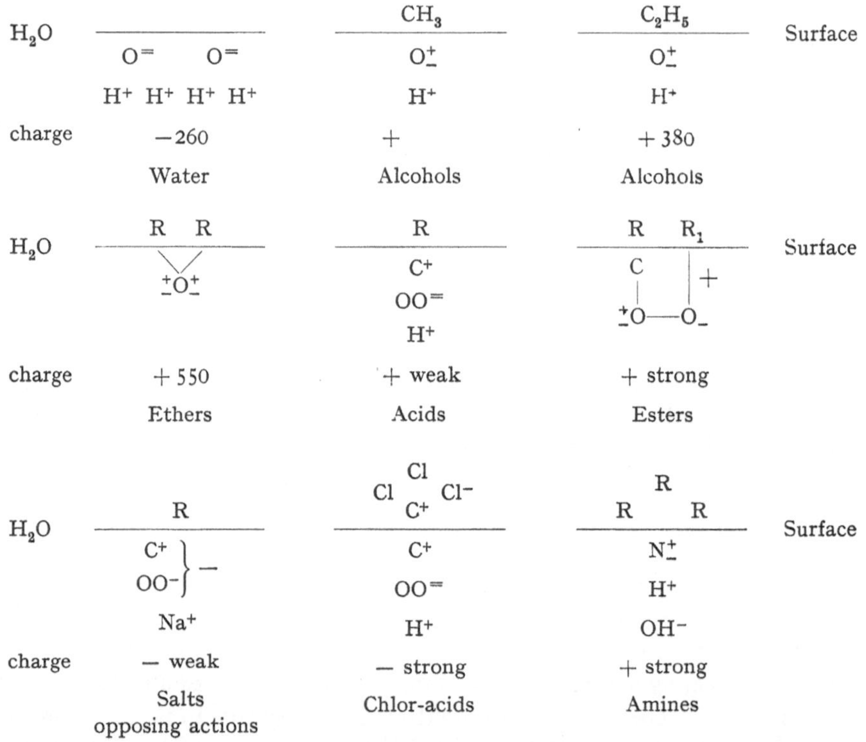

				CH$_3$		C$_2$H$_5$		
H$_2$O				—		—		Surface
	O=	O=		O$_\pm$		O$_\pm$		
	H$^+$ H$^+$ H$^+$ H$^+$			H$^+$		H$^+$		
charge		-260		$+$		$+380$		
	Water			Alcohols		Alcohols		

	R	R		R		R	R$_1$	
H$_2$O	—			—		—		Surface
	$_\pm$O$_\pm$			C$^+$		C	$+$	
				OO=		$_\pm$O—O$_-$		
				H$^+$				
charge	$+550$			$+$ weak		$+$ strong		
	Ethers			Acids		Esters		

				Cl			R	
			Cl	Cl$^-$		R	R	
H$_2$O	R			C$^+$		—		Surface
	C$^+$ $\}$			C$^+$		N$_\pm$		
	OO-$\}$ $-$			OO=		H$^+$		
	Na$^+$			H$^+$		OH$^-$		
charge	$-$ weak			$-$ strong		$+$ strong		
	Salts			Chlor-acids		Amines		
	opposing actions							

Fig. 14. FRUMKIN's diagrams illustrating orientation of various type solutions at an aqueous-air interface

the strong dipole moments of the hydrophillic segments of the molecules. No careful investigation of these have been made in the spray electrification studies, the nearest being the study of Na stearate. The very peculiar action shown by ammonium salts and fluorides in the Workman-Reynolds freezing potential studies is brought clearly into view in connection with this work. The NH$_4$ salts are strongly attracted into the water *interface* and give it a positive charge.

With this insight into the nature of the double layer and its properties it is possible to turn to the spray electrification studies.

D. The spray electrification phenomena

1. The studies of LENARD and his school

It is first best to review the early observations of LENARD's laboratory and their final rectification by his student BÜSSE in 1925[65]. As indicated LENARD[12] had in 1892 followed up the observations of ELSTER and GEITEL[54] of two years earlier on electrification in water falls. He had first observed the waterfalls and then carried out further experiments in a shower bath. There he confirmed the fact that the air in the neighborhood of finely dispersed water became negatively charged and that the residue of larger drops and water surface were predominantly positively charged. He observed that the effect was associated with breaking up of water. He found that the spray created near the water surface by falling drops had a positive charge but the finer spray in general was negative. He also observed that salts decreased the charging. Initially he ascribed the effect to an electrical double layer at the surface of the water with the air taking the negative sign and the water the positive sign. Various students carried the work forward in the intervening years—notably E. ASELMAN[66] and on the basis of this and other work as well as the work of COEHN and MOZER, LENARD published a rather lengthy summary in booklet form. Much of the substance of that book was later published as an extensive summary in a paper by LENARD[67] in 1915. LENARD's theoretical physical background was notoriously weak as his controversies on relativity and other matters amply showed. Thus the theoretical attempts to arrive at a distribution of molecules in surface layers of a liquid like water, based on the relative sizes of solute and solvent molecules, solute molecules being augmented by hydration appear naive today. In contrast, his experimental findings and his general conclusions were in the main sound. Unfortunately, neither he nor ASELMAN used an ion measuring device of sufficient resolving power or range to detect the various ion types and sizes of carriers created so that the later work of his student BÜSSE[65], in 1925, was needed to correct some of the picture. However, it took the advanced techniques of 1938 to get a more complete picture of the carriers produced in spray electrification. And again it is only today that studies of KUNKEL and DODD and others have thrown light on the homogeneous charging which overlaps the legitimate electrification process of LENARD.

Finally, the failure of non-polar liquids to reveal any sign of charging also required the modern techniques as applied recently by HARPER[61].

The conclusions of LENARD in the 1915 paper may be summed up as follows:

(a) The electrical double layer resides wholly below the liquid surface.

(b) The interfacial tension and thus the influence of gas on charging are unaltered by the nature of clean gases.

(c) There is no frictional electrification between gas and water, i.e., gas blown along a liquid surface that does not tear off droplets or carry off spray from bubbling, etc., is uncharged.

(d) The *collapse* of a liquid-air surface causes no electrification unless it disrupts the water surface.

(e) Water may be divided and disturbed with no electrification unless the surface is broken up into fine drops, such as by impact on surfaces, atomization by a sprayer, or bubbling.

(f) The electrification is produced by the detachment of the smallest particles from the surface, and these particles must come from the thinnest surface layers. This action was corroborated by "schlieren" photographs taken with spark flashes. Then visible spray droplets from 0.17 to 0.00067 mm diameter could be observed. He indicated that this occurs with bubbling but that the very largest bubbles appeared least efficient.

(g) He had started investigating the spray phenomenon in connection with lighthning discharges and was greatly stimulated by SIMPSON's studies in this direction. He studied the effects of updrafts and found shattering of drops at relative velocities in excess of 8 m/sec for drops the size of raindrops. Spark flash photographs showed how these drops broke up. The peculiar ring type of break up in uniform non-turbulent air caused *no* electrification by spraying. He thus had to assume that spray electrification in thunderstorms could not occur by coalescence or break up unless turbulence was present, a fact later corroborated by CHAPMAN [68].

(h) He concluded that waterfall carriers have large nuclei and are not air ions. This was owing to his inadequate measuring techniques.

(i) In agreement with COEHN and MOZER [57], he found that different gases only differed in electrifying effects in proportion to their mechanical interaction with water surfaces, e.g., their density. This confirmed him in the belief that electrification was a mechanical effect dealing with break up of the water surface.

(j) He considered the character of the double layer in relation to his surface force theory but the considerations lead to no conclusion.

(k) He found that the most prolific source of charge generation was with very pure water and by use of the 90° air blast atomizer. In this fashion, he was able to separate 7×10^{-10} coulomb per gram of water atomized as a fine negative mist in the air, the equivalent positive charge remaining on the atomizer or falling into the basin as larger droplets.

(l) From the diameter of the negative droplets which he claimed averaged about 8×10^{-7} cm and the fact that he got droplets to the extent 1% having a diameter 15×10^{-7} cm, he assumed that the water surface was stratified as follows:

1. Outer layer $\sim 8 \times 10^{-7}$ cm excess negative charge.
2. An intervening layer containing cations of an electrolytic solute.
3. The positive ions of the aqueous dipole layer.
4. The anions of the electrolytic solute.

Here he was in error as ALTY's considerations show.

(m) He attempted to prove the negative character of the outer layer through the photoelectric emission studies of his student, W. OBOLENSKY [69] who studied changes in photoemission for clean H_2O and for solutions. These are obviously not very enlightening.

(n) He indicated that the smallest particles were negative. That the next sized water droplets were positive and in electrical fields could be extracted showing the Na flame test if NaCl were present in solution while the negative spray did not show this. This confirmed the presence of an excess of Na^+ ions in the inner portion of the double layer. Larger droplets showed no net charge. These experiments mostly those of ASELMAN, were not too clear cut or decisive. K. DYK observed the phenomena in the author's laboratory in about 1931.

(o) LENARD also noted that for each salt the sign of the net charge of the spray changed at some given high concentration. The reversal was attributed to the inclusion of cations of electrolyte in the double layer. This would represent an excess of covering cations in consequence of penetrations of those ions into the layer. The inversion concentration depended on the vigour of disruption in the air blast, being higher the smaller the spray particles produced. The inversion depended on the ion type. COEHN and MOZER had observed that this varied with the valence of the cation. Actually, it depends on the action of the ions in increasing the surface tension as FRUMKIN has shown.

(p) Based on the inversion concentration and assuming that this required two cations for each negative ion of the outer layer, LENARD calculated that there must be 1 extra electron or negative ion at the surface of the double layer for 12000 H_2O molecules. At 10^{15} water molecules/cm^2 surface, this means 8×10^{10} ions per cm^2. This would give a surface charge density 40 esu or 1.3×10^{-8} coulombs per cm^2. It is noted that the number of anions in the surface layer of pure water as computed by ALTY was 1.9×10^6 per cm^2.

The concept of LENARD is here at fault in that he assumed that inversion meant bringing in excess cations to neutralize the electrons in the double layer. Actually, inversion in terms of ALTY's theory lies in increasing the external cation concentration until all anions are covered and possibly altering the surface layer of oriented dipoles by invasion of neutral salt particles to achieve reversal of charge. Hence, density of free anions available for spray electrification measurements cannot be calculated in this fashion.

According to LENARD, separation by spraying led to separation of only 2.1×10^{-12} coulomb/cm^2 surface or 1.3×10^7 electrons per cm^2. Thus if this estimate is correctly based on the amount of surface disrupted in finer droplets, the uncovered surface anion density for impure water is 50 times that calculated by ALTY from cataphoresis of gas bubbles in very pure water. This is a reasonable result, though LENARD's estimate of surface rupture involved in spraying 1 cm^3 of water could be badly in error.

In 1925, BÜSSE[65] carefully repeated the earlier work putting greater emphasis on using a good technique for getting the mobility spectra of all carriers. In this work, he copied the coaxial cylindrical tube of O. BLACKWOOD[70], whose thesis was carried out in Chicago under MILLIKAN with the author's immediate supervision. With his careful control and, for that time, superior technique, BLACKWOOD had found that spray electrified ions did not appear to have groups of distinct mobilities as reported by the NOLANS. BÜSSE carefully repeated the studies and found in agreement with BLACKWOOD for the slower ions which he could measure, *that there were no groups*, only a continuous distribution with maxima at a certain size both for positive and for negative ions. He used really pure water, having 1×10^{-6} mho conductivity. In this way, he got a total charge of one sign of the order of 2×10^{-9} coulomb/cm^3 of water using a sprayer with an integrated excess of $5-6 \times 10^{-10}$ coulombs/cm^3 of negative ions over positive ions in sensible agreement with LENARD's value of 7×10^{-10}. In this study, he rectified an error reported by LENARD, ASELMAN and COEHN and MOZER that with pure water and pure liquids, only negative charges were observed. This error was caused by their use of the earlier inferior ion mobility measuring techniques. This observation of BÜSSE's was confirmed in later studies, but merits considerable explanation only possible now.

BÜSSE's resolving power for mobilities was not adequate for the faster ions, and thus he failed to note the groups of ions later observed by CHAPMAN[60] in the writer's laboratory. Otherwise, his data are in good agreement with CHAPMAN's observations. BÜSSE further showed that negative carriers, whether produced either by impact on surfaces, spraying, or bubbling were always accompanied with *nearly* as many positive carriers, while others had only observed the net difference in the form of the excess negative ions in the fine spray, which for pure H_2O were 0.10 of the net carrier production. Faced with this fact and LENARD's theory, BÜSSE explained the presence of positive ions to droplets formed from fragments of the surface of greater depth after the fine negative particles had been torn off and before the double layer could reform. In support of this, he indicated that the time of relaxation in formation of a full surface tension field is of the order of 10^{-2} sec while the velocity of surface breakup in some of the spraying studies involved times of 10^{-4} sec. ALTY's calculations show this to be a correct inference. In how far it influenced BÜSSE's observations, one cannot say, for certainly with salt solutions and other than pure distilled water, the droplets observed by BÜSSE were influenced by symmetrical charging as well. He found that the positive carriers in stronger than 1% NaCl, ($\sim$0.3 N), were not mostly droplets carrying Na^+ ions. He also observed that Na was present equally in positive and negative sprays. This indicates that some Na was present as NaCl effectively undissociated in some of the droplets with uncovered Cl^- ions, in agreement with FRUMKIN. Below 0.2% of NaCl the Na^+ ion content of the negative ion spray was less than for the positive spray. At higher concentration, he correctly believed that there was more non-dissociable NaCl in all droplets.

BÜSSE also studied ion generation in a bubbler. At 0.2% NaCl the slowest carriers with closed bubbler has mobilities of 6.6×10^{-3} cm²/volt sec, while with the open bubbler, he got them to be 1.8×10^{-2} cm²/volt sec. Apparently in the latter cases, carriers were evaporating, see infra. The fastest ions in both cases were of molecular magnitude, i.e., had a mobility 0.4—0.5 cm²/volt sec though these could not be evaluated accurately. They were present from 1 to 3% of all carriers. The radii of the larger carriers could be estimated from the mobility by use of the Stokes-Cunningham law. Evaporation of spray droplets was found to be of importance in the 0.3 to 0.5 sec of age studied except that for radii greater than 6×10^{-7} cm the change was slow.

From his measurements, BÜSSE concluded that the layer in which the negative charge resided must be of the order of $7 - 8 \times 10^{-7}$ cm thick i.e., bound uncovered anions, while the equivalent positive layer was 1.4 to 2.2×10^{-6} cm below the surface. These figures are akin to LENARD's and derive from the estimated size of the average droplets of each sign.

2. The studies of CHAPMAN

The work was carried forward by CHAPMAN[60] in the author's laboratory, beginning in 1936. At this time, the excellent air blast method of ERICKSON[71] was available. In this, the ions were studied in a uniform plane parallel electrode field. The ions could be collected on a narrow strip electrode, the downstream distance of which could be varied over wide ranges. Air flow was laminar. The

apparatus is shown schematically in Fig. 15. Generating conditions could be
varied at will. The carriers could be studied within time intervals at least 0.1
that of previous workers. The humidity could be controlled if desired. While
absolute values of the mobilities derived were in doubt by perhaps 10% or so,
relative values were very accurate. The theoretically calculated form for the

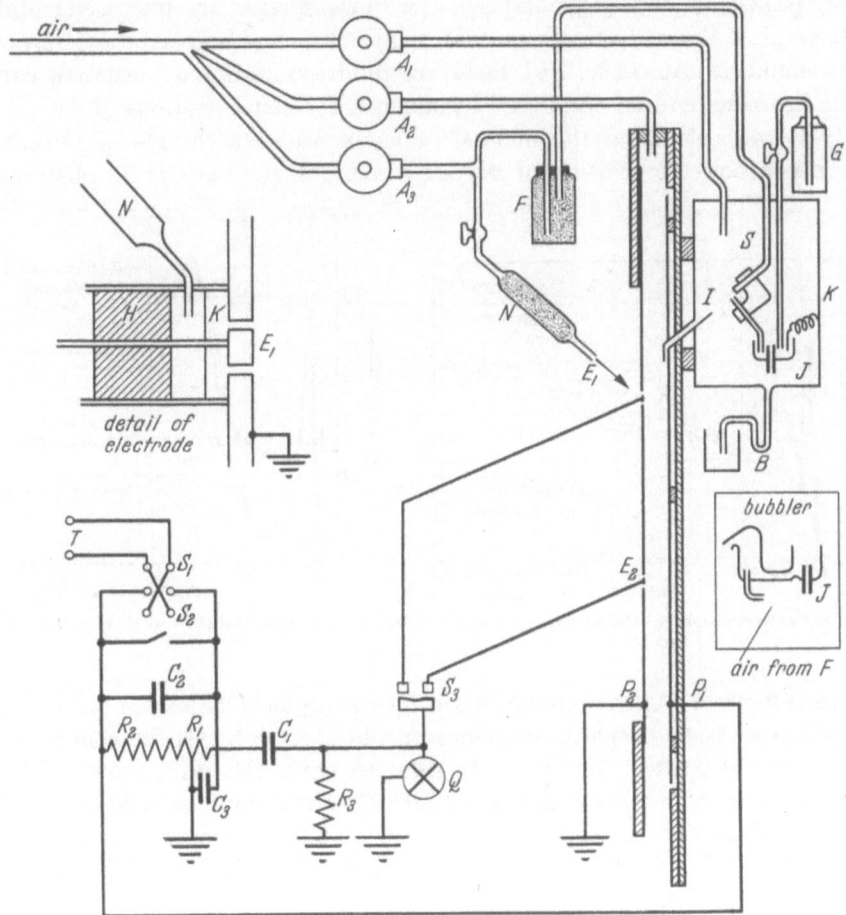

Fig. 15. Erickson ion mobility tube as adapted by CHAPMAN for the study of spray and bubble created mobility spectra

peaks of different mobilities was observed and in the last tube used mobilities
10% apart could be resolved. With proper plotting, the areas under the curves
gave the true relative numbers of carriers. The apparatus permitted mobilities
from 1.8 cm²/volt sec to be observed and intensities compared. Ions from 0.2 to
0.4 sec old could be observed. The humidity in the main blast in the measuring
tube ran from 35% up as required and that from the source varied. With the
90° sprayer, it was around 90% and with the bubbler could be reduced to around
50% or less. Temperatures ranged around 24° C. In the main tube humidity
was in general 35%.

Initial studies with a sprayer showed that *contrary* to *earlier studies* there
were *several sharp peaks representing ions of distinct mobilities at* the higher values.

Pure water showed a sharp rise of *negative* carriers at 1.8 cm²/volt sec reaching a peak at 1.6, other weaker negative peaks appeared at 0.41 and 0.26. Positive ions had peaks at 0.97 and 0.31. Such a curve is shown in Fig. 16a. Using a bubbler, the same peaks appeared much more distinctly, as seen in Fig. 16b. With the bubbling, the ratio of negative to positive carriers was in the ratio of 2.5 to 1. Addition of small amounts, 10^{-5} normal salt solutions enhanced the slower peaks and lowered the faster ones with the sprayer and increased mobilities of these ions. Around 4×10^{-4} normal, the positive and negative carriers were in equal number. At 0.2 M KCl there are more positive than negative carriers while the same effect is produced by much lower concentrations of $AlCl_3$.

CHAPMAN calculated the number of extra negative charges i.e., uncovered anions per molecule of water in the outer layer, on the basis of his observations

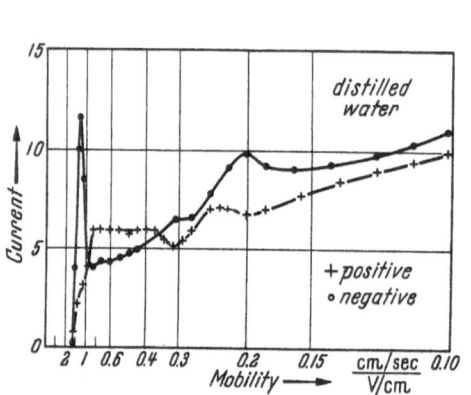

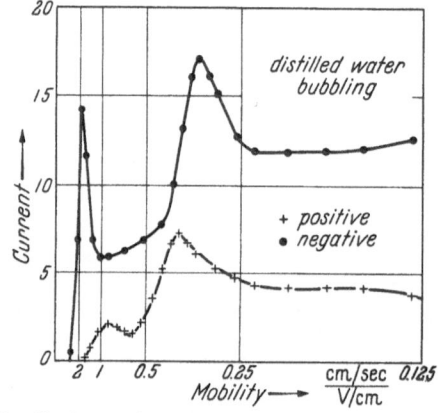

Fig. 16a. CHAPMAN's mobility curves for sprayed distilled water

Fig. 16b. CHAPMAN's mobility curves for bubbling distilled water

on the salt effect. The maximum height of the 0.5 cm²/volt sec i.e. peak for KCl for both positive and negative carriers occurs at 1×10^{-4} N. At this concentration, the number of positive ions from the salt just equal the extra negative charges torn from the outer layer. That is, at this concentration coverage of all anions is virtually complete for spraying purposes assuming of course, that this does not alter the double layer. Using MCTAGGART's[58] value of the electrokinetic potential, he calculated the separation of the double layer*. According to MCBAIN and SWAIN[72], the Gibbs adsorbtion for salt solutions, amount to a surface deficiency of but 2 molecular layers of water. This deficiency may, however, be in the vital areas as we known today. Thus the concentration of the salt in that region is that corresponding closely to that in the interior of the solution. CHAPMAN concluded that there was one extra electron (anion), for every 550000 H_2O molecules, and that the double layer separation was 2×10^{-6} cm. This figure of about 1 anion in 5×10^5 water molecules is roughly 0.02 that of LENARD's estimate and is perhaps more accurate. In terms of ALTY's theory, this involves 1.6×10^9 uncovered anions/cm². This is 10^4 times as great as ALTY's determination. It is clear that changes in salt concentration must alter the number of anions both covered and uncovered. In ALTY's and MCTAGGART's study, anion concen-

---

* This is not legitimate since that potential, while of the correct order of magnitude applied to pure water, and not to salt solutions. However, it is perhaps not far off.

trations lay around 10^8 per cm³. In the reversal of charge techniques, such as CHAPMAN's, anion densities were of the order of 10^{16} per cm³. This must *increase* the *number of bound anions* and likewise the *number of uncovered anions*. This could account for the discrepancy in the two estimates. Thus double layers are

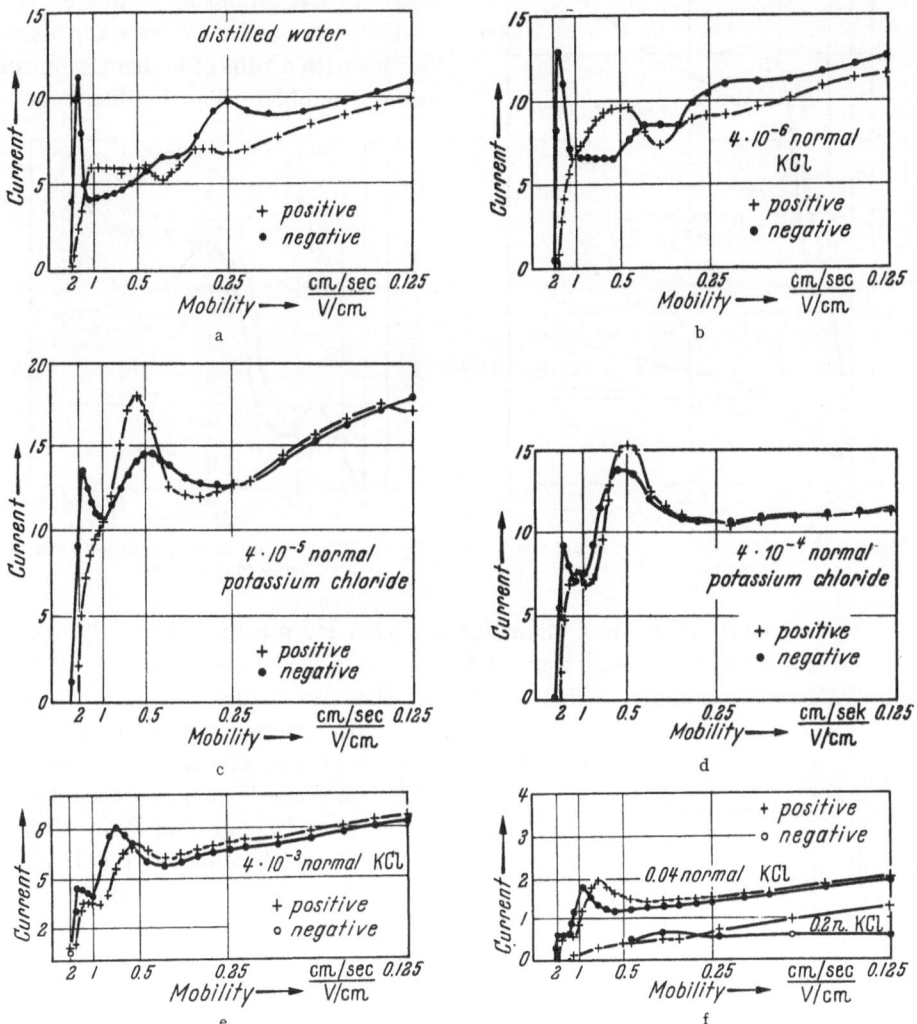

Fig. 17 a—f. CHAPMAN's curves showing the effect of different concentrations of KCl on spray electrification of water

thinner and more ions are involved. Here Coulomb forces become important, and diffusion may not be as free as in ALTY's theory.

The data on the effect of salt on the peaks with spraying is shown in the curves of Fig. 17 a—f for KCl. This figure clearly illustrates why earlier workers did not get reversal in sign of charge with salt until about 0.2 N solutions were reached, since they did not detect the faster ions and observed largely the slower ones. The effectiveness of AlCl₃ compared to KCl in reversing charge is shown in Fig. 18.

The effect of aging of ions on the spectrum of spray electrified water is seen in Fig. 19. The faster peaks have declined at the expense of the slower ones.

By reversing the directions of the sprayer so that carriers really took a long time to reach the measuring chamber, the fast peaks at 1.80 and 0.5 disappeared and mostly ions of 0.25 and greater were observed. This was partly loss by diffusion of faster carriers and partly owing to *growth* in the humid atmosphere.

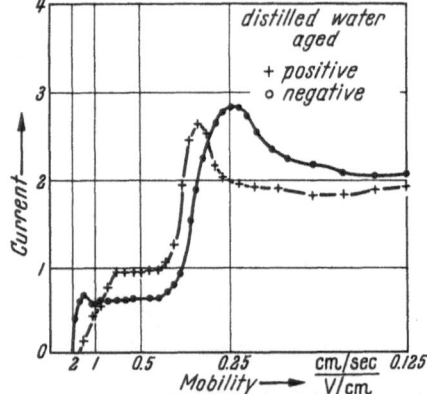

Far more significant were the observations with a tube of higher resolution using a bubbler. The bubbler mostly

Fig. 18. Effect of AlCl₃ on reversal of sign of charge compared to KCl

Fig. 19. Effect of aging on spray electrified water as observed by CHAPMAN

used had a nozzle of 0.7 mm diameter at 19 cm Hg pressure of air, though at times a tube 12 mm diameter was used. The abscissae of mobility were plotted to logarithmic scale under which conditions the area under the curves yields

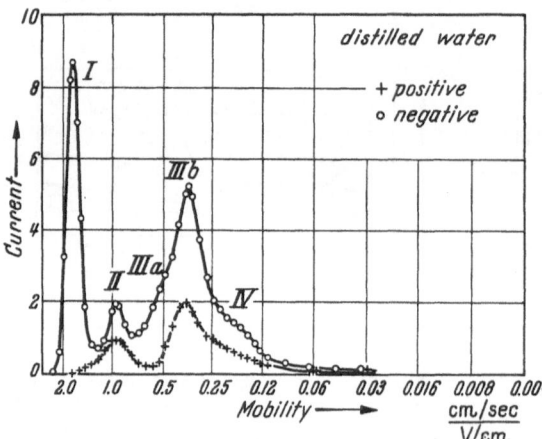

Fig. 20. CHAPMAN's curves for bubbling of pure water on improved apparatus. The logarithmic scale of plotting abscissae yield ion numbers proportional to area under the curves

the correct relative numbers of carriers. The age of the ions was of the order of 0.5 sec the relative humidity in the main stream was 35% while in the bubbler it was usually around 65%. Mostly evaporation of the ions occurred. Completely different curves were obtained as shown in Fig. 20. The spectrum consists *almost entirely of groups of carriers with very few larger ions.* The ions are designated by Roman numerals with the following mobilities all in cm² per volt sec. *Negative* I 1.8 ± 0.1, II 0.95, IIIa 0.45, IIIb 0.32, IV 0.20. *Positive* II 1.0, III 0.40. Here the negative charge is 4 times the positive charge. The effect of KCl is shown in the succession of figures represented by Fig. 21 a—c. With AlCl₃, the effect was the same except that there was more positive electrification. NaOH suppressed the II⁺ group more rapidly and enhanced II⁻ while HCl and HNO₃ did the reverse. Sugar added produced the series of changes indicated in Fig. 22a—d. The effect of violent and gentle bubbling is noted in Fig. 23. In the gentle bubbling, the negative charge predominates by 100 to 1. Sodium stearate, a soap which lowers γ and enters the surface layer, trebled group II⁻ in weak solutions. This is in keeping with FRUMKIN's observations on salts of organic acids. Stronger solutions, 1×10^{-3} molal virtually eliminated groups I and II, though not as effectively as salt. There was about twice as much electrification as with salt of the same high concentration. Heating the water reduced viscosity and changed the bubbling mechanism, the water spattered 5 times as much at 55° C as at 24° C. Group II⁻ was enhanced, II⁺, III⁺ and IIIb⁻ decrease in mobility and more negative charge is produced.

CHAPMAN's interpretation is illuminating. When a bubble expands the double surface stretches and grows thinner near the center top. Observations of JOHONNOT[73] studying the breaking of soap films by means of interference fringes observed films of 6×10^{-7}cm thickness just before rupture. This would form droplets of mobility 0.03 cm²/volt sec. What is more important, he observed the adjoining films *stratified* into zones of 5 uniform thicknesses or more. In the process of breaking, such films will give rise to droplets of

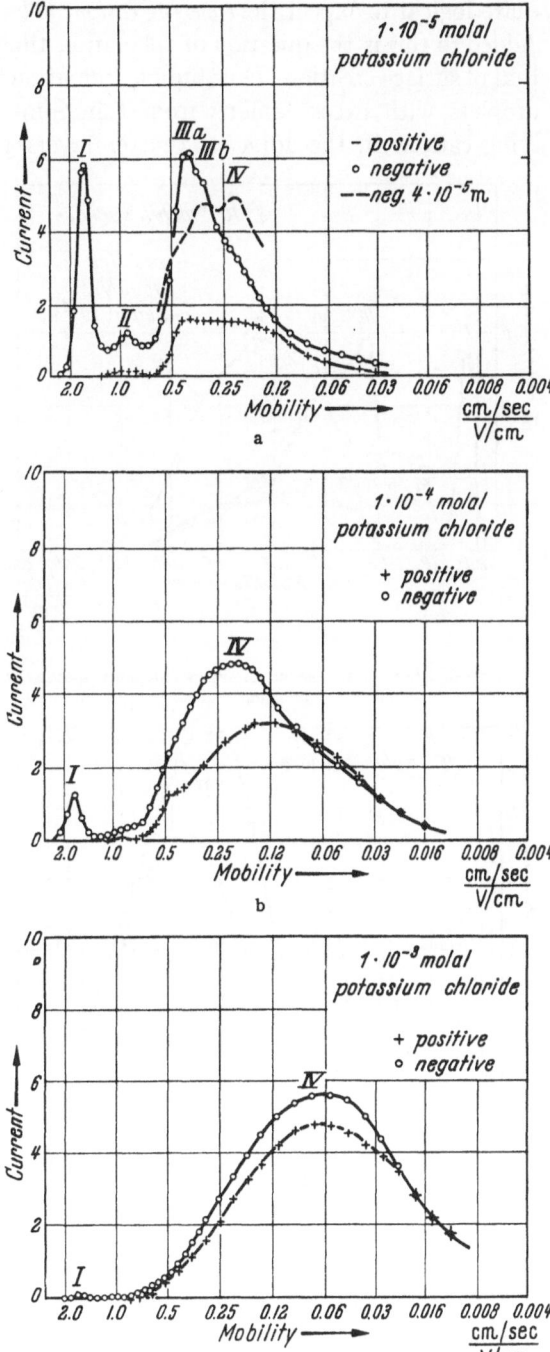

Fig. 21 a—c. Effect of KCl on CHAPMAN's mobility spectra from bubbling

varying and perhaps segregated sizes. If the bubbling is very violent, the turbulence of escaping air still further disrupts films. It should next be noted that as films get thinner so that the opposing double layers on opposite faces approach it is quite logical to expect the *covering cations to be squeezed back into the thicker layers*. Added to this is the question of relaxation time of layer adjustment compared to that of surface creation. Thus bubble formation is particularly favorable to creating droplets with excess anions from thin films and concentrating excess deeper lying cations in the adjacent thicker layers provided the relaxation times are

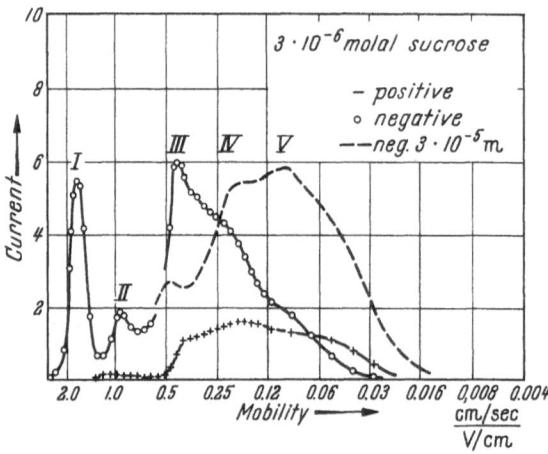

a

Fig. 22 a—d. Effect of sucrose on CHAPMAN's mobility spectra by bubbling

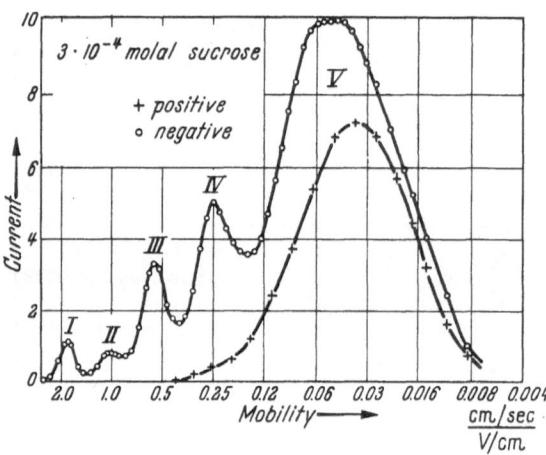

Fig. 22 b

adequate[59]. Formation times are longer in bubbling than in spraying aiding such action.

CHAPMAN next pointed out that J. BERNAL and R. H. FOWLER[74] indicate the existence of electropolar complexes in water surfaces as assumed by ALTY[59] and that these on rupture, can well break into molecular complexes with an excess H^+ or OH^- ion. There is also nothing to prevent one electron or uncovered anion in 10^5 molecules from being squeezed out into the surface. Thus whether it is partly rupture of the polarization charge of water droplets or partly due to excess bound uncovered anions, the rupturing of liquid water-air interfaces in thin layers, gives a rare droplet from the surface either with an excess solute anion, excess electron, an H^+, or an OH^- ion. CHAPMAN estimates from his charging currents that a small fraction $\sim 10^{-4}$ of the droplets of radii less than 5×10^{-7}cm are charged. This agrees well with an earlier doubtful estimate of LENARD'S.

He then advanced the hypothesis, borne out by his studies, that the fast negative I^- ions of $k = 1.8$ cm²/volt sec consist of possibly two to 3 molecules of water with an excess electron or possibly an OH^- or an uncovered anion. The larger I^+ positive ions in pure water of mobility 1.0 probably carry an H^+ from ruptured hydrogen bonds or dipoles and the II^- ions of $K = 0.95$ carry an OH^- ion or possibly an uncovered anion.

These ions I^- at $k_- = 1.8$ and I^+ at $k_+ = 1.5$ in the sprayer are probably the normal ions which in dry air have $k = 2.21$ and 1.6 respectively and add H_2O molecules in moist air to give $k_- = 1.8$ and $k_+ = 1.0$.

The larger ions such as II^+ and III^- of $k = 0.40$ and 0.45 respectively could be formed about electrolytic ions drawn into the double layer or squeezed out of the outer layers before rupture. The still larger droplets of 10^{-6} cm radius no doubt are stabilized with neutral salts and have a $k \sim 4 \times 10^{-3}$. CHAPMAN suspected they might carry excess charges owing to statistical fluctuations. The work of DODD[6] and others to be reported later on, make it very probable that in solutions some of these represent *symmetrical* charging produced by *statistical fluctuations* in ion distribution at the time of formation. If salt concentrations get too high, this will not occur (see infra) and larger droplets will be mostly uncharged because of Coulomb forces. These in concentrated solutions become so high that free independent diffusion ceases and is replaced by ambipolar diffusion.

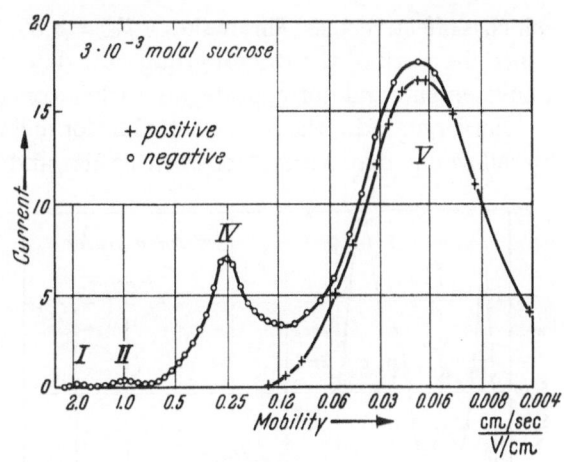

Fig. 22 c

It is expected that under the usual conditions of operation most droplet groups *evaporate* to an equilibrium size typical of the nucleii and size of the complexes with relatively little dependence on their original size except for the effect of original size in determining the nucleii included.

One strong support for assigning the II$^-$ and II$^+$ ions to adsorbed ions comes from the action of dilute acids which enhance II$^+$ and suppress II$^-$, i.e., cover the anions with H$^+$ effectively while dilute bases suppress II$^+$ and enhance II$^-$ i.e., remove covering H$^+$ ions and add excess uncovered OH$^-$ ions. It is interesting to note that the mobility

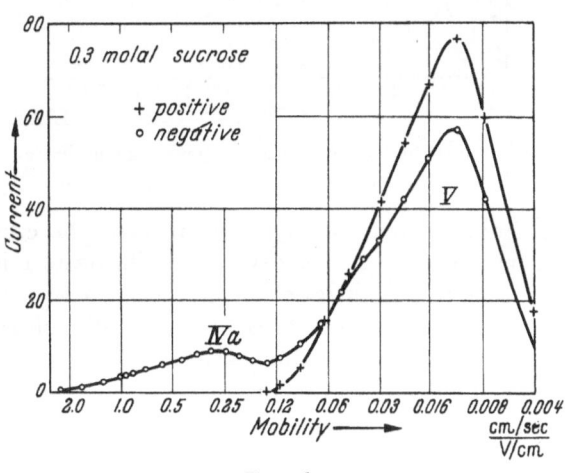

Fig. 22 d

k_{II^+} is greater than that of k_{II^-} indicating that the complex about OH$^-$ is larger than about H$^+$ as might be expected.

While much is known about mobilities in relation to the ionic nature for the so-called normal ions in air and other gases studied under controlled conditions, the nature and size of ions in moist air and in the range of size here involved is unknown. From the studies of the Bristol group, MUNSON, TYNDALL, DAVID, and HOSELITZ[75] on alkali ions in inert gases at various humidities, it is probable that the ions about H$^+$ and OH$^-$ in moist air would consist of perhaps 2 to 3 and, at most, 4, molecules of H$_2$O, certainly Li$^+$ ions in the inert gases with H$_2$O vapor are clustered to this extent. It is also possible that the difference in the ions in air noted here stems from the fact that both H$^+$ and OH$^-$ have the same number of clustered H$_2$O molecules perhaps 3, at most, and the difference in mobility stems from the slight difference in physical size and the increased mass ratio. Thus the molecular weight of H$^+$ (3 H$_2$O) is 55, while that for OH$^-$ (3 H$_2$O) is 71. Thus the molecular weight of H$^+$ (3 H$_2$O) is 55, while that for OH$^-$ (3 H$_2$O) is 71. The relative mobilities according to the mass dispersion law in air, is $\sqrt{\dfrac{M+m_+}{m_+}} \Big/ \sqrt{\dfrac{M+m_-}{m_-}}$ with M the molecular weight of air saturated with moisture which is effectively 28. This makes $k_+/k_- = 1.08$. The observed ratio of k_+/k_- is 1.06. Since the work of the Bristol group post dates CHAPMAN's paper, the conclusions above extend and corroborate his earlier conclusions.

For larger radii, the solid elastic law for mobilities is valid. For the discussion to follow one may assume ions of this size and generated under these conditions to be univalent although this does not hold for larger ions as will be seen in DODD's study. Thus to guide thinking in discussions to follow relating carrier mobilities, ion diameter or radii and numbers of carriers involved, Table 7 has been computed and assembled. This table is at best rough, but correct, in order of magnitude. For small or normal ions with molecular dimensions, mobilities of identified ions, are not accurately known certainly in air above 35% humidity. In fact, nothing

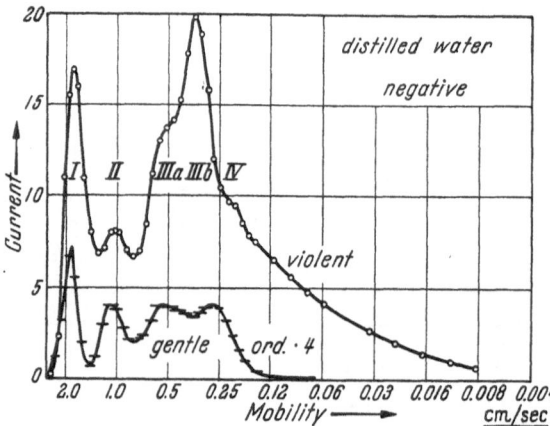

Fig. 23. Effect of violence of bubbling on CHAPMAN's mobility spectra

is known about the H$_2$O$^-$ ion in air. The effective radius of a water molecule is here chosen as 1.5×10^{-8} cm. The fastest ion observed with $k^- = 1.8$ could be ascribed to an electron attached to two water molecules, quite possibly it could be the queer O$_2$H$^-$ ion recently observed. The ions of $k^+ = 1.0$ and $k^- = 0.95$ have probably been correctly assigned as (H$_2$O)$_3$H$^+$ and (H$_2$O)$_3$OH$^-$ ions. Beyond this, the ions of intermediate speeds must consist of larger droplets carrying solute anions and solute cations. Formed from thicker layers of water, they are stabilized by the ions at about the mobilities of the groups IIIa$^-$, IIIb$^-$, and IV$^-$ at 0.45, 0.32, and 0.20 and III$^+$ at 0.40. In this intermediate region mobility theory is not too good. CHAPMAN used a solid elastic equation assuming a single charge given as $k = 7.6 \times 10^{-15}/r^2$, where r is the collision radius of ion and molecules. On this basis, he set the *radii* of the IIIa$^-$, IIIb$^-$ and IV$^-$ groups as 4, ~5, and 6 molecules respectively. In preparing the table, for carriers of radii ~10^{-6} cm and greater the relation used was the Stokes-Cunningham equation as given by MILLIKAN which reads

$$k = 4.7 \times 10^{-10} \left(1 + 0.87 \frac{10^{-5}}{r} \right) \Big/ r. \qquad (3.24)$$

For smaller carriers, a solid elastic equation derived from LANGEVIN with

$$k = 8.2 \times 10^{-15}/r^2 \qquad (3.25)$$

both in cm²/volt sec was used. Then CHAPMAN's groups should have radii of some 1.47×10^{-7} cm and 1.64×10^{-7} cm, implying 4.9 and 5.5 molecular diameters respectively, with about 950 and 1300 molecules respectively. Some of these ions could be doubly charged, but from the mode of generation, this is unlikely.

The number of molecules were calculated merely by dividing the volumes of the water molecules, chosen slightly large, into the volume of the droplet. This

Table 7. *Droplet Size, Number of Molecuces, Mobility*

Radius r in cm	Radius approx mol. diam.	Number of molecules approx.	Mobility calc. cm²/volt sec	Mobility obs. cm²/volt sec	Mobility equation used
1.5×10^{-8}	$\frac{1}{2}$	1	H₂O molecule	—	—
$<3 \times 10^{-8}$	$\frac{2}{3}$	$2\,H_2O + e$	—	1.8	—
$<3 \times 10^{-8}$	$\frac{3}{4}$	$3\,H_2O + H^+$	—	1.0	—
$<3 \times 10^{-8}$	$\frac{3}{4}$	$3\,H_2O + OH^-$	—	0.95	—
7.5×10^{-8}	2.5	125	—	—	—
9×10^{-8}	3.0	215	—	—	—
1×10^{-7}	3.3	322	0.81	—	Solid elastic
1.47×10^{-7}	4.9	950	—	0.4	Solid elastic
1.64×10^{-7}	5.5	1300	—	0.32	Solid elastic
2×10^{-7}	6.5	2350	0.21	—	Solid elastic
5×10^{-7}	17	3.7×10^4	0.033	—	Solid elastic
5×10^{-7}	17	3.7×10^4	0.015	—	STOKES, CUNNINGHAM
1×10^{-6}	33	3.0×10^5	0.0045	—	STOKES, CUNNINGHAM
1×10^{-6}	33	3.0×10^5	0.0082	—	Solid Elast.
2×10^{-6}	61	2.4×10^6	0.0013	—	STOKES, CUNNINGHAM
2×10^{-6}	61	2.4×10^6	0.0041	—	Solid Elast.
5×10^{-6}	170	3.7×10^7	0.00013	—	STOKES, CUNNINGHAM
10^{-5}	—	3×10^8	9×10^{-5}	—	STOKES, CUNNINGHAM
10^{-4}	—	3×10^{11}	4.7×10^{-6}	—	STOKES

STOKES-CUNNINGHAM $k = 4.7 \times 10^{-10} \left(1 + 0.87 \dfrac{10^{-5}}{r} \right) \Big/ r,$

Solid elastic $k = 8.2 \times 10^{-15}/r^2,$

Mass dispersion law says $k \propto \sqrt{\dfrac{m + M}{m}},$

m = mass ion, M = Mass molecule.

gives an overestimate for the number of molecules in the carrier since close packing is not possible. The greatest error comes for a single molecular layer about a single spherical central molecule. In that event, there are actually 13 molecules inside the sphere and the computation leads to 27. The error gets progressively less as the radius of the sphere increases relatively to that of the moleclues. In any event, the error is not serious for illustrative purposes.

Considering the group IIIb⁻ ions CHAPMAN indicates that they are the most stable. In fact, IIIa⁻ could be formed by evaporation of a layer of molecules from IIIb⁻ and IV⁻ by condensation of a layer of molecules onto IIIb⁻.

It is noted that with his bubbling there are very few droplets of k less than 0.1. On the contrary, with spraying most droplets have larger radii. Here the number of charged droplets extend in a continuous spectrum of equal numbers of positive and negative ions, even from pure water, to a value of k of the order of 0.003 cm²/volt second. Such ions have radii of the order of 1.2×10^{-6} cm and include some 10^6 molecules. Such carriers are the well known large Langevin ions of the atmosphere[78]. These ions are stable at room temperatures, except at high super-saturation and thus would form the upper limit in size to any spectrum of ions observed under these conditions. Larger droplets except in super-saturated air will evaporate rapidly down to 10^{-6} cm diameter even when multiply charged unless then contain much salt.

Concerning multiply charged ions, the process envisaged above will yield primarily singly charged ions and with droplets of 1000 molecules in short times multiple charges will not appear. The droplets of the Langevin ion size e.g., of 10^{-6} cm diameter and up will have anywhere from 1 to 10 electrons charge as shown by VASSAILS[76] and also as indicated for 10^{-4} cm diameter droplets in air by KUNKEL[77].

If the violence of bubbling is increased using a larger tube with more flow of air, more water surface is disrupted, more bubbles are torn by turbulent blast and the relative number of droplets with k less than 0.1 increases. Gentle bubbling reverses the distribution. The thinner films and turbulence in the bubble stream create nearly all small negative ions, the ratio of negative to positive here being nearly 100 to 1.

The bubbling with salts produces two changes, a reduction in the number of carriers in groups I and II, and the appearance of large numbers of carriers of both signs with mobilities less than 0.1. These effects are both caused by the appearance of ions in the double layer. This follows from the disturbance of the dipole layer of water by the increasing concentration of salt ions as indicated by FRUMKIN. The classes I⁻, II⁻, and II⁺ come from the water linkages, hydrogen bond, etc., while the classes III⁻ and III⁺ come from uncovered anions and the covering cations as well as from droplets containing neutral salts.

Concentration of the electrolytic ions in the thin films is probably irregular because of the statistical fluctuations of relatively small numbers. The ions at higher concentrations as indicated by FRUMKIN disturb the surface tension of the otherwise relatively uniform films in an irregular fashion increasing it in patches where they accumulate. Thus the uniform breaking of the surface film is disturbed. In consequence, the creation of uniform categories of drop sizes ceases. This yields the wide distribution of mobilities of drop sizes shown in Fig. 21 c.

In spray electrification, it takes 0.2 N KCl to obliterate the fast peaks and only 10^{-3} N to do the same with bubbling. This no doubt has to do with the relaxation times for ionic distribution relative to the rate of surface disruption.

However, as will later be indicated at the ion concentrations present the charging of the drops to nearly equal numbers of opposite sign may be due to the statistical fluctuations responsible for symmetrical charging as shown by DODD[6]. CHAPMAN correctly points out that with carriers of k less than 0.1, containing hundrets to tens of thousands of molecules there will be a considerable entrainment of solute molecules. The vapor pressures of such droplets, despite their small size, will be such that they would tend to *condense water vapor* rather than evaporate even at 60% relative humidity. He also points out that above 10^{-3} N the statistical fluctuations of charges in droplets even as large as 10^6 molecules of water begins to be such that relative charge differences of the two signs decrease. There is however, more negative charge produced on the large particles Group VI⁻ of $k<0.03$ than on positives as noted in Fig. 21 c which disappears for Langevin ions. These are the excess uncovered anions of ALTY.

It is to be noted in this connection that $AlCl_3$ and polyvalent cations are more effective, and at lower concentrations, in reversing the net charge from negative to positive than monovalent ions. This is to be expected since polyvalent cations covering the bound anions will have excess positive charge even in the bound layer. Thus drops of the size including this layer will have excess positive charge. There is, however, always care needed in the interpretation of the actions of specific salts, since these salts alter γ in different measure and thus influence bound and covered anions in unexpected ways, e.g., note the actions of KF, KCNS, $Th(NO_2)_3$, CaI_3 etc.

The action of sugar is very interesting in that it is inert as regards electrolyte production. It increases the surface tension and viscosity of the water. Thus suppression of I⁻ and II⁻ ions is less on addition of sugar than for salts at the same concentration. The increase in surface tension and increase in viscosity reduce the production of small droplets so that at 3×10^{-4} N sucrose, the I⁻ and II⁻ carriers are pretty well removed. The presence of large numbers of large molecules in the proximity of the double layer may act to increase its depth and will act to uncover the otherwise covered and bound anions. It thus results in an enhancement of negative ion groups III⁻, VI⁻ and yields a group V⁻ at $k\sim0.05$ cm² per volt sec. Positive ion group III⁺ is obliterated but a large positive ion group of $k\sim0.04$ is created, indicating that the former positive covering ions are deeper in the solution. Above 3×10^{-3} N the invasion of sugar is such as to reduce the group IV⁻ and the former covering positive ion charge appears about to equal the negative ion charge in low mobility V⁺ and V⁻ ions of $k\sim0.16$ cm²/volt sec. There are, however, still an excess of uncovered negative anions in the IV⁻ group, the equivalent positive charge presumably remaining in the bubble on larger droplets. The V⁺ and V⁻ peak does not appear to be entirely due to symmetrical charging as indicated for 0.3 N sucrose. No doubt break-up into smaller particles is precluded by increased viscosity and increased surface tension. However, increasingly larger droplets predominate approaching and exceeding Langevin ions in size at 0.3 N sucrose. Here the positive charge predominates with excess of cations for the larger droplets while the compensating negative charge persists in the uncovered anions of Group VI⁻. The low vapor pressure of the 0.3 N sucrose no doubt causes condensation and growth of the droplet size in the bubbler. It is suspected that viscosity may reduce the adjustment of ion concentrations as the surface breaks. Whether as indicated in Fig. 22d—there are really more positive ions than negative ions cannot be decided since areas over such ranges may not be accurately proportional to charge. Some of the observed asymmetry in favor of the larger droplets arises from loss of faster ones by diffusion enroute from bubble to measuring chamber.

The changes of humidity obviously will alter the carrier groups observed by causing evaporation or condensation. Since the ions age mostly in the bubbler, or spray chamber and on their way to the measuring chamber, because times spent *in* the measuring chamber are much less than in production, the effect of humidity is most conspicuous in these regions. Under normal conditions of

bubbler generation the relative humidity of the blasts used was around 35%
and evaporation of carriers was the general rule. This enhanced and emphasized
the very sharp division of carriers in pure water to the discrete groups shown in
Fig. 20. With nearly 100% humidity in the bubbler, the larger carriers do not
evaporate, and as these diffuse to the walls less in transit the spectrum of humidified gas strongly emphasizes these carriers. The high humidity smears out and
affects the mobilities of the negative carriers more than the positive ones. This
may be connected with the well known preference of negative ions for producing
condensation in the C. T. R. Wilson cloud chamber as noted by LOEB, KIP, and
EINARSSON[28]. If the auxiliary blast is super-saturated evaporation is retarded
to such an extent that carriers of all sizes are observed. By extending the time
for measurement of large ions in the dry blast of the measuring chamber by
increasing the downstream distance for measurement there is an increased amount
of evaporation of the large carriers so that the *average* mobility *is increased.*

Since the spraying measurements always occur in saturated air the mobilities
of the same groups in the spraying measurements are slightly lower than in the
bubbling measurements.

The spraying groups at $k_+ = 0.70$ and $k_+ = 0.049$ are the groups that shift
into the strong positive group observed when spraying salt solutions. They are
peculiar to the spraying mechanism and are not observed in bubbling.

The use of a soap, sodium stearate, gives results that are in some respects
analogous to those with sugar or salts. Conditions are however vastly different.
Sodium stearate *lowers* surface tension. The long chain hydrocarbon groups are
outside the surface. FRUMKIN has shown the structure. The net surface layer is
produced by the COO^- group with negative charge outward. The charging is weak
the Na^+ ions are inward. The dipole orientation of water molecules may be
disturbed. The lowering of surface tension and increased viscosity favor large
thin film formation. Actually, this substance should resemble a salt except for
its action on bubble formation.

Group I behaves as with salt solutions and the slowest group behaves like
Group V of sugar. The appearance of group II^- is almost exactly like that of
Group II^- with hot water.

Again sodium stearate hydrolyzes readily and yields sodium hydroxide.
Group I^- was decreased and group II^- was enhanced at 10^{-5} normal while 10^{-3}
normal virtually wiped out groups I^- and II^- but not as effectively as salt. The
action of the Na stearate appeared to obliterate the electronic type negative
ions, perhaps by altering the surface. The OH^- ions was enhanced by the OH^- ions
from hydrolyzed stearate. The effect on bound but uncovered anions and on
cations as indicated by FRUMKIN, should be analogous to salt action but less
strong. As with sugar, there was an increasing amount of electrification with
increasing concentration in the slower groups, less coverage, more deeply lying
cations.

Certainly the reduced surface tension leads to far greater surface production
and thus to greater fine droplet formation and charging. In fact, JOHONNOT's
film thickness of 6×10^{-7} cm should lead to many droplets of that diameter and
of mobility around 0.08 which is right in the center of the spectrum observed.

E. Spray electrification of water in relation to thunderstorm electrification

The final chapter on spray electrification of water and salt solution has not been written. CHAPMAN's study of electrification by spraying and bubbling was

Table 8. *Total amounts of negative and positive charge produced by bubbling solutions in arbitrary units*

Distilled Water . . .	− 4.0	+ 1.0	Sucrose 0.8 molal	− 25	+ 25
			Sucrose 2 molal	− 5.5	+ 6.5
KCl 1×10^{-5} molal . .	− 4.7	+ 1.7			
KCl 1×10^{-4} molal . .	− 5.3	+ 3.1	Water, violent, 200 cm³/sec, auxiliary		
KCl 1×10^{-3} molal . .	− 7.5	+ 7.2	off	− 25	+ 7.0
KCl 0.01 molal . . .	− 7.7	+ 7.7	Water, violent, 200 cm³/sec, auxiliary		
KCl 0.1 molal	− 5.0	+ 5.0	on	− 18	+ 5.5
KCl 1 molal	− 2.2	+ 2.2	Water, normal, 77 cm³/sec, auxiliary		
KCl 4 molal	− 1.5	+ 1.5	off	− 3.5	+ 0.95
			Water, gentle, 12 mm tube, auxiliary		
AlCl₃ 3×10^{-5} molal .	− 6.5	+ 3.5	on	− 0.7	+ 0.00
AlCl₃ 3×10^{-3} molal .	− 8.0	+12			
AlCl₃ 0.03 molal . . .	− 6.0	+ 8.0	Water, 99% humidity in auxiliary,		
AlCl₃ 0.3 molal	− 4.0	+ 4.0	bubbling temp. at equilibrium 6° C		
			below room temp.	− 6.0	+ 1.7
NaOH 1×10^{-5} molal .	− 4.4	+ 1.4	Water, 99% humidity in auxiliary,		
NaOH 1×10^{-4} molal .	− 6.3	+ 3.0	bubbling at room temp..	− 8.2	+ 2.6
HNO₃ 1×10^{-5} molal .	− 3.9	+ 1.4	Water, supersaturated in auxiliary . .	−10	+ 3.5
HNO₃ 1×10^{-4} molal .	− 3.2	+ 3.1			
HNO₃ 1×10^{-3} molal .	− 3.4	+ 3.4	Hot water 55° C, dry auxiliary. . . .	− 9.0	+ 1.2
			Hot water 80° C, dry auxiliary. . . .	−11	+ 0.6
Sucrose 3×10^{-6} molal	− 5.4	+ 1.7			
Sucrose 3×10^{-5} molal	− 7.5	+ 2.7	Sodium Stearate 3×10^{-6} molal . . .	− 4.5	+ 1.3
Sucrose 3×10^{-4} molal	− 11.7	+ 6.8	Sodium Stearate 1×10^{-5} molal . . .	− 7.0	+ 1.8
Sucrose 3×10^{-3} molal	−18	+ 14	Sodium Stearate 1×10^{-4} molal . . .	− 8.0	+ 4.5
Sucrose 0.03 molal . .	−28	+ 21	Sodium Stearate 1×10^{-3} molal . . .	−12	+12
Sucrose 0.3 molal . .	−35	+ 31			

undertaken among other things, to estimate the differential generation of electrical charge in order that estimates might be made of the rate of this process in thunderstorm generation. Data of this sort compiled by CHAPMAN showing the conditions of production of net positive or negative charge on bubbling and spraying are shown in Tables 8 and 9*. Absolute values are not given but the data are of interest and use in that they show how varied the phenomena can be. It should be recalled that LENARD in his spraying experiments observed measurable differences between negative and positive charge

Table 9. *Total amounts of negative and positive charge produced on spraying solutions in the same units used in Table 8*

Distilled water . . .	− 38	+ 38
KCl 1×10^{-5} molal .	− 40	+ 40
KCl 1×10^{-4} molal .	− 42	+ 43
KCl 0.01 molal . . .	− 11	+ 11
KCl 1 molal	− 2	+ 2

with negative charge predominating. LENARD however, did not observe any significant electrificant on the break up of drops in an air stream as would be

* These represent electrometer deflections since there is no meaning to absolute values which depend on water sprayed and percentage of charge sampled. Separate measurements are needed in a different device to get charges per gram of water sprayed.

expected in storm clouds unless there was marked turbulence. However, CHAPMAN [68], in a report on thunderstorm electrification, gave the results of later studies on the electrification of rain drops or falling water drops of equivalent size.

The procedure was to let droplets of water fall into a high velocity air stream in a tube which terminated with an opening into a larger space with a high potential and collecting electrode also with a high velocity air stream the system constituting an ERICKSON tube of low resolving power. Single drops disrupted gave carriers in the high mobility range of from 0.5 to 2 cm/sec per volt/cm. The negative charge predominated by about 10 to 20%. He found 20×10^{-12} coulombs of either sign per drop compared to the 1.6×10^{-12} coulombs per drop observed by SIMPSON.

To break up the drop, an obstacle was introduced downstream to create turbulence, or a second drop was introduced into the tube to strike the first one floating on the non-turbulent stream. These two drops floating eventually coalesced. On coalescence, the final drop has the size of one of the original ones, the equivalent water disappearing in spray too fine to observe. *No significant* charging was produced in this fashion. This is in agreement with LENARD's findings. If the drop falls in quiet air for some time before hitting the air blast, it disrupts in a more violent manner. In such turbulent disruption significantly more charge is produced than the common values given. Individual drops vary by a factor of 10 to 100 in the amount created on disruption. The charge created if there were 20 drops to the gram at 20×10^{-12} coulomb per drop would give about 4×10^{-10} coulomb/gram or about 1.2 esu per gram. With 10^{-4} normal salt solutions the electrification was somewhat enhanced. This agrees with BUSSE's figure of $5—6 \times 10^{-10}$ coulombs of negative ions per gram or cm³ on spraying. *This charge mechanism is not adequate by a factor of 10 to account for the main charge in thunderstorm cells.* It may be a mechanism active in the lower regions below the freezing isotherm in the turbulent throat of the storm.

F. Homogeneous or symmetrical charging of liquid droplets on dispersion

In the endeavor to understand spray electrification in all its details, the author assigned to CHAPMAN [79] as his first study, the investigation of spray electrification of droplets produced by an atomizer using the Millikan oil drop experiment. It had been noted by MILLIKAN in the atomizing of watch oil and mercury droplets, in the evaluation of the electron, that these droplets had high initial charges of both signs. Some study was made by MILLIKAN's students but CHAPMAN's study was an attempt to observe *any net charging* and by measuring charge and drop size to test COEHN's law if applicable. In fact, C. DYK [80], in the author's laboratory, earlier had looked for cataphoresis of gas bubbles in liquids of lower dielectric constant than water where definite movement had been observed. However, in order to observe the phenomena in all but water high potentials were required and these produced turbulence owing to ionic motion before cataphoresis could be observed. An ordinary brass atomizer was used by CHAPMAN on pure substances; Ameroil, (largely Octane), Nitrobenzene higly purified from Kerr cell studies, Aniline twice distilled, glycerine with water removed and distilled in vacuum, and distilled water. Drops of equal size appeared to be charged in equal amounts for both signs but the magnitude of charge increased with drop size roughly in proportion to the radius except for Ameroil. Drop radii ranged from 1 micron, (10^{-4} cm), to 5 microns in diameter and charges varied with the material. The results are indicated in Table 10.

Table 10. CHAPMAN's *data on spray charging of large drops relative to surface tension and dielectric constant*

Liquid	Dielectric constant	Surface tension in dynes/cm	Viscosity in dyne sec/cm²	Smallest drops		Largest drops	
				Radius in cm	Average charge in electrons	Radius in cm	Average charge in electrons
Ameroil. . . .	1.9	27.0	0.054	4.0×10^{-5}	2	1.0×10^{-4}	2
Nitrobenzene .	35.7	46.0	0.021	1.0×10^{-4}	150	4.0×10^{-4}	800
Aniline	7.3	44.0	0.044	1.5×10^{-4}	100	5.0×10^{-4}	450
Glycerine . . .	56.2	65.2	0.083	5.0×10^{-5}	40	2.5×10^{-4}	500
Water	80.0	72.8	0.010	1.5×10^{-4}	125	5.0×10^{-4}	600

CHAPMAN concluded (1) that there is no preference for either sign of charge.

(2) Larger drops of both signs have larger charges and this varies roughly in a linear fashion with diameter except for Ameroil.

(3) There is a rough parallelism between dielectric constant and charge for drops of the same size.

(4) Although there is no parallelism with viscosity, a parallelism exists between drop size and surface tension which is, however, difficult to differentiate from the influence of dielectric constant.

(5) Only Ameroil showed uncharged droplets and these were of all sizes.

It is clear that *this charging phenomena is not related to the spray electrification studies* and represents an other phenomenon which may, however, apply to the larger droplets collected in spraying.

The disagreement between these observations and those of the finer droplets is in no sense a contradiction of the data and interpretations which have preceded. The layers involved in spray electrification of the Lenard type are no more than 2×10^{-6} cm thick. The drops produced and studied had dimensions in excess of 2×10^{-4} and comprise 10^{10} or more molecules. If ionic impurities are present to the extent of one part in 10^6, there will then be 4×10^4 ions of both sign in each droplet.

The subject was not further investigated since the Millikan method was badly limited in scope for such study. The development by HOPPER and LABY[81] in 1938 of a much more effective device for the measurement of the electron, led the author to propose to HANSEN, then interested in a study of the electrification of dusts on dispersion, that he develop the method for such work. The skillful work of HANSEN[82] later improved by KUNKEL[82] led to a device which could carry out a rather complete analysis of the statistics of blown dusts yielding charge and mass for particles from 10^{-4} cm diameter to 3.0×10^{-3} cm diameter photographically recorded by a schedule of pictures taken at from 10 second to 20 minute intervals. In consequence, on completion of the dust studies, the investigation was carried over to the charging of liquid droplets produced by atomizing by

E. E. Dodd[6]. When Dodd under took this study, he had as precedent certain results and conclusions drawn by Kunkel[5] for dusts. With homogeneous dusts, e.g., S, SiO_2, etc., charging of both signs appeared in equal amounts for any particle size. It increased in a certain fashion with particle size. With the larger number of particles as well as range of size studied good statistics were available. It then proved that the cloud produced by separation with no contact between dissimilar surfaces was *electrically neutral as a whole*, and while many neutral particles were present, there were, however, many particles charged up to 3000 or more electrons and on down, but in equal numbers and of opposite sign. This was shown, as suggested by the author, to result from the chance that as various unit surface contact areas ruptured there were by chance excess negative or positive charges on one side or the other of the ruptured boundary. This theory statistically independently derived by Wijsman[5] and by Kunkel[5] proved to agree quite successfully with actual observation. It was with this background that Dodd approached the problem of liquids. The techniques, of Hansen, Kunkel and Dodd and the results, and conclusions drawn by Dodd on droplets are reproduced as written by them as their statements are as complete and as concise as possible, consistent with being serviceable to future user of the methods.

1. The spray and dust electrical analyzer of HANSEN and KUNKEL. Using the Hopper-Laby technique

The main feature of this method is to record photographically successive positions of the particles as they settle under the influence of gravity in a horizontal electric field. From the observed rate of fall, it is simple to calculate the Stokes' law diameter, and from the simultaneous horizontal deflection one can thus calculate the approximate charge of the particles. Such an arrangement, however, calls for a rather special optical system to obtain a strong dark field illumination. Clearly the light cannot be introduced at right angles to the direction of observation since the plates needed to set up the field have to be vertical and solid. Illumination at acute angles introduces great difficulties of construction. Thus, observation in the direction of propagation of the light is the only practical solution. The dark field is obtained by cutting out the central cone of the highly divergent light beam so that no direct light strikes the lens of the camera. Only light scattered into this dark cone will be caught by the camera lens and thus reach the film. The complete arrangement is perhaps best explaines by a simplified sketch (Fig. 24).

The optical system was designed to yield optimal contrast, i.e., maximum intensity with minimum scattered background, using the lenses at their disposal at the time of construction. It should be noted, however, that it seems possible to improve the performance of the instrument by proper changes in the dimensions indicated in Fig. 24. The following description, therefore, emphasizes the important features of the arrangement, rather than the exact dimensions. In this description items will be referred to by their letter designations in Fig. 24.

A high pressure mercury arc, water cooled, G. E. Type A-H6, serves as light source. It has to be operated in a horizontal position, so that a prism or a front silvered mirror is needed to tilt the virtual source into an upright position. The beam is first focused at a short distance by a large lens of short focal length. Near this first image of the source, a rotating disk interrupter which is driven by a synchronous motor is introduced. The interrupter, 12 inches in radius, was geared to run at 450 r.p.m. by a motor running at 1800 r.p.m. It carried four openings of 6° (B), two or three of which could be closed according to the time

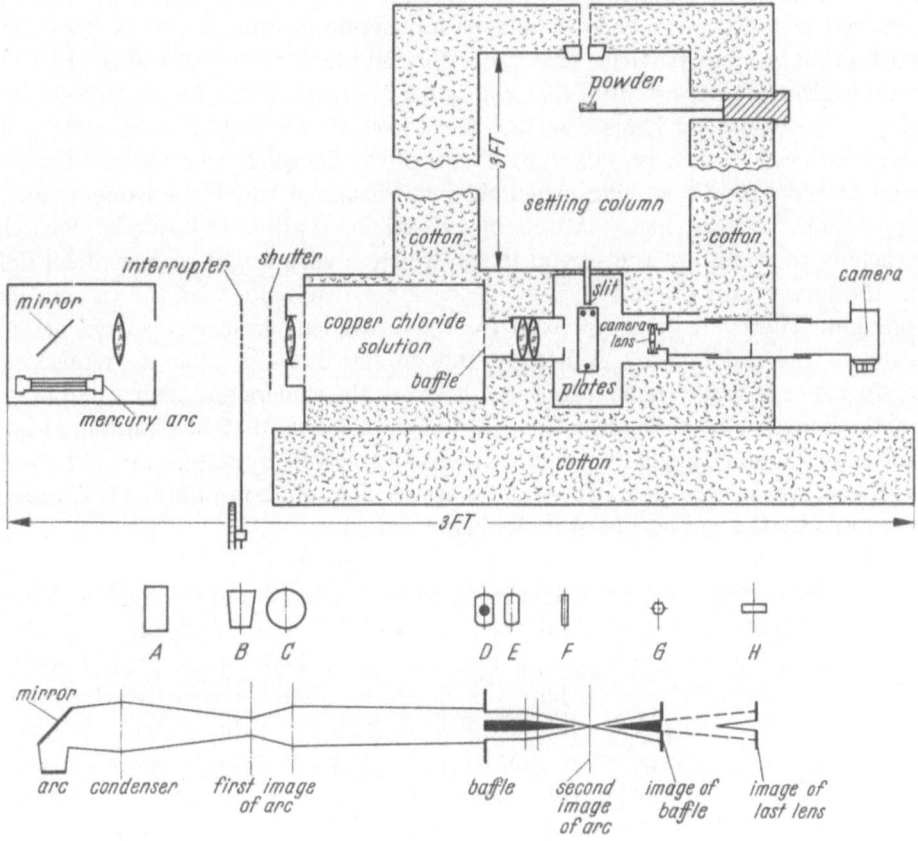

Fig. 24. Sketch of apparatus of KUNKEL and HANSEN for the statistical analysis of dusts and sprayed liquid droplets. Specifications of the optical elements used in the arrangement were (from left to right): mirror $d=57$ mm, $f=\infty$, first and second lens $d=56$ mm, $f=52$ mm, third and fourth lens (coated) $d=39$ mm, $f=63$ mm, fifth lens Leica objective $f=50$ mm set at $d=f/4.5$. Note: the third and fourth lens are considerably closer to the deflecting plates than shown in the diagram. All lenses are double convex

intervals to be observed. Since the intensity emitted by the arc fluctuates with 120 maxima per second, it is essential to line the shutter disk up so that maxima occur whenever an opening crosses the optical axis. Behind the disk a hand operated shutter is mounted. It is opened only during an actual observation. This shutter keeps all light and heat out of the system when no illumination is needed. A second large lens, similar to the first, condenses the light into a slightly converging beam which passes through a large volume of aqueous solution of $CuCl_2$, reducing the infra-red intensity by about 95% without appreciably affecting the strong green and blue light. Directly behind this heat filter, an opaque circular baffle (D) of $\frac{1}{2}$-inch diameter is interposed at the center of the beam. The light that passes, mainly above and below the baffle, is admitted by an oblong aperture (E) to the actual analyzing chamber. At the entrance of the analyzing chamber, a system of two lenses serves to focus the beam at the center between the vertical plates to form a second and reduced image of the source. The dimensions are so chosen that the image of the opaque spot has the size of and falls on the iris diaphragm (G) of the camera lens (set at f 4.5) which is detached from the camera body and brought close to the vertical plates. The position of the camera body should be adjustable in order to make it possible to bring into focus particles that are displaced along the optical axis. It is desirable although not essential that the camera operate with a focal plane shutter. They chiefly used a Leica camera. If no focal plane shutter is available, the exposures can be timed by hand by means of the flange shutter between the first and second lens. A copying attachment by means of which the camera can be replaced by a ground glass plate enables visual observations to be made. Particles reaching the region of the second image which is very bright scatter much light in the forward direction, i.e., into the dark cone subtended by the camera diaphragm. However, light scattered from the last lens surfaces also reaches the camera lens and appears as background on the film. In order to reduce this disturbance, a baffle (H) is introduced between the camera and its iris diaphragm at the image plane of the last lens. Although it is true that this additional limitation of the solid angle reduces the intensity received by the film in some parts of the focal plane, the gain in contrast is appreciable, since most of the background is caused by the last lens (despite its being coated).

The sketch is drawn to scale and so are the apertures and the outline of the beam indicated under it. The dust is introduced at the top of the settling column either by blowing it through a tube or by mounting a small thimble-like container, as shown in Fig. 32a and applying a measured short blast of air from above. In either case, the volume of air blown in should be small compared to the volume of the tower in order not to cause a violent disturbance in the system. The cloud thus produced will settle under gravity and partly enter the analyzing chamber through the slit right above the region of the second arc image. The slits used were made mostly of microscope cover glass edges which were 0.2 mm thick, 5 mm long and spaced approximately 1 mm apart. The dimensions and material of the slit are important as will be shown later.

For the observation of fine dusts, it proved advantageous to lower the slit as far down as possible to reduce the loss of particles by diffusion. The vertical plates are made of glass and coated with colloidal graphite. They are held very accurately 8 mm apart by three carefully ground quartz spacers and mounted electrically insulated to the vertical walls of the chamber. Thin wire leads connect the plates through insulators to a reversing and grounding switch outside the apparatus near the camera. Ordinarily both plates are grounded, the voltage supplied by B-batteries, chiefly 180 volts, being turned on only during the actual exposures. The chamber itself is made of heavy brass mounted on a bottom plate carrying set screws and a level indicator. All surfaces inside are coated with a highly absorbing flat black paint, and all parts touched by light at grazing incidence are covered with black velvet. Since in this arrangement it is impossible to avoid light striking certain areas inside the chamber, the total time of exposure to radiation has to be limited to the utmost minimum. Only during an actual observations is filtered light admitted at the rate of at most 30 flashes, of 1/450 sec each for an exposure of 1 sec duration, i.e., a total illumination of not more than 1/15 sec. If this illumination is not repeated after too short intervals, the absorbed energy will spread rapidly over the whole metal chamber so as not to cause any noticeable heating. After prolonged exposures, however, heating of the metal parts will result, followed by convection, rising at the walls and funneling down between the plates, increasing the apparent rate of fall of the particles. The apparatus without an elaborate thermostated heating system to maintain constant temperatures, therefore, has its limitations. If great accuracy and easy operation with long exposures are required it will be necessary to surround the entire settling tower and analyzing chamber by a thermostatic heat bath. In general, however, as in this case, wrapping the whole structure in two to three inches of cotton and keeping the room temperature constant to within a degree centigrade is all that is necessary. Under these conditions the temperature inside the system usually stayed constant within 1/100 of a degree centigrade for several hours, and the convection setting in near the end of the observation, at most, falsified the measurements on particles smaller than 1 micron in diameter. Great care should be taken to see that the camera is not shifted or tilted during a measurement. Since vibrations and shocks should be prevented from reaching the apparatus, the motor driving the rotating shutter is mounted on rubber supports resting on lead bricks, which in turn are imbedded in a large box of dry sand. No vibrations made themselves felt through this filter.

Procedure

Before the actual experiments begin, the vertical as well as the correct position of the camera and the associated magnification are determined by photographing first a weighted hair hanging freely through the slit and then a small calibrated scale held in the plane of observation. To accomplish this, the apparatus has to be opened. If the same slit is used repeatedly, and if neither the camera lens nor the camera holder are moved, these operations, of course, need not be repeated. After reassembling the apparatus, sufficient time must be allowed to let the system come to complete thermal equilibrium. They actually let at least one night pass before runs were made. A long stem thermometer graded in 1/20 of degrees, the bulb of which touched the metal of the chamber, served as a check on the inside temperature.

After equilibrium is reached and all convection presumably has died down, the cloud is generated in the manner desired inside the tower near the top or it is introduced through a hole at the top of the settling column. If rapidly falling particles are expected to be present, the arc is turned on and pictures are taken at once at short intervals of, say, 10 sec. The polarity of the horizontal *field is reversed between successive exposures to reduce the loss of particles by a steady drift toward one of the sides* caused by the bulging of the field beyond the ends of the plates. After about two minutes, when the largest particles have settled out, the intervals between pictures are gradually increased, roughly in inverse proportion to the average velocity of the particles. When after 10 to 30 min, the sizes and thus the rates of fall get very small, the time of exposure is increased to two, three, and finally four seconds, and the number of flashes per second is reduced by closing some of the shutter openings. This enables the measurement of particle sizes down to below two microns in diameter (depending on the density of the

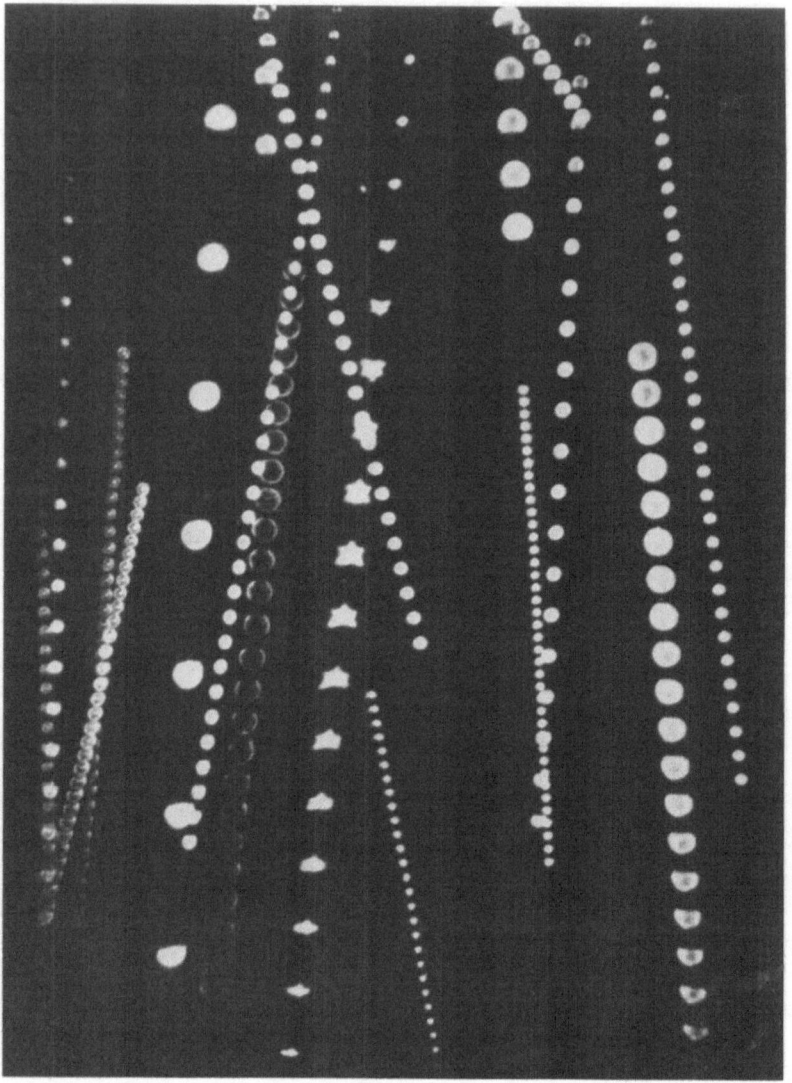

Fig. 25a and b. Tracks of dust particles falling vertically in horizontal electrical field as revealed for quartz dust with stroboscopic illumination (a) large (b) small. Stars and circles are diffraction images respectively too close or too distant from the focal plane

material). The tracks of the smallest particles of about one micron diameter and less cannot be resolved at this shutter speed. In this case the presence of completely resolved traces in the same picture are relied on to serve as a time scale. If no such traces are visible, then the scale is taken from the timing of the complete camera exposure. It would, of course, also be possible to arrange for a slower interrupter speed when these fine smoke particles are to be observed. Test runs must be made in order to determine and adjust the cloud density if possible, or to adjust the slit width so that the number of particles reaching the analyzer

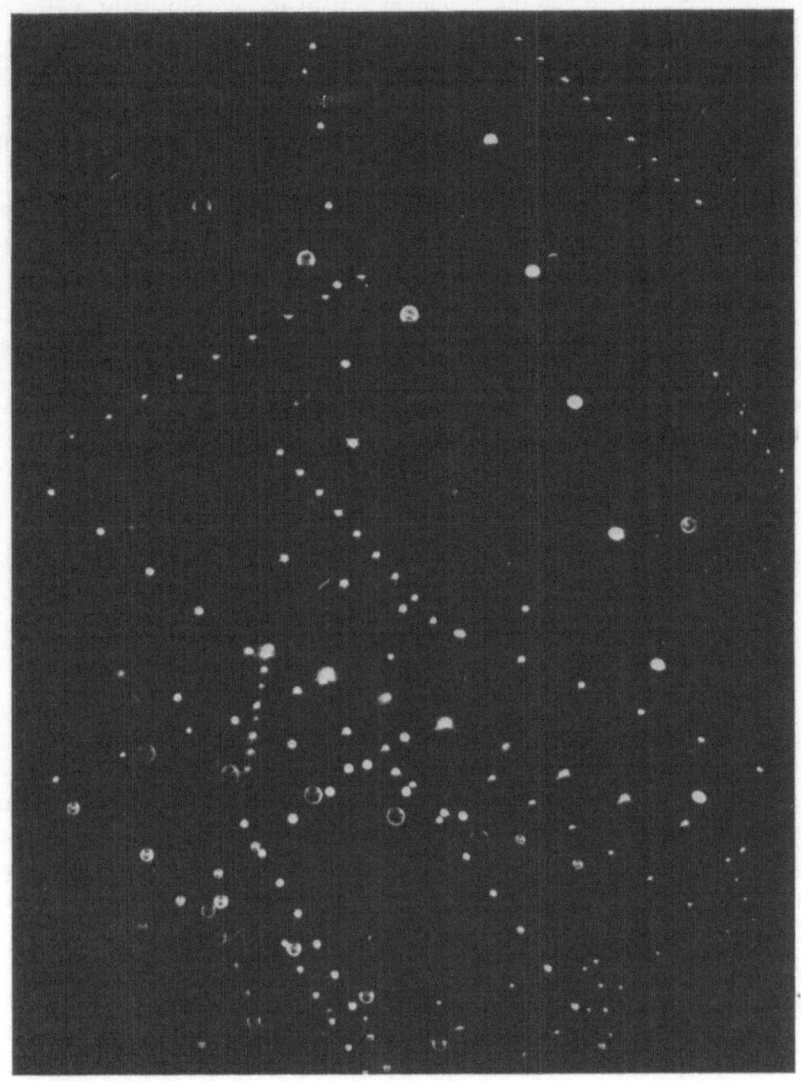

Fig. 25 b

is neither too large nor too small for a good evaluation of the cloud properties.

Examples of tracks obtained are shown in Figs. 25 a and 25 b. All of them are due to quartz particles. The first two pictures were taken without a field. They show that the trajectories of these particles are neither necessarily vertical nor straight. If they are very small, Brownian motion and unequal radiative heating causing photophoresis may interfere. If the particles are too heavy, their shape

may cause them to skid sideways*. Figs. 25 a and 25 b show particles of various
sizes and charges. The vertical dimension of the field of view is approximately
5 mm. Since they use the camera lens, which is corrected for distant objects, as a
projector lens, particles slightly too far away will appear as circles, the diameter of
which is proportional to the displacement out of focus. Similarly, point objects
too close to the camera appear as blurred stars or hexagons caused by the hexagonal iris diaphragm of the camera lens.

In order to calculate the sizes and charges of the particles photographed, the
negatives of the pictures are projected to a standard total magnification of, say,
50 onto a screen on which the vertical and horizontal displacements are easily
measured. If the vertical length of a certain trace is denoted by Y and the
horizontal displacement by X, the corresponding velocities are given by

$$V_x = \frac{v\,X}{M\,N} \quad \text{and} \quad V_y = \frac{v\,Y}{M\,N},$$

where M is the magnification, v the number of flashes per second, and N the
number of intervals between dots of the trace. Using STOKES' law to estimate the
particle size, we have for the diameter:

$$d = \left[\frac{18\,\eta\,V_y}{(\varrho_1 - \varrho_2)\,g}\right]^{\frac{1}{2}} = \left[\frac{18\,\eta\,v}{(\varrho_1 - \varrho_2)\,g\,M}\right]^{\frac{1}{2}} \cdot \left[\frac{Y}{N}\right]^{\frac{1}{2}} = A\left(\frac{Y}{N}\right)^{\frac{1}{2}}; \qquad (3.26)$$

similarly for the charge

$$\frac{d}{q} = \frac{3\,\pi\,\eta\,V_x}{E} = \frac{900\,\pi\,\eta\,l\,v}{V\,M} \cdot \frac{X}{N} = B\,\frac{X}{N}, \qquad (3.27)$$

where η is the coefficient of viscosity of the medium (air), $g = 981$, ϱ_1 and ϱ_2 the
densities of the material and the medium respectively, and V and l the voltage
(in volts) and the distance between the plates respectively. All quantities are
expressed in c.g.s. units. It should be kept in mind, however, that for solid particles these formulas in general are only very approximate, since mostly we are
not dealing with spheres. Both values will, at least as long as the analogy with
the viscous flow holds, appear smaller than they actually are. The magnitude of
the error will depend on the shape and may easily exceed 10%, but it will hardly
ever falsify the order of magnitude. This is also the reason why it is in general
not necessary or justified to apply the Cunningham correction. Only for particles
smaller than 1 micron, which often do not deviate appreciably from the spherical
shape, will any such correction have to be considered. It should also be pointed
out that the Cunningham correction opposes the shape correction, the first letting
the particles fall faster than STOKES' law predicts, and the latter slowing them
down again. For the same reason, the density of air can, in general, be neglected
relative to the density of the material as can the temperature variation and uncertainty in the coefficient of viscosity. There is little sense in attempting to
use accuracies better than 1% in any of the constants involved. For the objec-
tives of these applications greater accuracy is usually not required.

---

* In a previous article it was shown that wedge-shaped platelets will appear to drift
laterally as if charged. This is owing to a short time observation of the well-known helical
paths of such platelets under viscous free fall.

The constant B in Eq. (3.27) depends on the experimental conditions only, while A in Eq. (3.26) also varies with the density of the material involved. Both, however, will usually not change in the course of one experiment, except, perhaps, for the factor ν, the flash frequency, which might be intentionally decreased (preferably by a factor of 4) when the fast particles have settled out and only the slow small ones are being observed. The evaluation of the quantities q/d and d will then simply consist in multiplying the velocities and, respectively, the square root of the velocities, as given by X/N, and by $(Y/N)^{\frac{1}{2}}$, into these constants. This can be done by slide rule if N is counted and X and Y are measured in centimeters, but it is usually faster to employ a combined multiplying nomogram measuring tool. It is easy to see how such a device should be constructed. It should give the same reading for constant ratios Y/N etc.; e.g., by plotting N at right angles to the displacement and drawing in lines of constant slope each carrying the appropriate value of d. Taking into account the polarity on the plates and the sense of the deflection, a complete analysis of the charge and size distribution of the sample observed is obtained.

Limitations

It is essential to investigate some of the numerous instrumental limitations of the method. The most serious shortcoming of the arrangement described here is the tendency to develop convection currents inside the analyzer. If the convection has a horizontal component, it is easily detected and the experiment can be decleared invalid. But it so happens that the convection currents most often are vertically upward or downward in the region that is photographed because the walls of the chamber are either cooler or warmer than the plates. In the former, the small particles whose settling rate is less than the air velocity are swept out, while those larger particles that reach the field of vision are slowed down and appear smaller than they really are. Usually this difficulty quickly becomes apparent by the general dearth and rather sudden cut-off of particles of small size. The vertically downward convection however, is not only the most likely convection, as pointed out before, but also the one most easily overlooked. All that happens is that the particles settle faster than in still air, thus appearing too large, an effect counteracting the shape correction. When the average velocities of the particles seem to increase with time, increasing downward convection is the most probable cause.

The most definite check on the reliability of the size measurement and thus indirectly on the charge estimates is a comparison of the observed size distribution with a microscopically determined one. Especially convincing is a check against the dust that is collected on a cover glass placed below the region observed in the measurement. This should be done in every crucial experiment. When the size frequency curves for the photographed and collected samples show great similarity, the effect of convection can be considered negligible.

Another critical factor is the width of the slit. It will be desirable to make the slit as narrow as possible in order to admit the particles that are falling very close to the focal plane only. On the other hand, it is clear that highly charged small particles will have less chance of passing through a narrow slit than those with lower charge or larger diameter because of the attraction by the induced charges in the slit material. In order to keep this disturbing variation of the effective slit width below 10% for a reasonable upper limit of charge on the smallest particles to be studied, the slit will have to have a minimum width. A crude calculation giving a first approximation appears to be very simple. Consider a particle of charge q approaching the slit very much closer to one edge than to the other. Assume now that the slit materials is not only conducting but that the total image charge $-q$ is concentrated on the very egde at the point of closest approach of the particle trajectory. Then the effective collision radius of the charged particle provided q is large is given by:

$$a' = 2(C/V)^{\frac{1}{2}}, \tag{3.28}$$

where $C = q^2/3\pi\eta d$ is due to COULOMB's and STOKES' laws and $V = g\varrho d^2/18\eta$ is the settling rate of the particle. Thus

$$a' = \left(\frac{24}{\pi g} \cdot \frac{q^2}{\varrho d^3}\right)^{\frac{1}{2}}. \tag{3.29}$$

If this is required to be less than 5% of the slit width S, we have

$$S > 20 a' \cong 0.001\, \frac{q}{\varrho^{\frac{1}{2}} d^{\frac{3}{2}}}, \tag{3.30}$$

where q is now to be expressed in electrons and d in microns. For $q = 200$ and $d = 1$

$$S > (0.2/\varrho^{\frac{1}{2}})\ \text{cm},$$

i.e., in general of the order of 1 mm. From this it is also clear how important it is to use a material of low dielectric constant to form the slit, since in that case the attraction will be much smaller than here calculated. In addition, it sets an upper limit for the density of the clouds to be studied. Since it would not be desirable to obtain more than N tracks per picture and since the volume visible at any time is AS where A is the area of the field of vision, we must require a density

$$n \leq N/(AS). \tag{3.31}$$

In one case, $A = 0.2\ \text{cm}^2$, $S = 0.1$ cm, and taking $n = 50$, we find $N \leq 2500\ \text{cm}^{-2}$, a number which is at the same time small enough to make any agglomeration during settling highly improbable.

The third limitation to be considered is due to the effect of the electric field which makes itself felt at some distance above the plates. In order not to lose small highly charged particles which will drift sideways under influence of the field above the plates during an exposure, the polarity is reversed for successive observations and the time between pictures is chosen long enough to select the particles coming from a region relatively unaffected. This minimum interval T will depend on the charge q and the size of the particles, i.e., their settling rate V_0, the duration of the previous exposure τ, and the dimensions involved. As a first approximation, we find for large q and a permitted displacement $l/4$ of particles of diameter d

$$T = \frac{l_0}{V_0} + \frac{l}{2V_0} \cot \frac{3\pi^2 \eta\, l^2 d}{2V\tau q} \tag{3.32}$$

where l_0 is the distance between the observed region and the upper edges of the conducting plates. l is again the distance between plates and V their potential difference. As an example for one of the cases, $l = 0.8$ cm, $l_0 = 0.5$ cm, $V = 180$ volts, $\tau = 4$ sec, $q = 200$ electrons, $d = 1$ micron, and for quartz, $V_0 = 0.008$ cm/sec, which yield $T = 110$ sec. This means that 4-sec exposures at 2-min intervals will include all particles with $d \geq 1$ micron and $q \leq 200$ electrons, but may lose particles exceeding one of the limits. The relation between the critical values of q and d that will be just admitted is implied in (3.32) but is rather complicated as V_0 is proportional to d^2.

2. The measurements of Dodd on liquids

In the Hopper and Laby method, the aerosol particles are introduced into an analyzing chamber where they settle in still air between two vertical plane parallel electrodes. Under the influence of gravity and a constant horizontal electric field, each particle has a Stokes' law velocity whose vertical component depends upon the particle size and whose horizontal component depends upon the ratio of particle charge to its size. Intermittent dark field illumination of the particles allows their paths to be photographed as a line of equally spaced spots, from which the velocity components and thus the simultaneous size and charge of the particles can be obtained.

The apparatus was adapted by minor modifications from the one built and described by Kunkel and Hansen and was operated in essentially the same manner.

The particles analyzed obeyed Stokes' law, the upper diameter limit taken to be that given by Arnold. Cunningham's correction was not needed since particles of diameter less than about one micron were not analyzed in order to avoid Brownian motion, undue length of time for particles to settle into the analyzing region, and increased susceptibility to errors caused by minute air currents.

Instrumental errors in the determination of the diameter and charge of an individual aerosol particle amounted to about 1%. However, rather than carefully measuring the film recorded particle paths, nomographic devices were used for simplicity. Although this increased the diameter error to about 5% and the charge error to about 10%, the error was of random sign and thus was not so large for the average values of a large number of particles.

This method of analysis actually reveals the initial particle diameters and charges only in so far as changes in these quantities or selective removal of certain particles does not occur between the times of aerosol origin and analysis. Various processes which might cause such error were carefully considered in order to determine their effect during the particle settling period.

Coagulation of particles resulting from contacts caused by Brownian motion or electrostatic interaction and charging of particles by ion diffusion to the particles were found to be negligible for the low particle charges, concentrations, and settling times encountered.

The stray field above the deflection plates may cause lateral removal of settling particles having a high ratio of charge to diameter so that this class of particles escapes analysis. This effect was avoided by successively alternating the polarity of the applied field from one periodic observation to the next and by adjusting the field strength, duration, and field interval. Values of the ratio charge to diameter as high as were needed were easily obtained for any diameter particle analyzed.

A prominent source of error in the analysis of droplet diameters was the rapid evaporation of small droplets. This process can be approximated by

$$-dS/dt = 8\pi p\, MD/\varrho\, RT = C, \qquad (3.33)$$

where S = droplet surface, t = time, p = droplet vapor [82A] pressure, M = droplet vapor molecular weight, D = droplet vapor diffusion coefficient in air, ϱ = droplet density, R = molar gas constant, and T = absolute temperature.

Evaporation rates calculated form (3.33) have negligible deviation from complete evaporation theory for diameters above about 20 microns and are only slightly high for diameters down to the order of one micron. However, experimentally observed evaporation rates often disagree with (3.33) because of surface impurities, etc., and hence should not be used for evaporation corrections. Nevertheless (3.33) can serve as a guide in selecting slowly evaporating liquids, e.g., for water droplets in air (3.33) give $C = 10^{-4}$ cm²/sec, so that a 100-micron diameter water droplet might evaporate into a one-micron diameter droplet in the order of a few seconds. In fact, sprayed water droplets evaporated completely before reaching the analyzing region.

Using STOKES' law and $-dS/dt = C$, it follows that an evaporating droplet undergoes a constant change in the fourth power of its diameter while settling through a fixed distance H,

$$\Delta d^4 = \text{const.} = 36\eta\, CH/\pi\, \varrho\, g, \qquad (3.34)$$

where η = coefficient of air viscosity and g = gravitational acceleration. Although H was not strictly fixed owing to the initially disperse nature of the aerosol, an approximate value for the apparatus used in this investigation was adopted.

Partial neutralization of charged particles can be expected if they contact the inner surfaces of the grounded settling chamber in which the aerosol is formed. Care was taken to eliminate this effect, and it was found that most particles, especially the smaller ones, were carried past surfaces in the stream lines of moving air or shielded from surfaces by the adjacent static air cushion. A dust particle or oily droplet which does contact some surface generally remains attached there and hence does not contribute error to the analysis.

Statistical errors occur when only a finite number of particles are observed. In the case of a Gaussian distribution in charge among particles of a certain diameter with an average charge Q and a standard deviation S, there is an even chance that the average $\bar{q}$ of a sample consisting of a number n of these particles differs from Q by as much as $0.674 - S/n^{\frac{1}{2}}$ and for a large sample (say $n > 10$) that the standard deviation s differs from S by as much as $0.674 - S/(2n)^{\frac{1}{2}}$.

It was initially apparent that the distribution in charge among the particles included a wide range of charge and depended upon the particle diameter. In order to obtain good sampling, several hundred particle observations should be made in each of many small diameter ranges. Since insufficient data were obtained from the analysis of a single spray or dust dispersal, many runs were made, each in the same manner, and all data for each substance were combined. Data from individual runs were satifsactorily consistent in view of the disagreement to be expected between such small samples. In general, as the number of particles of a sample increases, its average charge and standard deviation approach the aerosol values, and in this investigation the sample values were used as the estimate of the aerosol values with the recognition that these values were subject to purely statistical inaccuracies.

Results

Size-charge observations of sprayed droplets were made for the following liquids: 1. Di-n-butyl sebacate—Eastman White Lable Grade. 2. Dibutyl phthalate—reagent grade further purified by multidistillation. 3. Oleic acid—reagent grade. 4. Paraffin oil—narrow fraction, aromatic and unsaturate impurities removed by sulfuric acid treatment, double distilled (mostly normal dodecane). 5. Nitrobenzene—reagent grade. 6. Mercury—highly purified.

The dibutyl esters and oliec acid were chosen in anticipation of negligible evaporation effects ($P \sim 5 \cdot 10^{-6}$, $3 \cdot 10^{-5}$, and 10^{-6} mm Hg, respectively). Paraffin oil was used to determine the charging of a nonpolar liquid free from molecular double bonding. Since these liquids all have low dielectric constants, nitrobenzene was tried in order to have some information about a high dielectric constant liquid in the hope that it would be possible to avoid evaporation effects. Mercury was chosen as a typical conducting liquid.

A glass sprayer was used (Parke Davis and Company "Glaseptic Nebulizer") with a spraying duration of 30 sec at reproducible pressure for each experimental run.

The data for the nonconducting liquids analyzed were segregated into the following diameter ranges in microns: 1.0–1.4, 1.5–1.9, 2.0–2.8, 2.9–3.9, 4.0–5.6, 5.7–7.9, 8.0–11.2, 11.3–15.8, 15.9–22.3, 22.4–31.6, and 31.7–44.6. It was found that the numbers of positively and negatively charged droplets in each diameter range were approximately equal, as were their average charges, and the net charge in each diameter range was approximately zero. This indicated the spray electrification distribution in charge among the droplets to be *symmetrical and net neutral*. The following description of the di-n-butyl sebacate droplet analysis is typical of the treatment of data for the other liquids.

The distribution in charge among the droplets in each diameter range (normalized to unit probability) was indicated by histograms to be closely Gaussian (except as mentioned later). Gaussian curves having zero average charge and dispersions equal to the mean square charge of the droplets in each diameter range were superimposed upon the histograms. This comparison is shown in Fig. 26 for the three most populated diameter ranges.

Further indication that the distribution is Gaussian is the close agreement between the average absolute charge $\langle |q| \rangle$ of the droplets in each diameter range and the value derived from their mean square charge on the assumption of a Gaussian charge distribution, $\langle |q| \rangle = (2\bar{q}^2/\pi)^{\frac{1}{2}}$.

Because of the Gaussian shape of the data histograms and the good agreement between the average absolute charges and their values calculated from the mean square charges, it was assumed that the charge distribution among droplets in each diameter range was actually Gaussian without further "goodness of fit" tests.

The droplet analysis is given in Table 11, which lists the ranges of diameter d in microns, the numbers of droplets observed n, the average absolute charge in electrons as observed and as calculated from the mean square charges, and the statistical probable errors p.e. $\sim 0.674 - \langle |q| \rangle / (2n)^{\frac{1}{2}}$. The numbers of droplets in diameter ranges not included in Table 11 are less than in those included and are considered too meager for satisfactory sampling.

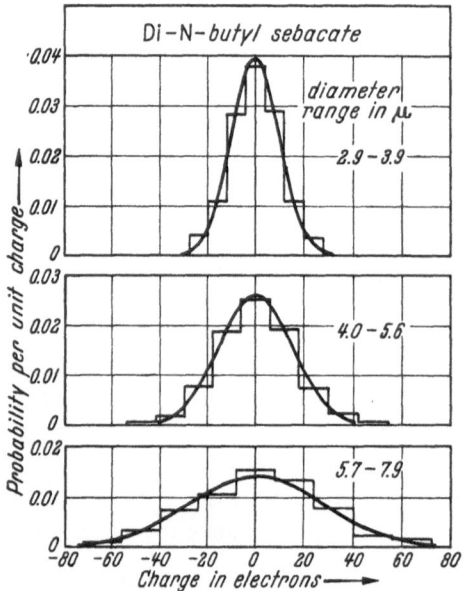

Fig. 26. Probability per unit charge against charge in electrons on droplets of Di-n-butyl sebacate for three diameters showing Gaussian character of the charges

The average absolute charge was found to be proportional to the $\frac{3}{2}$ power of droplet diameter except for a deviation in the smallest diameter range listed in Table 11. Here a discrepancy between the computed and the observed average absolute charges suggested that the analyzed charge distribution might not be Gaussian. In fact, the histogram of data in this range had a definite depression at zero charge, which was not present in the larger diameter ranges.

If we assume the true distribution to be Gaussian, a deficiency of neutral droplets in the smallest diameter range was occurring before their analysis in some manner as yet unexplained, resulting in an observed average absolute charge which was too large. It was found that the first half of the total number of droplets which settled into the analyzing region did not exhibit so pronounced an effect as did the second half. Therefore this effect was considered to be a time dependent sampling error rather than a true effect of the spray electrification mechanisms, although the sampling errors previously considered did not explain this effect.

The true average absolute charge was estimated by approximating the true charge distribution, and this estimate was inserted (parenthetically) into Table 11. Then a relation of the form $\langle |q| \rangle = A d^{\frac{3}{2}}$ fits the corrected data, given $A \sim 1.26$ electrons/micron$^{\frac{3}{2}}$.

Table 11. *Showing statistics on Di-n-butyl sebacate droplets as observed by* DODD

d	n	$\langle |q| \rangle$	$(2\bar{q}^2/\pi)^{\frac{1}{2}}$	p.e.
2.0–2.8	77	7.5	7.0	(± 0.28)
		(5.1)		
2.9–3.9	181	8.2	8.1	± 0.29
4.0–5.6	396	12.2	12.4	± 0.29
5.7–7.9	138	22.1	22.4	± 0.88
8.0–11.2	76	33.3	33.0	± 1.82
11.3–15.8	30	72.7	74.4	± 6.33
15.9–22.3	12	120.6	111.4	± 16.6
22.4–31.6	8	153.0	151.3	± 25.8

Analogous data were obtained for the dibutylphthalate. The smallest diameter range had to be corrected for evaporation by (3.34). Then for this liquid $\langle |q| \rangle = A d^{\frac{3}{2}}$ with $A = 4.16$ electrons/micron$^{\frac{3}{2}}$. Evaporation and deficiency in neutral droplets reduced accuracy for the smallest droplet diameter in oleic acid but the larger diameter gave $A = 1.31$ electrons/micron$^{\frac{3}{2}}$. Parrafin oil caused trouble in spraying because of its high viscosity and loss of small droplets by evaporation; data limited to a smaller range of droplets yielded $A = 0.32$ electrons per micron$^{\frac{3}{2}}$. Data for nitrobenzene were very meager and required extensive evaporation correction. The $\langle |q| \rangle = A\, d^a$ law did not fit *all* nitrobenzene data *well* for any a. When $a = \frac{3}{2}$ was used, the two most populated diameter ranges gave $A = 0.97$ electrons/micron$^{\frac{3}{2}}$. Aniline and glycerol were tried and found unsuitable for analysis by this apparatus, and no attempt could be made to evaluate A for these liquids.

The presence of relatively large charges with a net neutral Gaussian distribution among the droplets indicates that the Lenard liquid-air interface spray electrification is *not dominant* in the electrification by atomization of the microscopic droplets of liquids here observed. This result must be interpreted as follows:

No doubt all sizes of droplets are found and some particles down to 10^{-7} to 10^{-6} cm diameter are created as Chapman noted. These must exhibit the spray charging with finer particles negative and larger ones positive. Such relatively fine particles may be present floating around in the settling chamber but are not detected. In spray electrification the net negative charge in the air, i.e., the total negative charge in the finest spray less the total positive charge on the somewhat coarser, but still fine particles, must be compensated by a net positive charge on the sprayer liquid and the very large drops such as observed here. The net negative charge may be relatively small with these liquids and the compensating positive charge could reside largely on the sprayer and thus leak to ground. The fine negative charge eventually reaches the settling chamber walls, and can be studied only whith a device like the Erickson tube. These observations showing no effective net positive charge on the droplets studied, i.e., symmetry, means that the net negative spray charge may be small for these substances, and that relatively the number of large particles of net neutral charge here studied is so great that the residual positive charge distributed over the mass plus that on the sprayer may not be noticed.

To account for the homogeneous charging which in this instance is a true volume effect with $\langle |q| \rangle \propto V^{\frac{1}{2}}$ or $q^2 \propto V$, where V is the droplet volume, Loeb suggested that this charging would be expected to result *from statistical fluctuations of the electrolytic ion concentrations of both signs present in relatively low concentrations compared to those in aquous solutions*. Statistics described by Margenau and Murphy[83] are readily adapted to the analysis. Let the positive and negative ions be univalent and have equal concentrations in the liquid. Then the probability that a droplet with n ions has m positive ions is $P_n(m) = \binom{n}{m} \big/ (2^m \cdot 2^{n-m}$. As n becomes large, $P(m)$ tends towards $P(m) = \text{const. exp.} [-(m - \bar{m}^2/m]$, where $\bar{m}$ is the mean value of m. The droplet charge is $q = 2m - n$ so that $m = (q + n)/2$ and $P(q) = \text{const. exp} [-(q - \bar{q})^2/4m]$.

A droplet of volume V may contain any number of ions n, but the average number will by $\bar{n} = 2VN$, where $N =$ the ion pair concentration $=$ concentration of dissociated molecules. Thus $\bar{m} = n/2 = VN$ and $\bar{q} = (2\bar{m} - \bar{n}) = 0$. Thus, it is clear that $P(q) = \text{const. exp} (-q^2/4VN)$.

The droplet charge probability is a net neutral Gaussian distribution with a standard deviation of $S = (2VN)^{\frac{1}{2}}$. Accordingly $\bar{q}^{-2} = 2VN$ and $\langle |q| \rangle = (2/\pi)^{\frac{1}{2}}(S) = (4VN/\pi)^{\frac{1}{2}}$. Empirically it was seen that $\langle |q| \rangle = A d^{\frac{3}{2}}$ so that $N = \frac{3}{2}A^2$. For dibutyl sebacate the value of A was 1.26 electrons/micron$^{\frac{3}{2}}$ yielding $N = 2.3 \times 10^{12}$ ion pairs per cm^3. This is of the order of the number of ion pairs to be expected in a relatively pure substance of low dielectric constant.

It was logical following the success of this approach to consider studying these with known but varied ion concentrations. Before this was undertaken it was discovered that in 1949 NATANSON[84] working in the U.S.S.R. had made a rough study of sprayed transformer oil droplets by means of a sinusoidal alternating potential between vertical plates. With droplet diameters of 1 to 4.2 microns the value of q^{-2}/V was found to vary linearly with the conductivity of the transformer oil. The references cited by NATANSON indicate that at various times publications not listed in *Physics Abstracts* had theoretically considered this type of charging largely, however, in relation to electrolytic solutions. Actually NATANSON was the only one to apply this experimentally to homogeneous charging of sprayed droplets. As early 1911 H. BATEMAN[85], in an endeavor to illustrate the usefulness of certain statistical methods developed by him, had arrived at the quantitative relations above and had indicated that they might be applied to the analysis of charges produced when *electrolytic solutions* were dispersed. In 1946 LEONTOVICH[86] in the U.S.S.R. applied the theory of fluctuations to the free energy of electrolytic solutions, and NATANSON[81] in 1946 applied the Boltzmann probability to the energy of formation of a charged spherical region within an electrolytic solution in the presence of thermal motions. They also obtained a net neutral Gaussian charge probability with $\bar{q}^2 = 2VN$. It was thus believed that the work of NATANSON on transformer oil strongly indicated the accuracy of the inferences made independently here when considered in the light of the relatively accurate data obtained from the Hopper and Laby method.

Table 12. *Correlation of charge and dielectric constant as observed by* DODD *for a number of liquids*

Liquid	Dielectric constant	A in electrons per micron$^{\frac{3}{2}}$
Paraffin oil	2.1	0.32
Di-n-butyl sebacate .	4.4	1.26
Dibutyl phthalate .	6.2	4.16
Oleic acid	2.5	1.31
Nitrobenzene . . .	35.7	0.97

It must be pointed out that the statistical analyses apply only to "droplets to be" within the bulk liquid. If they are to apply to separated droplets, the act of spraying must not change the concentration probability by liquid conductivity during the finite separation time. Such changes were not apparent in this investigation.

It has been indicated by CHAPMAN's early studies that the greater the dielectric constant of pure sprayed liquids, the higher the charge for a given diameter droplet. Table 12 indicates the dielectric constant of the substances here studied in relation to the constant A. It is seen that the highly purified paraffin oil and dibutyl esters do show charge increase with increasing dielectric constant. In contrast oleic acid with a low dielectric constant gave a large A, indicating that it has a larger concentration of ions. This is what is to be expected, since the oleic acid was reagent grade and not very pure. The position of nitrobenzene is also highly improper. It was reagent grade with a high dielectric constant. Much impurity thus should have gived high concentrations of ions and a large A. The causes for its deviation are perhaps many. Actually the data really apply to too few substances for any proper conclusions to be drawn. It appears as noted by CHAPMAN that for the pure substances the value of A increases as the dielectric constant increases. If this is correct, it can only mean that for a generally constant concentration of certain types of ionizable impurities in the first three liquids the concentration N of dissociated molecules is larger the higher the dielectric constant. Such action should follow on the principle that dielectric constant is largely responsible for electrolytic dissociation by reducing the electrostatic forces between ions in molecules. This interpretation of the general role of dielectric constant presents a naive view. High polarizability, meaning large molecular dipole moments of bulky molecules, does not suffice to reduce ionic binding forces in molecules surrounded by these, since much stronger linkages, such as the hydrogen bond, are the effective dissociating agents in substances like water. This most likely explains the behavior of nitrobenzene and is accepted as such by the chemists. Any relation between dielectric constant and A in homogeneous charging if present can only depend on the effect of the dielectric constant on N, if any.

3. Significance of DODD's results

DODD's results clearly indicate that aside from carriers derived from the thin surface layers involved in the surface structure and the uncovered ions of the endosmotic layer mostly lying below 10^{-6} cm in diameter there is produced on dispersal of ion containing liquids into droplets an *overlying symmetrical* charging process. This cloud of particles of various sizes has a net neutral charge yet contains a continuous Gaussian distribution of equal numbers of charged particles of opposite sign in each size group. This *symmetrical* charging arises from pure statistical considerations depending on the chance that one particle on separation has more ions of one sign than another. DODD at the author's suggestion along with others has calculated the charge distribution as a function of drop size and ion density and has confirmed the law. This symmetrical charging will overlie the asymmetrical charging produced by the break up of thin sections of the surface layers of all liquids containing electrolytic ions.

It is therefore of interest to note at what *ion concentrations* and *drop sizes* the *symmetrical charging* will *appear* and *mask* the data resulting from the more interesting waterfall, or spray, effect. Actually what is of interest would be to note for what drop size and ion concentration the chance of observing singly charged droplets by the symmetrical charging process is good. This calculation would be tedious. However, Table 13 calculated in terms of DODD's theory and observations will provide a very helpful guide.

This table is calculated on a purely statistical basis depending on number of ions assuming a non-conducting liquid. If the liquid is conducting the charge separated will be limited by return currents during separation. Again high ion densities will by Coulomb fields limit such transfer. These actions are not considered. The calculation is based on a Poisson distribution obtained by STIRLING's approximation. For small charges, the more accurate Gaussian distribution given by the original series should be used.

In column 1 are listed the molar concentration of the ionic solvent, in moles per liter. This represents normal solutions for univalent ions. The second column

8*

Table 13. *Values of ion concentration, droplet sizes and average charge calculated on the basis of* DODD's *statistical theory*. This indicates regions where symmetrical charging may be anticipated as a complicating phenomenon

1 Molar conc.	2 N	3 $\sqrt{\tfrac{2}{3}N}$	4 Diam. d cm	5 $d^{\frac{3}{2}}$	6 $\langle|q|\rangle$	7 Comment
1	6×10^{20}	2×10^{10}	10^{-4}	10^{-6}	2×10^{4}	
1	6×10^{20}	2×10^{10}	10^{-5}	$10^{-7.5}$	630	
1	6×10^{20}	2×10^{10}	10^{-6}	10^{-9}	20	Ten H_2O diameters
1	6×10^{20}	2×10^{10}	4×10^{-7}	$6\times10^{-10.5}$	3.8	2×10^{3} molecules
10^{-2}	6×10^{18}	2×10^{9}	10^{-4}	10^{-6}	2×10^{3}	
10^{-2}	6×10^{18}	2×10^{9}	10^{-5}	$10^{-7.5}$	63	Langevin ions 10^{6} mole-
10^{-2}	6×10^{18}	2×10^{9}	10^{-6}	10^{-9}	2.0	cules
10^{-4}	6×10^{16}	2×10^{8}	10^{-4}	10^{-6}	200	
10^{-4}	6×10^{16}	2×10^{8}	10^{-5}	$10^{-7.5}$	6.3	Langevin ions 10^{6} mole-
10^{-4}	6×10^{16}	2×10^{8}	10^{-6}	10^{-9}	2×10^{-1}	cules
2.1×10^{-6}	1.25×10^{15}	2.9×10^{7}	10^{-4}	10^{-6}	29	Conductivity water high
2.1×10^{-6}	1.25×10^{15}	2.9×10^{7}	10^{-5}	$10^{-7.5}$	0.91	purity
10^{-8}	6×10^{12}	2×10^{6}	10^{-3}	$10^{-4.5}$	63	Organic liquids
10^{-8}	6×10^{12}	2×10^{6}	10^{-4}	10^{-6}	2	di-n-Butyl sebacate
2.5×10^{-10}	1.5×10^{11}	3.1×10^{5}	10^{-3}	$10^{-4.5}$	10	Ameroil
10^{-10}	6×10^{10}	2.5×10^{5}	10^{-3}	$10^{-4.5}$	0.03	

gives the number N of ions of both sign per cm³ expected to the first significant figure. The next column give $\sqrt{\tfrac{2}{3}N}$. The fourth column gives the various drop diameters d encountered in spraying and bubbling in cm. Column 5 gives $d^{\frac{3}{2}}$ and column 6 gives the computed value of $\langle|q|\rangle$ the representative number of excess charges of one sign, while column 7 presents certain comments.

It is at once noted that the molar concentration yields average charges for micron, (10^{-4} cm), diameter drops which are quite large. More important it is to note that even for droplets of the diameter of the thickness of a soap solution film before it breaks, the average carried by a drop is 3 electrons. It should be noted that while the conductivity of the molar solution might preclude the occurrence of 2×10^{4} charges on 10^{-4} cm droplets, if the separation process allowed time enough for ion movement, the potentials and return currents for the excess charges of 2 or 3 electrons on smaller drops cannot equalize as readily so that conductivity would not interfere with charging in this region. This means that with molar solutions symmetrical charging could effectively mask most of the spectrum of charge carriers produced by asymmetrical charging. What is worse, it will complicate matters by having small droplets of the same size with more than one electron of charge. This could even introduce *new* groups of ions. However, with a *spread of drop sizes* such solutions should virtually obliterate all ion groups and give *continuous spectra* of charge from mobilities extending upward of perhaps 0.3 cm²/volt sec. At 10^{-2} molar concentration symmetrical charging extends down to the 10^{-6} cm diameter particles belonging to the class of Langevin ions containing roughly 10^{6} molecules with mobilities of the order of 0.003 cm² per volt sec. However, it is noticed that the 2 electrons per drop charge are of fairly frequent occurrence on these drops. When the low efficiency of the asymmetrical charging is noted with only one in 10^4 of the created droplets charged, there can still be some masking of the lower mobility carriers of the spectrum by this process.

At 10^{-4} molar solutions the chance of serious masking for smaller droplets ceases, despite the small fraction of droplets containing uncovered anions. When the concentrations of carriers responsible for the conductivity of really good conductivity water, according to Alty[59], (largely dissociated H^+ and OH^- ions), are reached it becomes unlikely that much symmetrical charging will occur and the mobility spectrum of carriers from pure bubbled water will result chiefly from electrons, OH^-, H^+ and the few impurity ions located in the segregated portions of the surface structure. Below the concentrations represented by conductivity water aqueous solutions need not to be of concern. However, the charging observed in spraying various liquids as carried out by Dodd is symmetrical and the sizes of the droplets for manifestations of this charging with different low ion concentrations should be noted.

One feature of the statistical theory of Dodd must disturb the reader as it did the author. This is the apparently indefinite increase in $\langle |q| \rangle$ as drop size and ion densities increase. If this were possible why would not extraordinarily large charges of this type be observed by spraying a liquid like Hg in which there are 10^{22} electrons/cm^3 of metal. It is true that charging occurs with Hg metal as shown by Dodd. It is here that care in study and observation is needed. Hg is an excellent conductor. It also derives charge in contact with glass and other surfaces. Careful control indicated that Dodd's *whole Hg surface was charged equipotentially* and this charge was created by loss of electrons by Hg to the glass container as seen in the Chap. IV on *Contact charging between insulators and metals*. Thus if his drops come from the conducting surface, each drop carried away a charge of electricity at separation equivalent to its capacity times the potential of the surface. The 10^{22} electrons per cm^3 it is true, are bound by a surface potential barrier and there is an electron cloud outside the surface held to the surface by image forces. No concentration of mechanical energy on thin enough films of the mercury is adequate to overcome the work function and remove the electrons from the double layer. For this reason *there is no true spray electrification of molten metal surfaces* and *there should be no symmetrical charging unless forces and/or mechanisms occur for tearing single charged atoms from the surface*. Evaporation of atoms will, in fact, remove single atoms but ions cannot escape in this fashion as the heat of vaporization of ions against image forces is very much higher than that for neutral atoms.

Returning from the extreme case of an electronic conductor such as a metal to that of a molar solution the question of charge limitation in the process of symmetrical charging will depend on the field produced at the surface of the droplet say of 2×10^{-4} cm diameter when it receives, for example, 2×10^4 electrons. The capacity of the sphere of radius r is r cm. The charge amounts to 10^{-5} esu which gives a potential of 10^{-1} esu or 30 volts. The surface field of the drop will be 3×10^5 volts/cm. Such a field is sufficient to create ionization by collision of occasional electrons in the air, it will sweep in existing air ions and thus seek to neutralize itself if the drop ever could acquire that charge. The fields which can produce a discharge at the surface of water drops > 0.06 cm in diameter have been shown by ZELENY, MACKY and by ENGLISH[88] to be able to disrupt the water surface below discharge values. It is not known if this holds for small drops. Thus physical limitations of a drastic sort are imposed on the net charge that can be acquired by any droplets.

Assume however that the charge accumulated was just 0.1 of that above. At the instant of separation from the main body of liquid, the field would then be less than but commensurate with 3 000 volts/cm. One might calculate the time taken with such a field acting across a junction area of 10^{-12} cm² during separation to neutralize the charge producing the field by current flow. This time is of the order of a microsecond. Such a calculation really makes little sense since the field would not build up to that value under these conditions. Thus no charges of a symmetrical *type of this order of magnitude will be observed with solutions* of such conductivity. However, many larger droplets will charge up with a *number* of *unit charges*. The phenomenon of larger charges is observed in poorly conducting substances and in better conductors for smaller particles. In this event, mechanical forces are adequate to break up the liquid into appropriately small particles. The few single charges created, though on droplets of small radius, are separated from other charges sufficiently so that fields do not cause discharge in the micro volumes. Small droplets of 10^{-6} cm or less in radius for example while having very intense surface fields, even with an electron of charge, cannot cause electrical

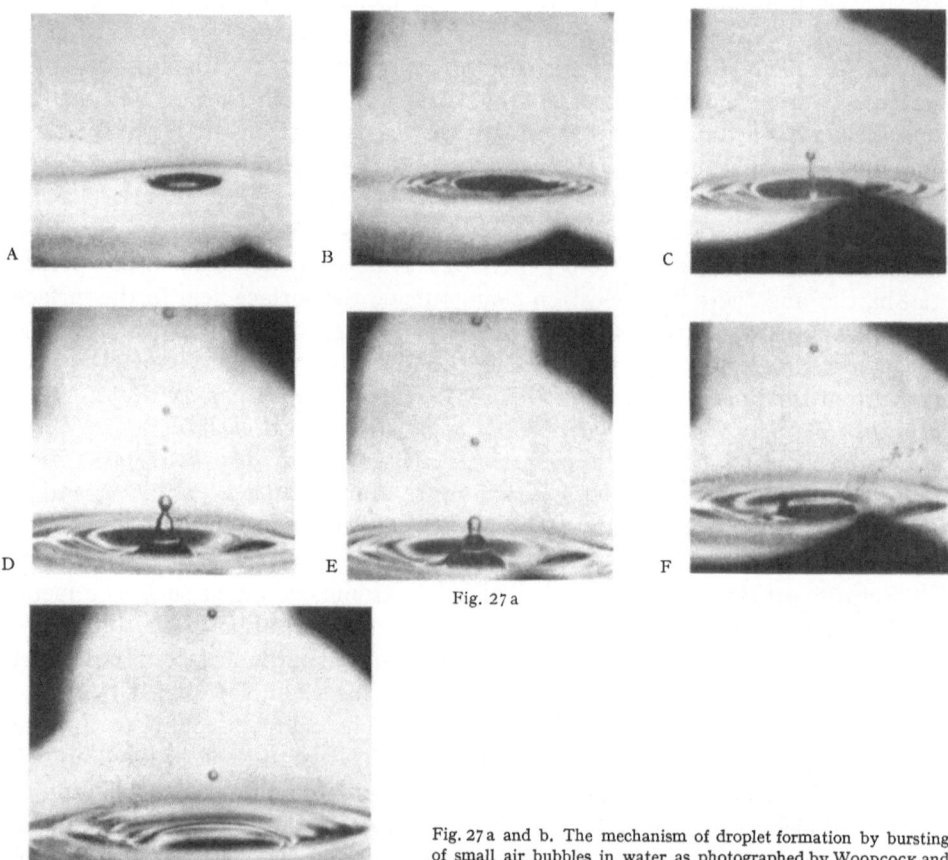

Fig. 27 a

Fig. 27 a and b. The mechanism of droplet formation by bursting of small air bubbles in water as photographed by WOODCOCK and associates and reproduced from their article in Tellus[90]

breakdown of the gas as the high fields extend over distances so short compared to electron free paths that the electrons cannot gain ionizing energy. Electrostatic forces between ions and molecules in the droplet are adequate to prevent evaporation or disruption. Thus the symmetrical statistical charging can and does appear in those regions, with aqueous solutions and for larger droplets with more insulating liquids. In view of this, care must be used in extrapolating and applying DODD's law, while at the same time, its influence on spray electrification data cannot always be ignored.

G. Asymmetric charge distribution of droplets of micron size from bursting of small air bubbles in concentrated solutions

In recent years, the appearance of positively charged droplets projected upward from the surface of sea water on the bursting of fine bubbles has been reported. These charged droplets occuring in the visible range of 10^{-3} cm diameter and upward, present an enigma in terms of the mechanism and nature of spray electrification.

Meteorological studies of the salt nuclei from the sea air, effective in causing corrosion of metal structures and in connection with rain drops by A. H. WOOD-COCK[89], at the Woods Hole Institute of Oceanography, led to further studies of the origin of these nuclei in the sea. The study of the size distribution of the sea salt particles observed in the atmosphere provided a clue to the origin of the salt

Fig. 27 b

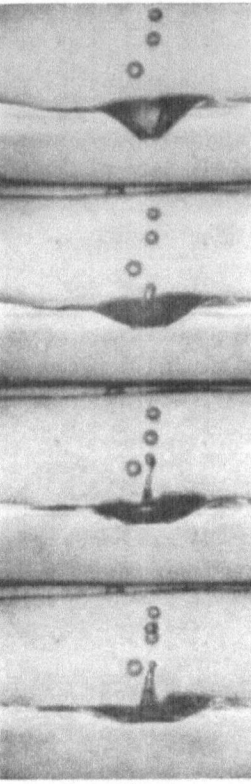
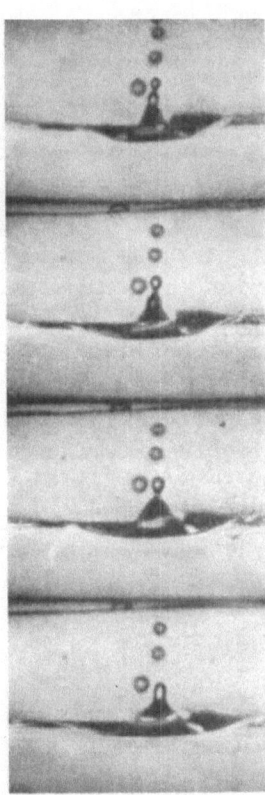

in the production of fine water droplets as arising from the ordinary turbulence
of sea water in the breaking white caps at sea. A study of these indicated that
the projection of the droplets of water into the atmosphere arose through the
bursting of minute air bubbles entrapped well below the surface in the wave
turbulence. This was demonstrated by WOODCOCK, KIENTZLER, ARONS, and
BLANCHARD[90], through high speed stroboscopic motion pictures of the bursting
of bubbles in salt water, using speeds of more than 3000 frames/sec. The bubbles
varied in diameter from 1.7 mm down to 0.04 mm. The ensuing droplets projected from the bubble ranged in size from 0.174 to 0.0038 mm diameter. Figs. 27a
and 27b are taken from the paper in Tellus and show the mechanism of breaking.
Table 14 gives the relation between the weight of salt, the diameter of the droplet,
and the approximate bubble diameter needed to produce the droplets. The
weight of the sea salt in micro-micrograms corre-
sponds well with the weight of salt particles
collected in sea air or in rain droplets. This
bubble producing process was studied further by
BLANCHARD[91]. He found that by forcing air
through finely drawn glass capillaries at air pressures ~ 30 psi, some cm below a water surface,
he could produce a uniform stream of bubbles,
each of which yielded some 5 or 6 droplets projecting them into the air various heights depending on bubble size. A previous study of spray
production on pure water preceding BLANCHARD'S
work had been done by O. STUHLMAN[92]. In
studying the size and distribution of the droplets
projected in this fashion by means of fall velocity
and STOKES' law, BLANCHARD[93] discovered that

Table 14. *The weight of salt, diameter of sea water droplet and
approximate diameter of the air
bubble associated with these in
ocean spray*

Weight of sea salt nucleus in 10^{-12} gm	Diameter of sea water droplet from which nucleus came in 10^{-4} cm	Approx. diameter of bubble yielding droplet size. In 10^{-4} cm
1	3.8	40
10	8.1	80
10^2	17.4	170
10^3	37.6	380
10^4	91.0	910
10^5	174.0	1700

the droplets were charged. The size of the droplet and its charge could be
determined first by applying a vertical field countering the action of gravity
until the droplet remained stationary, and then reducing the field such that the
speed of the particle under reduced field plus gravity could be measured. In
this fashion, using STOKES' law, the charge and the radii of the droplets could
be ascertained. In Fig. 28 there are assembled all the data to date. The drop
sizes ranged in radius from around 10^{-3} mm up to 5×10^{-2} mm. The straight line
A represents charged drops created by induction as the droplet separated from
the bubble. For a given radius, the induced charge was constant for fields of
from 50 to 300 volts/cm and the charge increased up to 10^6 electrons. With the
high conductivity of the salt solution, this indicated that the charge segregated
was limited by some back discharge or current flow mechanism and not determined entirely by polarization in the field. The curves B, C, and D represent
charges on bubbles in the absence of any but the contact potential fields.
Here the small droplets less than 3×10^{-3} mm in radius follow a law of their
own with charge increasing nearly as r^5, while the second group for drops between 3 and 6×10^{-3} mm follow about an r^2 law. All of these positively charged
droplets rise to less than 2 mm above the water surface.

The unique charging of such large droplets does not properly fit into the type
of spray mechanisms considered, where the charged droplets came from the very

small drops. Of course, larger isolated sections of water from which small negative or positive tips were torn off, should show residual positive or negative charge. Since the data were not taken under conditions of field free space and because of the difficulties in accounting for the observations on kown theory, the author asked BLANCHARD to repeat his mea-surements. Since BLANCHARD was engaged in other work, requiring his absence at sea, he had little chance to do the work as intensively as desired.

The generation studies reported to the author in a private communi-cation, dated May 8, 1956, were pur-sued by bubbling in a shielded field free space in which CHALMERS and PASQUIL's filter paper technique was used to present only pure water sur-faces to the breaking droplets, with no metal surfaces exposed. Thus the maximum fields were less than 0.26 volts over some cm. The bubb-les were also allowed to break in fields of ∼4 volts/cm. There was no charging or change of charge observed due to this field.

The observations first were applied to drops projected some 4 to 12 cm above the water surface. All of these were *negatively* charged. Then drop-lets were studied from smaller bubb-les rising only 3 to 4mm. Here again the droplets were *negative*. The drops in this size range had previously been positively charged. The mystery was clarified when it was found that small bubbles rising from a tip and passing through *about 1 cm of* H_2O

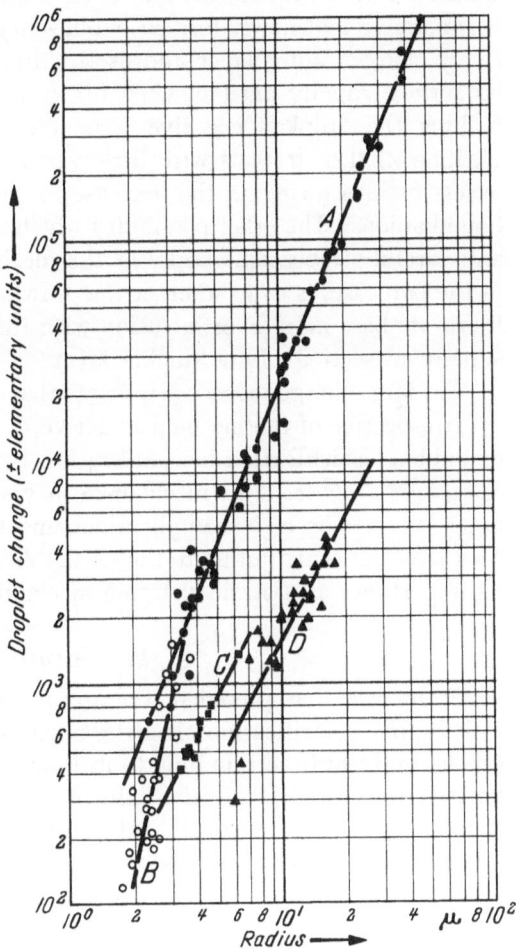

Fig. 28. Curves observed by BLANCHARD between droplet charge and radius for droplets produced by breaking of small bubbles. Curves B, C, and D are for natural charge on droplets. Curve A is for bubbles allowed to break in an electric field, hence are charged by induction

yielded a top drop that had negative charge. If the bubbles rose for several cm through sea water before reaching the surface, the droplets produced were positive as before. The glass tip was not concerned. Bubbles that were created by pouring sea water into sea water rising up from depths of several cm and forming thus slowly, all yielded small droplets rising less than 2 mm above the surface and all positively charged.

The phenomenon, which may have far reaching implications in the mainten-ance of the earth's electric charge relative to the atmosphere and other meteoro-logical phenomena, as salt particles are found far inland in the atmosphere and in rain, is thus a new phenomenon.

The clue to the explanation may come from two possible causes. As indicated by ALTY and McTAGGART for fairly dilute solutions and pure water, the relaxation times for the achievement of equilibria of ions at surface layers in dealing with small bubbles were relatively long. In the concentrated, 0.3 normal salt solution that is sea water, the times may be less than the hundreds of seconds noted. Thus it is possible that newly formed bubbles may have an initial more strongly negatively charged outer layer and as the droplet forms from this layer expanding its surface rapidly until pinched off, it may have many anions in the surface making the droplets negative. The more slowly rising and forming bubbles yielding smaller droplets with little vertical velocity may produce droplets with excess cations owing to the decrease in surface concentration produced by the bound anions. The other possibility which needs investigation by more extensive and careful stroboscopic study of the mechanics of breaking off of the droplets in the two cases is as to whether the drawing out of the fine filament as the drop leaves its base may favor inclusion with the drop of more or less of the negative superficial section of the double layer.

The large charges observed indicate that the first mechanism is the more likely.

Irrespective of the mechanism active, it is clear that a new asymmetric spray or bubble electrification mechanism has been found applying to *strong solutions and larger droplets*. This mechanism in view of the vast ocean areas may have extensive meteorological implications and requires much more extensive study. As a factor in contributing to the earth's electrical field maintainance it manifests its importance as a static charging mechanism in its own right.

H. Conclusions

From what has preceded, there appears to be a clearly defined mechanism called spray electrification. It arises in consequence of the orientation of electrical dipoles inherent in certain liquid substances at a gas-liquid interface with negative polarity outward and positive inward, if the measurements of COEHN have any significance. In any event, the situation holds in water where owing to its nature a large majority of the surface molecular dipoles are oriented according to CHALMERS and PASQUILL. Such oriented dipoles bind anions of the liquid, or those dissolved in the liquid, rather tightly to their inner positive ends. The weaker forces in the interior between attached anions and their corresponding cations allow only a fraction of the anions to be "*covered*" by their opposite cations, the remainder of the cations being in dynamic equilibrium in the liquid in consequence of thermal motion. The amount of anion adsorbed and the fraction of these covered vary with the nature of the anion, its influence on surface tension, the cation and the ion densities in solution. The *uncovered anions* and *their loosely bound partner cations form* the *mobile cataphoretic*, and *endosmotic double layer leading to the Helmholtz potential* ζ at liquid-gas interfaces. For distilled water this lies at around 0.06 volt while the total interfacial potential is 0.26 volts. Certain substances notably organics, can alter these potentials and even reverse them. Electrolytic ions also modify the covered and uncovered proportions. Depending on concentrations the formation of these layers have characteristic relaxation times which may influence spray electrification depending on mechanical processes involved.

Electrification results whenever the surfaces containing such double layers are disrupted in such a fashion as to create fine droplets involving various thicknesses of the double layer. Certainly with water the thinnest films on rupture appear to show carriers consisting in relativey dry air of possibly $2(H_2O)^-$ or O_2H^-, of $(3\,H_2O)H^+$ and $(3\,H_2O)OH^-$, and of some hundreds to a thousand molecules, including solute or impurity ions, e.g. Cl^-, Na^+ from the inner bound layer and perhaps a few molecules of undissociated salt in concentrated solutions. Beyond that, more nearly symmetrical charging with carriers of both signs in drops of up to 10^{-6} cm diameter having 10^6 molecules constituting the so called Langevin ions, appear in a more or less continuous spectrum. Larger charged droplets disappear in unsaturated air for the Langevin ions mark the limit of stability of droplets short of supersaturation, unless much salt is dissolved.

The number and distribution of the carriers produced depends critically on the mechanical processes of dispersion ion concentrations and relaxation times. The most successful for charging purposes is probably the breaking of large thin bubbles. More violent bubbling breaks up water into larger masses. Spraying results in many more larger particles as does impact on a surface. One more factor in charging which cannot be ignored is the time factor in disruption compared to the relaxation times needed to form the surfaces. With low ion content times may be in the order of seconds. With adequate ion content, it may fall to 10^{-2} sec or less and be comparable with the rate of surface creation. The humidity in which the carriers are produced and exist alters the charge spectrum by enhancing certain groups by condensation or evaporation relative to others. The efficiency of charge generation is not high. Many of the larger stable droplets of both signs of charge in equal number accompany most dispersion processes. Some of these with adequate ions concentrations stem from symmetrical or statistical charging usually with smaller charges per particle. Something like 1 in 10^4 of 10^5 of the charged anions uncovered in the surface layer appear as charged drops in spraying. The total amount of positive or negative charge per gram of pure water sprayed or bubbled is around 7×10^{-9} coulomb of each sign and a differential excess of some 6×10^{-10} coulomb per gram is separated out and carried away by the air blast as fine negatively charged spray. This is the separation accounting for the waterfall effect. With suitable arrangements for catching sprayed liquid and the segregated fine mist considerable charges can be accumulated and stored so that suitably high potentials can be built up. If 100 grams of water are dispersed per second and the charge is segregated, a charging current of 6×10^{-8} ampere is achieved. Thus a capacity of 10^4 $\mu\mu F$ or 10^{-8} F will be charged to 6 esu or 1800 volts in a second. This sort of charging could readily be achieved by a jet of wet steam.

Care must be used in insulation owing to the effect of spray and humidity in causing leakage. Apparently in certain waterfalls where air currents and geometry are adequate, considerable sparks are drawn. These indicate primarily low capacities and much water sprayed. The presence of water droplets lowers sparking thresholds and encourages spark breakdown[130]. Probably the most effective generators of this type are connected with jets of wet steam at high pressure. Salts in very low concentrations may increase electrification, higher concentrations usually decrease asymmetrical charging and increase symmetrical charging and in some cases, salts at higher concentrations, may reverse the sign of charge. The process appears to be a *possible mechanism of charge generation in the turbulent wet region of thunderstorms below the freezing isotherm*. It is certainly one of the important mechanisms in the precipitation charging of aircraft by liquid precipitation in flight, but *may be altered by electrolytic mechanisms*. In industrial processes, the conditions under which it would occur for aqueous media provide themselves with discharging conditions such that static generation except by steam jets rarely presents any serious hazards or annoyances. In fact, precautions are needed to provide insulation required to render spray electrification of water observable. It must also be recalled that if fields created by charging of this type exist they may produce very much heavier charging currents by inductive charging of water drops which might be confused with true spray electrification, and quite generally spray electrification to be studied must be generated in field free spaces.

The question whether other liquids which are less conducting than water can produce true spray electrification is still an open one. It is in theory possible that this can occur with such polar substances as ether. Whether it can occur with true non polar substances of low dielectric constant has not been completely established though HARPER's results yield a negative answer. Symmetrical charging of all such liquids do occur on larger droplets if they contain ions enough as CHAPMAN and DODD have shown. Most studies or observations on the charging of such substances have been performed under conditions particularly propitious to electrolytic effects produced by flowing over metal surfaces, etc. The charging of these inflammable liquids in industrial and other uses present great hazards. Certain ways of reducing the hazards might be suggested.

1. Where such liquids are handled in more or less closed systems static charging can be allowed to occur without harm by using gases that are inert, N_2, or CO_2. In one plant cooled flue gases deficient in oxygen were used. In this way sparks produced cause no inconvenience.

2. The liquid can be rendered conducting in greater measure. Magnesium oleate has been used in cleaning fluids but is gradually removed in the cleaning process. Addition of water and ions to some liquids may help. The degree of conductivity required for prevention of large statistic charge accumulation is not high. The charging currents produced are usually not great. They differ with circumstances, but currents of the order of 10^{-8} ampere are frequently cited. With bulk resistivities under 10^{10} ohm cm in liquids static charging is unlikely. The demarcation line appears to be a specific resistivity of 10^{11} ohm cm.

3. Adequate grounding of all surfaces involved, in essence the metal container and the vessels receiving both fine drops and the larger drops removes all chances of serious sparking. Surface conductivities of many insulating substances in contact with sprays can be increased. Glasses and enamels can be rendered conducting by coating such as SnCl. Various thin metal and carbon coatings are useful. All conduits, pipes, especially their orifices and adjacent receivers should be well bonded electrically and grounded together.

4. Use of radioactive ionization of the air in prevention of static accumulation can be resorted to. However, use of such devices, if adequate, is accompanied by far more serious and insidious industrial hazard than the static generated by processes above cited if reasonable precautions are not taken.

IV. Mechanisms of electriciation in solid-solid contact

A. Introduction

It is precisely in this area that the greatest confusion exists. Aside from the complications produced by the action of films of moisture, electrolytic effects and/or the formation of electrolytic Helmholtz double layers, or by films of impurity, phenomena are complicated by other factors. These include the manner in which measurements are made, the failure to avoid external electrical fields, the character of the contacts, e.g., sliding, rubbing, or rolling contact or just normal contact with no lateral motion. In some cases, there was a disregard of the electrostatic circuit arrangements and failure properly to analyze the circuit as a whole including, especially, the electrometer capacity. The effect of heating either at local contact points or asymmetrical heating of the same substances has been ignored. Heating may involve local fusion, etc. Again, the nature and state of surfaces, possibly exposure of crystal planes and the removal of oriented layers of ions as a result of mechanical or heat stresses. Finally, assuming that the measurements are conducted under controlled conditions designed to avoid all unnecessary complicating factors when one regards the electrification on contact of the whole gamut of inorganic, organic, insulating and conducting solids known and used in study, the various possible interchanges of charged carriers leads to such an extensive range of interactions that it is virtually impossible to derive any basic understanding of the mechanisms at work. This is especially true when one turns to such complex substances as vulcanite, (hard rubber), the various plastics, sealing wax, amber, cat's fur, silk, glass and wool commonly associated with the phenomena of static electrification.

Since something is known of the nature of metals and of their contact potentials, work functions and static charging by their means, it would seem reasonable to use such as one side of a system of charging studies. These substances can be obtained in a clean state and can yield reproducible surfaces. It is next logical to consider as the other partners in contact charging studies, substances of known simple chemical composition yielding reasonably clean reproducible and smooth surfaces, if possible of single crystalline habitat. It will be essential to study these under conditions where disturbing aqueous films, and external electrical fields cannot interfere and to yield quantitatively meaningful results capable of analysis and interpretation.

In reviewing the literature of the past, little attention was paid to such systems. A great variety of studies were made with two types of systems, (1) the dispersion of dusts or observing the impact of dust-air jets on surfaces and (2) either the rubbing, rolling, or pressure contact and separation with no rolling, between solid surfaces or liquid Hg surfaces and solids, the amount of charge being measured in one fashion or another. Much work was done in air, a considerable proportion of it was done in humid air, say, of 60% or less, relative humidity at 20° C. Surface cleanliness was never certain; the widest assortment of cleaning methods were employed extending from the use of chromic acid, or caustic soda and distilled water on surfaces like glass to washing with acetone, absolute alcohol, 95% alcohol, water, ether, carbon tetrachloride, etc., etc. It must be said that virtually all cleansing using organic solvents, such as "chemically pure" alcohol, acetone, ether, etc.*, tend to leave behind even visible films of organic residues that are highly undesirable. Acetone and alcohol being water soluble are the least objectionable *if followed while moist by washing with distilled water*. It is, however, probable that if contact charging is done with any appreciable pressure between surfaces, these films may break down and contribute little unless they act as lubricants in sliding friction or contribute by their surface conductivity in which case they can be very disturbing.

In consequence, numerous observers have come out with various contact electrification series, none of which, though consistent within themselves, agree with any other series achieved under different conditions. With certain preliminary comments and precautions, this chapter will present the attempt by using the simpler metal—insulator systems, under increasingly rigorously controlled conditions, in terms of studies largely in the author's laboratory as well as other significant studies and interpretations, to develop basic data and principles of static charging with attempts at interpretation based on current knowledge of surface interaction and atomic structure.

B. Quantitative measuring techniques and the influence of electrical fields

1. Simple studies

The earliest technique for observing static electrification consisted in rubbing two surfaces of the substances to be tested, (or bringing them in contact), subsequently separating the surfaces and bringing one of the surfaces provided with an insulating handle into a Faraday cage, or deep cup, connected to an electrostatic potential indicating device. This was either some form of gold leaf electroscope or for lower potentials an electrometer. If the charged body was a conductor with an insulating handle then contact of the body with the cage and subsequent removal was indicated. If an insulator, then measurement was made while it was in the cage and if needed, the cage could have been made complete by covering with a top.

In this process, the situation is as follows: Contact between the two surfaces of geometrical area A results in actual contact between much smaller areas fA where f may be $\sim 10^{-7}$ or so on single contact. The contact points across which charge is transferred are within some small distance of the surface $\delta \approx 10$ Å. The exchange of charge on account of some sort of, intrinsic "effective contact potentials", V_c, results in the transfer of an actual charge Q. It may also result from a movement of ions from one surface to the other through differences in energy, concentration or through specific chemi adsorbtion affinities. In these cases, transfer takes place until equilibrium is achieved, leaving the surfaces with a potential difference V_c effective. The virtual capacity of the condenser charged is $C_1 = \dfrac{A}{4\pi\delta}$. Actually, in separating the surfaces, if V_c is high at distances $\sim \delta$ considerable charge will leak back in the form of effective field emission or other conduction currents so that the actual charge collected will be $q = V_c^1 C_1'$ at separa-

---

* It is probably not known to physicists who are unacquainted with organic preparation procedures that chemical purity in this field is far from that in the inorganic field.

tion where $C_1' = \dfrac{A}{4\pi d}$, with d beyond the field emission, or return current, range so that only a small fraction of q/Q of the possible charge will be collected. As the surfaces separate to large distances, the capacity C_1' is reduced to the mere electrostatic capacity of the isolated surface C_1'. Thus q is constant and V rises to many times V_c^1; in fact, to $V_c^1 \dfrac{C_1'}{C_i'}$.

Placed in the Faraday cylinder, the quantity q is now placed on a system consisting of the capacity of the Faraday cage plus that of the associated capacity of the electrometer and its leads, which can readily be evaluated by use of a parallel standard condenser and the method of mixtures. If the electrometer and Faraday cage system have a capacity C_E. Then the potential read is $V_E = q/C_E$ and q can be measured.

The measurement of charging of dusts by blowing or impact can be carried on in the same fashion. Thus in a *field free space*, best a Faraday cage, a blast of clean dry gas interacts with the dust surface and the dust may either be caught in a separated shielded Faraday cage connected to an electrometer, or it can pass through a duster with a series of baffles where it impacts on the surfaces and then it enters the Faraday cage. The two shielding systems are grounded. Electrometers had best be attached to both the generator, (dust chamber and baffles), and to the Faraday collector to insure that equal and opposite charges occur.

The simple systems measure quantity of charge generated per square cm of geometrical surface, or per gram of dust dispersed. If the size of dust particles are known, then charge per particle can be estimated and average surface charge density and can be calculated for either case. If it is desired to measure the rate of charge generation, a quantity somewhat more significant, then the same system can be used in that the test material in the form of spheres or cylinders can be released at different points on an inclined plane and allowed to roll different distances to collection in a receiving Faraday cylinder. This method is mechanically awkward and more convenient systems can be devised.

The other approach is to have the test sample either roll continuously, or slide, relative to a surface and to measure the state of electrification through the charge induced by the charged insulator on a test probe or plane connected to the electrometer. Such probes or test planes are generally very useful at recording the rate of charging, indicating saturation charge, etc., but in most methods the actual charge q acquired is not registered. However, granting surface uniformity of charge distribution and adequate probe area for proper sampling relative values proportional to q are yielded. With proper geometry in some instances, the probes can be calibrated to read absolute charge. This would be achieved by taking the charged test sample indicating a certain induced charge on the probe and immersing it in a Faraday cylinder without discharging the sample. Dusts may be dispersed and statistical studies made of charge and associated size distribution for the cloud using MILLIKAN's oil drop method, the Hopper-Laby modification thereof, or horizontally applied alternating potentials stroboscopic illumination and rate of settling of the cloud in cruder measurements. Studies of finer particles could use the ERIKSON[71] tube with an air blast method.

There is one danger in the use of the probe technique. Assume that in an external field the test sample was subjected to an induced dipolar charge separation with long relaxation time* in virtue of its dielectric constant (e.g. minutes). The probe shielded from the inducing field would sense and measure the unrecombined, or polarized charge nearest it and register this, falsely as an acquired charge. However, if the sample were rotated 90° the dipolar nature of the charge on the sample is quickly revealed. Similarly any non uniform charge disposition can lead to erroneous conclusions. Such situations must be looked for and be guarded against in the interpretation of data under any system.

2. Action of external fields on charging

In a series of studies of static charging, GILL and ALFREY[8], working under conditions using room air investigated the effect of external electrical fields on static charging. Particles of vulcanite or sand slid down an inclined metal plane in an electrical field. They were caught in a Faraday cage and the charge was measured. The charging observed was given by an equation of the form $q = -AX - q_0$. Here q_0 is the charge in the absence of external field and happened to be negative while A was positive when the upper inducing electrode was positive and the earthed plane was relatively negative. If the inducing plane was negative A was negative and if X was large enough, then the charge reversed and q became positive. Reversal occurred with $q = 0$ and under these conditions $+AX - q_0 = 0$ and $X = q_0/A$.

It was clear that in this event, the external field was inducing a charge separation on quasi conducting, or conducting, particles to the extent that the charge of the same sign as the inducing electrode leaked off to ground and that on separation from the plane in falling into the Faraday cage the particles retained the sign of charge bound by the inducing plane on their upper sides. In this case, the air was an insulator so that no current was drawn and the charge separated was directly proportional to the field X through the constant A. The value of A depended on the actual field X across the particles and the amount of charge separation to reduce the internal field across the particles to zero.

GILL and ALFREY[8] then extrapolated these results to indicate that possibly *all static charging* could be led back to induction charging produced in some fashion presumably by surface fields of some nature at the interface metal-insulator. This would require that the range of the surface fields be of such character as to polarize the whole particles if the analogy were to be strictly carried out. However, such a general statement could be applied to refer to the contact potential fields between metals when the Fermi levels adjusted on contact. Aside from being very general and also quite ambiguous this statement is of little help in understanding charge transfer between surfaces which may depend on such complicated in electrical processes as chemi-adsorbtion of ions of one surface by the other and dipole formation in consequence of oxide layer formation, a process which cannot be neglected when contact charging is considered.

Leaving their theoretical considerations, it is of interest to consider an extension of these measurements made by J. W. PETERSON[7] in rolling dielectric spheres of a borosilicate glass along an inclined Ni plane. The apparatus was such

---

* Many dielectrics of composite nature such as papers in oil etc. on the basis of the Maxwell-Wagner mechanism as well as dielectrics having ions that cannot leave the surface show such action.

that the rolling could be studied in very dry air, or vacuum and with varying degrees of humidified air present. The system could be heated to 210° C to remove some of the water vapor. Complete removal of aqueous films to a mono layer calculated for unit geometrical surface area is achieved for this glass only above 300° C. Charges were measured by letting the pyrex spheres fall into a Faraday cage after rolling. An external field between the Ni rolling track and an opposed Ni electrode was applied. Using outgassed dry spheres and chamber, the charge q_0 caused by *true contact charging* of the glass by the Ni was mostly collected. The charge was very slightly modified by an applied field even then. That is, with $X \sim +1200$ volts/cm, q_0 differed from that with $X \sim -1200$ volts/cm and the quantity Δq represented this observed charge difference. With water vapor present in the air there was strong charging in which the observed charge $q = -AX - q_0$ was observed to be linear with applied field as noted by GILL and ALFREY. In view of this, PETERSON, fixed the fields at some value, e.g., that for an applied potential ± 1200 volts, the charge change $\Delta q = q_- - q_+ = 2AX$ was observed for different values of humidity. This was necessary as q_0 varied with humidity decreasing progressively as humidity

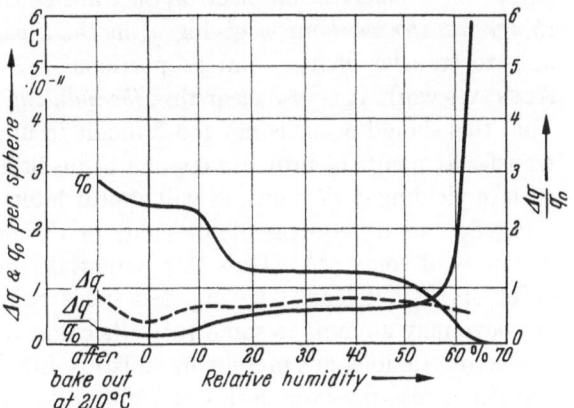

Fig. 29. PETERSON's curves for induction charging of borosilicate glass spheres as a function of relative humidity

increased. The variations of Δq and q_0 as a function of humidity are shown in Fig. 29 $\Delta q = 2AX$ is sensibly constant being lowest at low humidity but *appearing to rise after bakeout*. The *natural charge* q_0 decreased slowly up to 20% humidity then fell about one half remaining constant up to 60% where it sharply fell to zero. This action of *humidity* which *increases surface conductivity* will materially reduce the surface charge q_0 from contact electrification owing to the high potential on the sides and top of the sphere and the conducting film of water. The significant quantity is the ratio $\Delta q/q_0$ which is very small at low humidities, but increases very sharply as q_0 declines while $2AX$ is not materially affected. This is in perfect keeping with the explanation of the field action as being such as to polarize a conducting medium so that on separation of conducting glass and metal, the induced charge is measured, with A representing the quantity that must be induced to create zero fields within the sphere, a quantity readily calculable for an accurately figured geometry. Note that the degree of conductivity is not essential unless the relaxation time for charge separation becomes comparable with rolling time.

It will be noted that $\Delta q/q_0$ was not 0 at 0% humidity and, in fact, it was even *slightly larger after bakeout* of quartz and metal to 200° C. This appears paradoxical if the charging is to be attributed to surface conductivity of an aqueous layer as the humidity measurements indicate. The increase in $\Delta q/q_0$ on heating could indicate a contamination of the glass by moisture, etc., from the heated metal and walls, for, indeed, its conductivity was found to be increased by such heating in later studies. This increased conductivity would depress q_0 and increase Δq leading to the increase observed.

Be that as it may, PETERSON points out, however, that in an electrical field, the dielectric spheres are polarized by the field. Thus close to the surface where electrons, or whatever agency transfers to the sphere surface in contact with the metal there is, *in addition to the natural surface potentials causing transfer*, a very highly localized field at the lower surface of the sphere which reverses sign with X.

Under PETERSON's operating conditions this field by action on the layer of humidity or dirt on the surface of the spheres and metal was causing electrolytic charge transfer to the surface in the direction of the imposed static field as indicated. This polarization electrolytic transfer charging *superposed on the natural charge transfer mechanism* giving q_0 in the measure that the inducing field was able to transfer charge, and proportional to it. In this case, as in GILL and ALFREY's work, it is also clear that *the inducing field did not alter the nature of q_0*. That this should occur is not too difficult to understand. The electron or charge transfer at points of intimate contact of insulator asperities, indenting the metal surface yielding q_0 depends, as will appear later, on an effective contact potential difference, or its equivalent, of value of the order of a volt. This acts across distances of some 5 Å. Thus this potential corresponds to electric fields of the order of 10^7 volts/cm. The imposed fields causing polarization charge transfer as above may amount to some 10^4 volts/cm at most and film thickness is 1000 Å. If electrolytic ions are present in moisture films quite large charge densities can transfer across the boundaries at points other than those of intimate insulator-metal contact.

In the case of PETERSON's supposedly clean dry sphere there was a charging Δq which amounted to 1/4 the maximum polarizing surface charge density to be expected at the contact area of the glass and metal sphere in uniform field. Thus in the time of contact and with the moderately dirty surface of the relatively dry sphere, (baked out at 200° C), enough charge transferred to neutralize 25 % of the polarization charge.

Thus even supposedly clean surfaces have enough ionic contamination in dry air or vacuum. The higher the humidity the stronger and more complete the polarization charging observed. For with more moisture larger areas of such films of 10^{-6} to 10^{-5} cm thickness for charge transfer in the field were available. Since charging was proportional to impressed field in the rolling spheres it is clear that the ionic separation was adequate and fast enough to yield saturation charge transfer to the polarized layers as the sphere rolled.

This now raises the question as to whether the charge transfer needed for achieving polarizations charging will occur with perfectly clean surfaces of dielectric in contact with a metal. Such surfaces have been prepared in later studies of WAGNER.

This charging in an external electrical field, independent of the contact charging yielding q_0 *can only occur if there are mobile or transferable charges in the metal or on the dry solid in contact that are bound by forces so weak that the polarization fields across the boundary can effect transfer*. In the case of electrolytic conducting films such as in PETERSON's and GILL and ALFREY's work, both types of ions were present and mobile and fields were adequate to effectuate transfer. *Usually there are not free carriers* at the surfaces of *clean* dry substances in contact. Electrons or ions are tightly bound in varying degrees and unless the polarization fields across the boundaries are stronger than the effective contact potential divided by the surface separation distance involved no transfer may be expected.

Where such fields are stronger, then *if the charge transferred is electronic* as from metal to insulator, the polarization charging by the external field *will be asymmetric in regard to field direction*. That is, the charging will proceed in one direction only, i.e., when the sense of the imposed field can force electrons from metal to insulator. If transfer depends on a unique sign of ion transfer, e.g., Cl^-, from insulator to metal, it will likewise be asymmetrical, going only when the field can send that ion across. Since fields needed to remove even loosely bound electrons with 0.01 volt work function across 5 Å will be 2×10^5 volts/cm the chance of removing metal ions from the clean metal lattice requiring 10^9 volts/cm is nil. Thus actual true polarization charging of clean insulator-metal contacts will hardly be expected and if observed will be asymmetrical. Charges just generally do not traverse metal insulator interfaces at easily realizable boundary fields.

There is, however, one falsifying mechanism which might in some studies lead to an apparent polarization charge where the charging is observed in a static system, i.e., in one where spheres or insulator do not rotate while in the field. For the dielectric need not have charges transferred from metals *into it* or vice versa to show such apparent charge displacement. A very large number of insulators placed in electrical fields will undergo polarization with long relaxation times. In an alternating fields this action yields the well known dielectric hysteresis loops. Three mechanisms of dielectric polarization can cause an insulator in a field to become polarized and remain so for long times. (1) Dielectrics having dipoles of long chain character and so bound that it takes considerable time in a field to orient them and to disorient them. (2) Dielectrics having conduction carriers, e.g., ionic or otherwise that *can move inside the insulator but cannot transfer these across the barrier to electrodes*. These will suffer displacement polarization *annihilating the field in the interior* and producing strong surface polarization charging. In fact, many insulators of specific resistance 10^{11} ohms cm on up show this behaviour. Notable in this connection are the borosilicate glasses designated under the trade name Pyrex. Here Alkali ions are mobile. They will polarize the glass under high fields and retain their polarization for hours. (3) A very large number of heterogeneous plastics and substances, having constituents of different dielectric constants undergo the famous Maxwell-Wagner polarization. This accounts for the dangerously high residual charges on certain high potential condensers.

Such polarizing substances exposed to a high field will suffer charge segregation in an inducing field which may persist for hours even after the field is removed. The electrometer will record such a polarization as if it were produced by actual charge transfer onto the dielectric in static studies. In dielectric studies such polarization produced by a field on a stationary sphere will be disclosed by its bipolar induction when the sphere is rotated. The action however differs from the polarization and symmetrical charge transfer in that the sign of the charge induced will be opposite to that for charge transfer. This is not likely to be noticed unless it accompanies a unidirectional static charge transfer in which transfer for one field polarity will yield $+q_i-q_0$ and the other $-q_i-q_0$. For this analysis it is best to consider a static study carried out by MEDLEY[94] who charged insulators on contact with metals with and without superposed fields. He used wafers of dielectric such as nylon, alkathene, mica, cellophane, keratin, and filter paper. The wafer closed off an upper glass tube containing clean mercury. The lower side could be touched by a flattened meniscus of Hg which could be raised or lowered by rotating the apparatus about a horizontal axis. Under normal conditions, the two mercury surfaces were earthed. Then the apparatus was rotated until contact was made over the surface. On reversing the motion, the lower Hg surface separated off and the charge left on the underside of the wafer by the receding Hg could be measured on an electrometer. This appears to have been done using a condenser, apparently of capacity large compared to that of the wafer across a Lindemann electrometer. Sheets of plastic dried by flaming or warming in the air were placed in a resonable vacuum. The electrometer was next disconnected, the two Hg pools grounded and contact was made with the plastic. During contact a high potential of some 1500 volts was placed across the two surfaces.

The apparatus was then rotated, removing Hg from the lower surface and the high tension was removed, the lower Hg pool being earthed. At this point, the electrometer was connected to the upper pool and earth and the potential was read. Observed was a certain charge transfer q which was evaluated. This charge transfer consisted of two parts, the natural charge q_0 normally given the plastic by contact with mercury *plus an induced charge*. A plot of charge with applied potential, opposing the natural charging gave a linear relation between observed charge of q as applied potential decreased which reversed at appropriately high potentials. It had the form $q=q_0-C_w V_a$ where C_w the slope of the line, equalled the capacity of the cell and V_a is the applied potential. In principle, the situation as reported by MEDLEY may be described in terms of the diagram of Fig. 30.

In this Fig. 30a shows the situation with no outside field imposed. A contact of Hg on both sides as in Fig. 30a—1 with $-q_0$ transferring by contact potential charge across both top and bottom boundaries and both Hg pools grounded. The high boundary capacities compared to the low capacity of C_E and grounding lead to no charge on C_E. In Fig. 30a—2 the lower surface is removed after electrometer ground is broken as in MEDLEY's study reducing the lower boundary capacity thus raising potential across C_c to V_0 and placing $-q_0$ on C_E at a low potential V_E.

Fig. 30b—1 shows MEDLEY's concept of the process in an inducing field assuming that the applied potential $-V_a$ opposing the action of $-q_0$ on the lower plate, *causes charge transfer of induced charge q_i across both metal boundaries* onto the insulator without affecting q_0. This occurred for PETERSON owing to moisture films. It doubtless occurred in MEDLEY's case, possibly for the same reason, as will be seen. However, in this event, there is *no surface field across the boundaries* insulator-metal to alter the q_0 process (not shown) so that it proceeds as before. This is the argument MEDLEY makes for assuming charge transfer to occur. Fig. 30b—2 shows what happens on separations of the lower surface and grounding the lower plate. The net charge $q = +q_i - q_0$ is left on the lower plate. MEDLEY sets $q_i = C_w V_a$ where V_a is the potential applied and C_w is the capacity of the wafer. Then as observed $q = V_a C_w - q_0$ and from the slope of q plotted against $V_a C_w$ can be evaluated. Its value agrees reasonably well with the capacity of the Hg —wafer—Hg condenser as measured directly. In this case

$$C_c V_{0i} = q_i - q_0 = C_E V_E.$$

Note here that the sign of the charges on the system add up to $q_i - q_0$.

If there is *polarization* with *no charge transfer* the situation described in Fig. 30c—1 appears when potential is applied. In this q_i has the same sign as q_0 at the lower separating boundary. However, q_i will not necessarily equal $V_a C_w$ as before, unless the applied field is on sufficiently long to cause polarization to be complete.

Separation leads to the charging shown in Fig. 30c—2. Here

$$C_c V_{0i} = q_i + q_0 = C_E V_E$$

and unless the polarizing potentials on top and bottom plates are reversed $q = -(q_0 + q_i)$ increases with V_a.

Polarization charge then as indicated *gives a sign opposite* to that *for transfer charging.* It should also *show a time dependence* on the *duration of application* of V_a and it should show a decrease of V_E with time.

In MEDLEY's observations the charging was clearly one of transfer charging. This need not cause concern as MEDLEY used surfaces *less clean than Peterson,* despite his vacuum precautions. His cleaning procedure consisted of *flaming* the surface—about as nice a method of placing a contaminating film on the surface as one could wish. For the products of combustion of gas include SO_2 and SO_3 as well as H_2O. Despite evacuation MEDELY'S

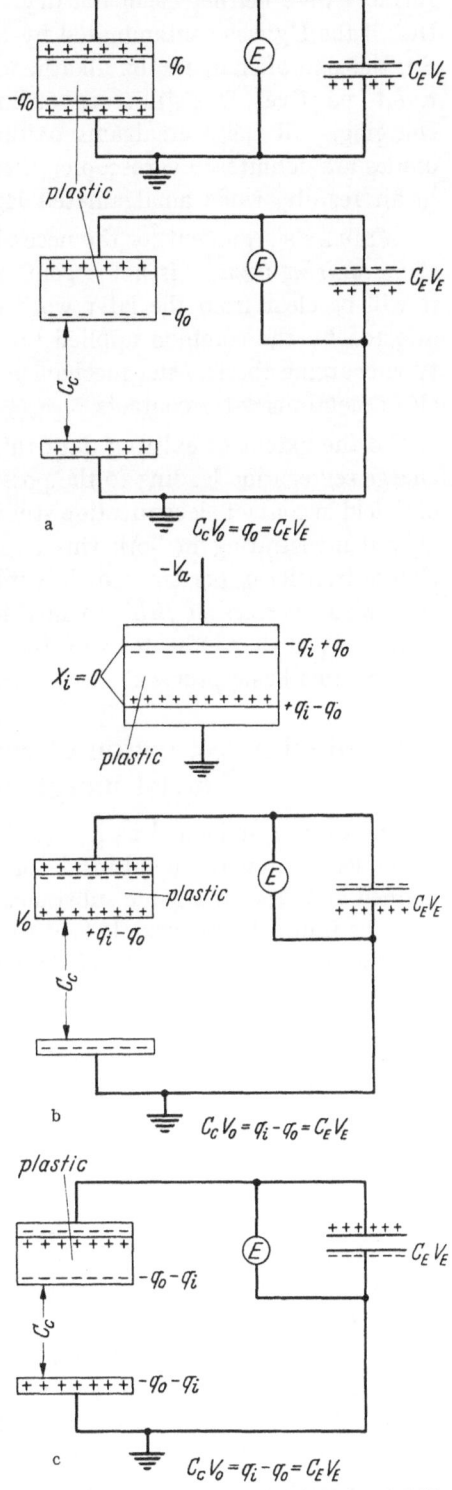

Fig. 30a—c. Diagram of conditions in MEDLEY's field induced charging process using plastic wafers and Hg contacts

surfaces were neither clean nor dry. In his direct charging studies he observed that if the Hg was contaminated by amalgamation with traces of Na, Zn, Pb or Sn, the sign of charge q_0 on alkathene, Nylon, and Mica was reversed from negative to positive. This still more confirms the electrolytic nature of, some of the charging. All these amalgams oxidize readily and rapidly in moist air and the oxides are definitely hygroscopic. Clean Hg remains bright and does not oxidize in air readily, while amalgamated Hg is rapidly soiled.

MEDLEY's argument for the necessity of *zero applied field across the boundaries thus requiring transfer* is not cogent as far as its effect on q_0 is concerned. For it will be clear from the later work of WAGNER that the q_0 process cannot be affected by the common applied boundary fields. Thus except for electrolytic type charging the transfer mechanism in applied fields is not to be looked for at clean metal-insulator contacts.

To the extent of existing data this discussion does clarify, the nature of the charge segregation leading to the possible and observed inductive charging effect of a field in contact electrification studies. It goes further, however, in unquestionably demonstrating in both this and PETERSON's study, that there is a large charge transfer q_0 per cm² which is inherent in the surfaces in contact *and is not influenced by external fields* in any major fashion despite GILL and ALFREY's beliefs. It further indicates why the external fields usually applied are unlikely to alter this basic process.

C. Initial investigation of controlled contact electrification of metal-inorganic insulator systems

It has been indicated as highly desirable that contact electrification studies be undertaken with metal-insulator systems where the insulator was a well-defined and clean inorganic substance. It was to be performed in the absence of aqueous films. It was also desired to know whether the presence of gases altered the charging. Towards this end, D. DEBEAU [95], in the author's laboratory, undertook studies in a simple system. Two conical Ni funnels with 5 mm steams were separated by a short length of quartz tube that surrounded them on the outside. The inner diameter of the stems of the funnels was 0.5 cm. The tops of the funnels were closed with fine mesh Ni screening. These funnels were each electrically connected by leads through the walls of the Pyrex glass envelope. The envelope could be rotated about an axis at right angles to the longitudinal axis of the two funnels. Thus the one funnel was filled with particles of the insulator to be studied, sized by differential screening. Then by rotating the envelope, the particles could be made to roll, slide, and fall through the Ni funnels without making contact with the quartz insulator and the charges on the particles and on the Ni funnel from which they emerged could be measured by electrometer. Gases used were air, O_2, N_2, and H_2 cleaned by removal of H_2O vapor, Hg vapor, oil and dust by liquid air. From the average size of the particles, the density and the mass used the number of particles could roughly be estimated. All particles came from pure substances crushed in a mortar, passed through metal sieves and dried.

Heating to 200° C in vacuum, ($\sim 10^{-6}$ mm), for several days gave Ni, SiO_2 and NaCl surfaces that were properly reproducible. Gas pressures were varied from 7×10^{-5} mm to 760 mm Hg. Measured was the potential acquired on a reversal of the funnels and at any pressure 5 to 8 potential measurements were made. Actually, since the capacities were constant, this measured quantity of charge separated. Reproducibility of charge was of the order 1 in 25 to 1 in 30 in successive reversals, e.g., ~ 2 to 4%. Quartz charged negatively relative to Ni and NaCl charged positively. Potential-gas pressure curves were reproducible and the same for any pressure irrespective of the gas used. The number of particles used ran around 10^5 and from 4 to 7 grams of SiO_2 or NaCl were used. It is doubtful if every particle charged in a single inversion though the masses of sand used were such as to allow all particles a fair chance to contact the Ni surface. The charge separation per particle observed was a function of pressure. It seemed to saturate near 36×10^5 electrons per particle at 700 mm pressure, and ran down to 12×10^5 at 0.97 mm pressure for quartz. Charges were about the same for NaCl.

The charging observed decreased as pressure decreased, reached a minimum at somewhere near a mm of pressure and then rose again. DEBEAU raised as a chemist, attributed the effect of the gas to adsorbed gas layers on the Ni surface and logically attributed the pressure-charging curves as representing adsorbtion isotherms. The agreement obtained by plotting reciprocal of charge against reciprocal of pressure fitted the expression for an adsorbtion isotherm with remarkable fidelity. If x is the fraction of the surface covered by the gas layer, then $x = a P/(1 + a P)$. If the charge separation depends on adsorbed charge $V = b x$ so that

$$V = \frac{b a P}{1 + a P} \quad \text{or} \quad \frac{1}{V} = \frac{1}{a b P} + \frac{1}{b}.$$

This indicated that perhaps one adsorbed layer was needed. If that disappeared, there was little charge and then at lower pressures a second layer came off either the initial surface or from the surface of the other substance yielding a bare metal or quartz that charged again. DEBEAU had found out that H_2 gas was particularly bad with quartz and was difficult to remove at the low temperatures permitted for outgassing. He attributed this to water vapor produced by the H_2. In this, he was correct. It appears that SiO_2 is exceptionally susceptible to reduction by H_2, especially if electrical discharges are active. This, in fact, was occurring, the H_2 was forming H_2O by reducing SiO_2 as a result of sparking of which DEBEAU was unaware.

This investigation of DEBEAU brought out several important points having to do with planning for future work. However, the gas adsorbtion effect appeared startling and impressive. In 1947, GILL[96], had been looking at the pressure variation of the charge observed by DEBEAU and thought that the replotted curves looked very much like the variation of sparking potential at fixed gap lengths with pressure. He then computed the surface fields of DEBEAU's particles and concluded that the variation of q with pressure observed by DEBEAU must represent a limitation of the charge permitted the particles as a result of their discharging to the Ni cylinder through the gas if potentials got too high. This surmise proved correct, as will be seen. Actually, the average radii of the quartz particles assumed spherical were $\sim 2 \times 10^{-2}$ cm with 2×10^8 electrons per cm² yielding surface fields of ~ 375 volts/cm. This is the sparking threshold field for air at 15 mm pressure. Since particle sizes varied and the sand grains were irregular reduction of charge by electrical break down could well begin at higher pressures and increase with pressure decrease.

One difficulty with the work of charging the particles under dry and reproducible conditions was the time taken to discharge the particles after a charging run. However, it did appear that *charging was reproducible if clean metal-insulator systems* were used. It indicated that the sign of charge transfer was not the same for different substances. It indicated that gas type did not alter charging as long as water vapor was excluded.

The studies of CHAPMAN[79] on spraying droplets using the Millikan oil drop-techniques indicated a chance for detailed investigation of charged particles as individuals. Thus the development of the Hopper-Laby[81] version of the oil drop measurements appeared to offer unique opportunities to study that aspect of the problem. It also appeared that it was desirable to get more controlled studies of the type initiated by DEBEAU. Thanks to a grant from the Office of Naval Research for investigation of these possibilities and for finding among the post World War II student group a number of good men who were interested in pursuing these studies in the author's laboratory, the two lines of investigation were simultaneously begun. The first study by J. W. HANSEN[82], who was a veteran in studies of dust electrification in the prewar years, set up the Hopper-Laby electrostatic dust analyzer. This piece of equipment is described in detail in Chap. III. His investigations were carried further by W. B. KUNKEL[5, 82]. In the meanwhile, J. W. PETERSON[97], using a modification of DEBEAU's scheme, undertook to determine whether amorphous quartz particles sliding over stressed piezo electrified quartz surfaces would be able to carry with them the charges of the oriented dipoles. This investigation will be reported on later. However, it gave PETERSON experience in the techniques required and led to the extension of DEBEAU's work on contact charging. It happened that KUNKEL's study of dusts, completed before PETERSON's work, so clarified certain issues that better to interpret the later work, it is worthwhile to report the dust electrification studies before proceeding on to the contact work of PETERSON.

D. Electrification of dusts on dispersion and impact on surfaces

1. Experimental techniques

After working with and perfecting the Hopper-Laby technique initiated by HANSEN, KUNKEL investigated a number of factors essential for interpretation of results. One of the problems in the dispersion of dusts involved the effect of various shape factors on the rate of fall as given by STOKES law used in computing mass and charge of the particles. Toward this end, KUNKEL[100], studied the fall of various arrangements of particles, such as the microscope had revealed to have been present in his various dust samples, e.g. as spheres, chains of spheres, planes made up of small spheres, clusters or clumps, etc. in a large cylinder with a viscous liquid. In all cases, the fall was *slower* than for the equivalent sphere and the particles fell in liquids presenting their largest area normal to the direction of fall. Asymmetries in plate-like particles caused a slow lateral drift of helical character making the particles appear to have a charge if plates were present. These asymmetrical particles could also be detected in the actual photograsph of the fall of particles in the absence of the horizontal field. Correction for the percentage of such particles could thus be made.

The slowing of the odd shaped particles produced the effect of making the masses appear too small. The greatest difference observed between the Stokes law calculated and true size was 56%. Since in dusts, the extreme shapes like plates and needles giving the most serious deviations are not present in great numbers the maximum error in size estimates was under 50%. KUNKEL[101] also investigated the change in particle size due to possible agglomeration as the cloud settled. Aggregation was found to be negligible if the cloud particle density was less than 10^6 per cm^3 and if the average charge of one sign is well below 10^3 electrons per particle.

In this method, a cloud dispersed in a controlled fashion in clean air of controlled humidity was allowed to fall through a settling tower of suitable length and enter the slit of the Hopper-Laby analyzing vertical condenser plates. By a study of the rate of appearance of various sized particles as a function of time after blowing a series of properly timed photographs of the stroboscopically illuminated dust particles enabled very good statistics of the size and charge structure of the cloud as a whole to be had. For large ranges in size photographs were taken at 30 second intervals to begin with for large particles and at 20 minute intervals 4 hours later for the micron (10^{-4} cm diameter) sized particles.

The dispersal was controlled so that it was by gentle blast with little turbulence of by violent blast with much turbulence. If needed, blowing was controlled in a nearly *quantitative* fashion. Dusts were blown with minimum contact with surfaces or else deliberately blown through a system of baffles to give a maximum surface contact. Surfaces similar to the dusts but also radically different were used. Sometimes homogeneous dusts were used. At other times the dusts were mixed. The surfaces were as carefully prepared and controlled as room air would permit. The size distribution inferred from the stokes law measurements could be checked by catching the samples from the same cloud on a cover glass slide without field and making a microscopic analysis of size against frequency of occurrence. The origin of the cloud above the slit of the measuring chamber could be controlled in order to improve the statistics as needed. Particles from about 0.5×10^{-4} cm (microns) to 30 microns could be satisfactorily measured. Larger particles fell too rapidly and those less than a micron were too slow and subject to falsification by the unavoidable slow convection currents that could not be eliminated. The charges observed varied from 0 to 3×10^4 electrons. The samples photographed in a given cloud ranged from 200 to 2700 particles. Photo micrographs and electron microscope photographs of sample particles were made. Figs. 31a and b show typical examples of the crushed quartz powder particles at different magnification.

The discussion of the statistics involved in the measurements will be left for the reader to look up in the original article. The methods used presupposed that the cloud did not have its particles change in size and charge with time. The change in size did not occur and the change in charge was generally negligible[77]. In this connection, some interesting data for micron sized particles over many hours are reported by KUNKEL[77] showing that in time such particles acquire an equilibrium Gaussian charge distribution by atmospheric ions whether initially charged or uncharged.

No significant difference in distribution was observed for clouds dispersed with no turbulence, or with much turbulence. This means that aggregates observed in the clouds were in the clouds initially and did not grow.

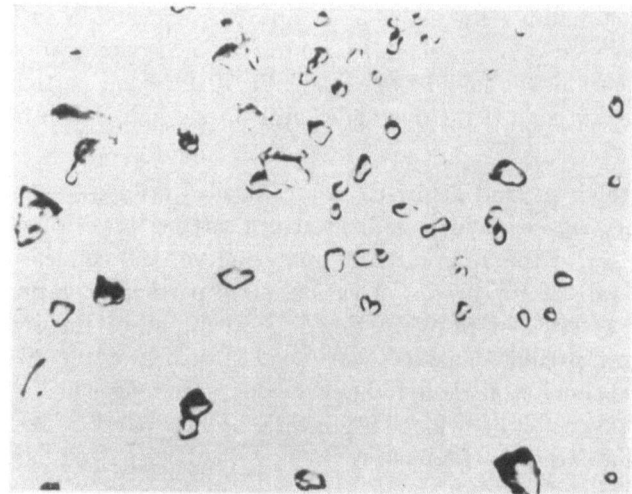

Fig. 31a. Photomicrograph of samples of crushed quartz used in KUNKEL's study

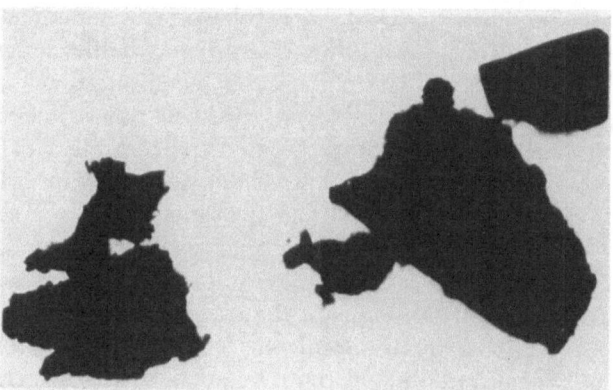

Fig. 31b. Electron Microscope studies of KUNKEL's crushed quartz samples. Note the very fuzzy unresolved surface

The choice of powders was limited by various factors. They had to have powder form from 1 to 30 microns diameter; they had to be easily dispersible with little lumping and they had to be reasonably clean. Neither hygroscopic or chemically active substances could be used. Purity and chemical homogeneity was paramount and led to use of other than the desirable spherical shapes.

At first, two substances were used—quartz and sulfur. The quartz was obtained by crushing and grinding. The sulfur was prepared by precipitation or crushing. Size ranges could be obtained by sifting or by elutriation methods. Containers could be made out of the same materials to avoid contact charging. Both substances suffered from irregular shapes and the fact that larger particles were coated with finer dust. The strong electrostatic charging of sulfur causes this sticking. Commercially sulfur fordusting pur poses is admixed with an agency preventing lumping. Here pure sulfur was desired. The amount of falsification of data by helical motion, retarded fall, etc., as a result of such roughening was relatively unimportant compared to the real charges and rates of fall. The sulfur containers were allowed to crystallize before use. Crystalline and fused quartz behaved in the same fashion so that this caused no trouble.

Commercial grade crushed quartz was used. It was treated with HCl and thoroughly rinsed in distilled H_2O. It was dried for several hours at $120°$ C and stored in a closed glass jar. It was found that despite precautions quartz would acquire from the air a layer of 2 molecules of adsorbed H_2O calculated in terms of grain size surface. This could only be removed by prolonged heating at $500°$ C. This is in agreement with findings of BOWDEN and THROSSELL[4].

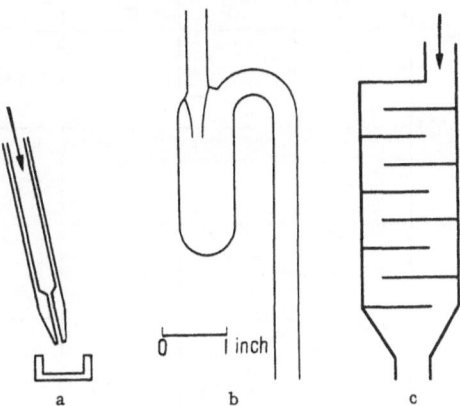

Fig. 32 a—c. Represent the devices for dispersing dusts so as to get as little or as much contact between particles on dispersal. The device (a) disperses with minimum contact, (b) with more but less with walls and (c) gives maximum wall contact

The quartz acquired this layer even standing in a jar dessicated by P_2O_5. Increasing humidity to nearly 80% at $22°$ C did not alter the moisture content. Sulfur powder did not appear to be altered by the humidity of the air. It was, of course, impossible to bake out S. Attempts to dry sulfur at normal temperature in vacuum resulted in sublimation of sulfur rather than removal of water layers. Supersaturated water vapor condensed on the quartz and sulfur making the powders wet and lumpy. There was no indication that the monomolecular water films exerted any influence on charging. Presumably the molecules occupied certain special sites on surfaces whose actual areas were at least three to four times the geometrically calculated areas.

Another powder used was a very pure natural rice starch, the granules of which were very nearly spherical and uniform in size (between 3 and 5 microns diameter). As work progressed, other substances were used, one being a very dry talc powder. Ni powder was prepared by abrading Ni with sand paper. The Ni was removed by magnet. Then it was demagnetized and reduced in H_2. Towards the end a sample of nearly pure carbonyl iron powder was obtained. This was fairly heavily oxidized and its surface was thus not reliable.

Dispersion had to be of such character that it created no turbulent air currents. The dusters used are seen in Figs. 32a and b. The duster in Fig. 32b gave a chance for a maximum turbulence and contact while yielding a quiescent stream outside. Fig. 32a shows the other extreme where the powder was gently dispersed out of the dish by a puff or air with minimum contact with the vessel walls. To achieve a maximum contact between dust and walls, the baffled tubes of Pt or Ni indicated by Fig. 32c were used. A special gun was designed to yield puffs of uniform assorted and controlled magnitudes. Various dried gases as well as air could be used for dispersion. In all, more than 3000 exposures were made photographing some 50000 particle trajectories.

2. Observations

1. Every single cloud has *positive and negative* particles.

2. When contact of dissimilar substances was avoided, e.g., blowing S out of S, or quartz out of quartz, the *number of positive and negative particles* was about equal.

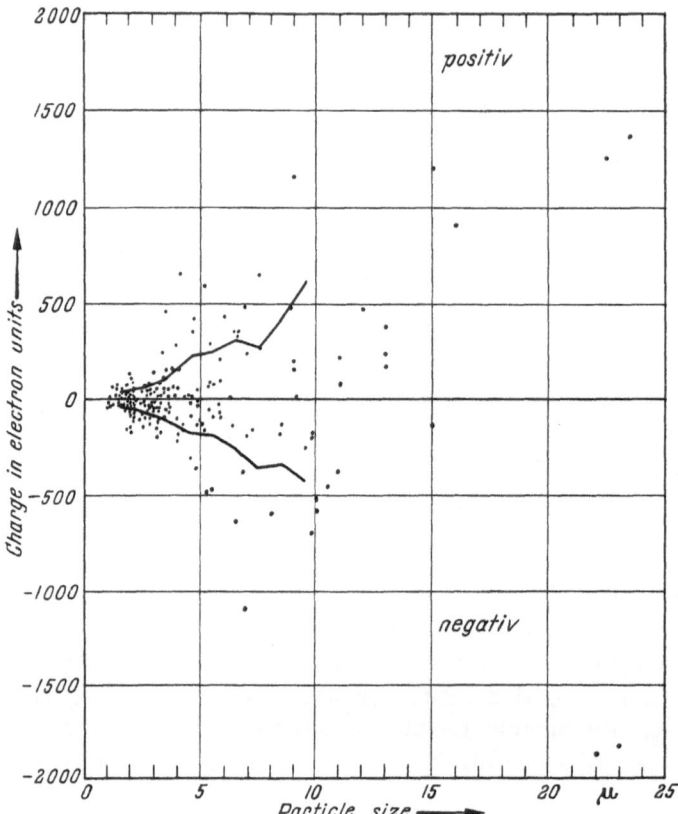

3. Dividing particles into size groups, counting positive and negative particles and calculating average charge per particle for various sizes showed that the charges were *equal* for the two polarities.

4. The average charge per particle increased in general somewhat more *slowly than the square of the diameter*. It increased somewhat more *rapidly than linearly with the diameter*.

5. Very few of the particles in the ranges studied were completely uncharged.

Fig. 33. KUNKEL's charging statistics: charge in electrons vs particle size in 10^{-4} cm for symmetrical charging of quartz. Only 1 in every 5 particles is plotted

To visualize these relations, the distribution of droplets in charge and size are shown for a cloud of quartz as single dots in Fig. 33. In this cloud, 1500 particles were observed. However, only one in every five is depicted for clarity. The averages shown as solid lines are calculated for the 1500.

6. This result was typical for homogeneous dusts and was remarkably independent of the form of dusting process used. Typical data illustrating this are given in Table 15.

7. It may be concluded that contacts *in the* turbulent cloud were of little consequence and that *most of the charging occurred on separation of contiguous particles*. Whether charges were segregated before blowing or whether charging occurred at the moment of separation could not be determined.

8. The relative humidity produced no observable effect on charging unless the powders became too moist to separate. This is to be expected on the basis of the direct studies of moisture content as a function of humidity.

Table 15. KUNKEL'S *statistics of charge size distribution for quartz from a quartz duster and a quartz cup*

SiO₂ size micron diameter	SiO₂ duster					SiO₂ cup				
	Number of particles			Average charge		Number of particles			Average charge	
	+	0*	−	+	−	+	0*	−	+	−
0–1	147	4	161	15.3	18.2	158	0	157	13.9	14.1
1–2	446	32	474	22.2	23.2	340	8	409	23.4	25.2
2–3	84	5	96	45.8	49.3	127	6	125	52.9	59.8
3–4	29	3	30	72.4	67.3	36	1	35	103.8	86.8
4–5	7	0	8	112	99.2	21	3	23	134	172
5–6	5	0	4	105	117	15	1	9	155	159
6–7	0	0	2		342	7	2	6	141	381
	Total numbers			Total charges		Total numbers			Total charges	
	718	44	775	20000	23000	704	21	764	27000	31300

* Here particles were considered neutral when their charges were too small to be determined; i.e. $q < 3$ for $d \approx 1\,\mu$ increasing to $q < 30$ for $d \approx 10\,\mu$.

Table 16. KUNKEL'S *table of results of charging of homogeneous and inhomogeneous systems.* Showing symmetrical and asymmetrical charging

Powder	Symmetric				Powder	Asymmetric			
	Container	Sign	Devi-ation***	St. dev.***		Container	Sign	Devi-ation***	St. dev.***
SiO₂ ...	SiO⁵	--	0.035	0.05	SiO₂ ..	S	−	0.21	0.11
S	S	−	0.036	0.04					
Starch ..	Pyrex	--	0.020	0.12	Ni ...	SiO₂	+	0.21	0.07
Talc ...	Pyrex	+	0.015	0.08	S** ...	Pt	−	0.60	0.21
SiO₂ ...	Pyrex	−	0.10	0.12	Ni ...	Pt	+	0.34	0.10
Talc ...	Ni*	--	0.06	0.11	Ni ...	Ni	+	0.32	0.09
Slightly asymmetric					Starch .	Ni	−	0.33	0.11
S	SiO₂	+	0.13	0.11	SiO₂ ..	Ni	−	0.39	0.07
S	Pyrex	--	0.13	0.11	SiO₂ ..	Pt	−	0.50	0.11
S** ...	Ni	−	0.19	0.16					

* Ni duster found coated with a layer of talc powder.
** Sulphur powder could not be well dispersed.
*** The deviations from symmetry are expressed by the ratio $x = \dfrac{N_+ - N_-}{N_+ + N_-}$.

A table of results using different dusts including inhomogeneous systems is shown in Table 16.

9. One of the strange things is that so many of the *inhomogeneous* systems showed nearly the symmetrical charging observed in Fig. 33. When quartz, talc or starch was blown out of the pyrex duster, the charges on particles balanced as before. A similar result was obtained by HANSEN with $CaCO_3$ powder in a SiO_2 duster. When S was blown out of a quartz, Pyrex or Ni tube, the predominance of one sign was only very slight.

10. Interpretation appears to be relatively simple. a. Most charging is produced by separation and is of the homogeneous symmetrical type. b. A relatively small percentage of the cloud particles strike the dissimilar surfaces. c. Turbulence was not on a fine enough scale to insure contact. d. The air stream cushions the particles in it from contact with surfaces. e. Only the more massive and denser particles have inertia enough to strike the walls. f. In many cases, the finer dust particles are charged and adhere to the surfaces in question, especially for the good insulators that retain their charge. Thus talc so coated the Pt duster that only impact of talc on talc occurred and no asymmetry was observed. This was verified on opening the duster which was found to be heavily coated with talc.

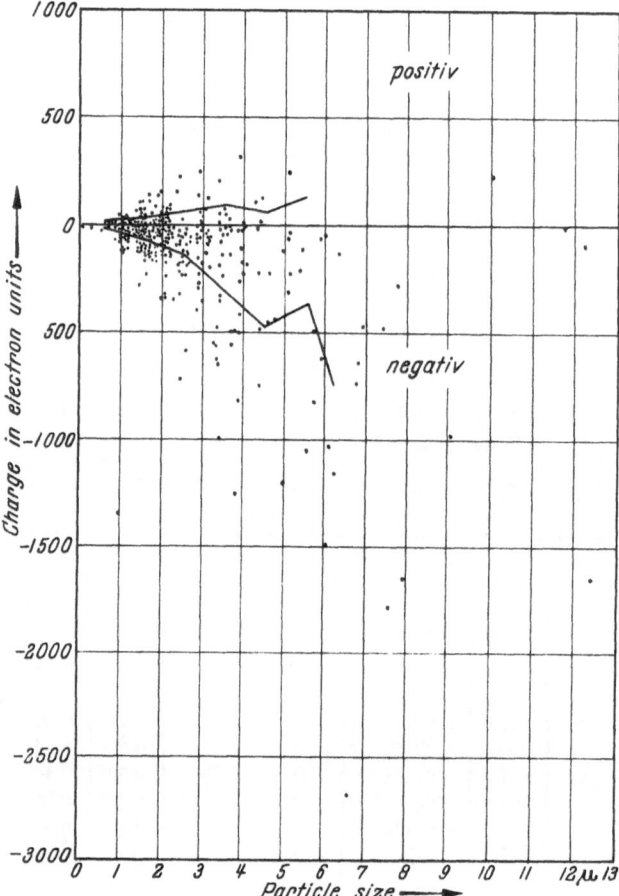

Fig. 34. KUNKEL's charged-particle size distribution for quartz dust out of a Pt duster showing asymmetrical charging

11. The most pronounced case of asymmetrical charging was for quartz blown out of a S duster. Here the quartz retained a negative charge. Where S was blown out of quartz, there was a preponderance of positive charge on the dust.

12. The asymmetry was greater for the larger particles.

13. All other strongly asymmetric charging cases involved metallic elements.

14. Insulating materials against metal showed net negative charge, and metal dust against insulating materials showed positive asymmetry.

15. Fig. 34 shows the distribution of quartz out of a Pt duster, with every other particle plotted, it reveals the degree of asymmetry.

16. A crude measurement of *total charge asymmetry* in the cloud and the net charge on the metal duster showed them to be approximately equal.

17. Ni powder out of a quartz tube showed a predominatingly positive charge. The density of the Ni particles insured better contact with the walls.

18. Ni powder out of the Ni duster did not act like a homogeneous system. This was inevitable because of the difficulty in creating Ni surfaces in equivalent states of oxidation in powder and sheet. The powder gained a positive charge

since it had been reduced and thus yielded up electrons to the more oxide coated duster. Likewise Ni charged up positively against Pt since the latter has a higher work function than Ni.

19. The asymmetry of charging was expressed by the ratio $x = \dfrac{N_+ - N_-}{N_+ + N_-}$. Symmetrical charging was taken to exist when the ratio x was less than the statistical standard deviation. The degree of asymmetry is represented by the relation of x to the statistical standard distribution.

20. To indicate that charging takes place on separation easily lumping sulfur was mixed with quartz dust. They were blown very very gently out of the cup by a vertical stream of air. The large particles were fairly randomly charged with only a slight excess of negative sign. The smaller particles that arrived towards the end of the run were overwhelmingly negative. These were the unclumped quartz particles.

21. Such a mixture of powders with charge and size segregation presents all the elements of static machine-like separation which could ultimately lead to electrical breakdown.

These results led to certain important conclusions. These are:

a. The charging results from separation of surfaces in contact or in the making and breaking of contacts of surfaces, *apparently not from friction.*

b. In contrast to the results and conclusion of Debeau gas films and water films have relatively little influence on charging for all but hygroscopic substances up to nearly 90% humidity at 20° C.

c. Two types of charging appear (1) *homogeneous or symmetrical* charging largely between particles of the same substance in the same state in which the difference in the number of particles of charge of each sign divided by the sum of the numbers charged is about the standard statistical derivation and (2) *asymmetrical* charging in which the number of charged particles of one sign consisting of substance A exceed the number of oppositely charged particles consisting of particles of substance B, or a surface of some different substance B, by more than the standard statistical derivation.

d. The summarized facts in conclusions a, b, and c indicate strongly that the separation of homogeneous surfaces in intimate contact, perhaps even fracturing of pressure welded surfaces and the separation of surfaces of dissimilar substances can lead to charge transfers. The electrolytic processes of Chap. I cannot be active here.

e. The average charge increases somewhat more slowly than the square of the radii of the particles that is not quite in proportion to the geometrically computed surfaces assuming spheres. However, it increases faster than linearly with diameter. This indicates that the charge transfer per unit geometrically calculated surface available slowly decreased as size increased. It indicated, however, that the actual surface areas in contact causing charge transfer were probably less than the actual geometrical surfaces and decreased somewhat more rapidly as geometrical surface areas increased. In the light of Bowden and Tabor's work and of the electron microscope studies showing the extreme roughness of the surfaces the latter conclusion is logical.

f. In view of this circumstance, the author suggested to KUNKEL and to R. J. WIJSMAN that the homogeneous charging could be accounted for in the fashion to be outlined below. On this basis, KUNKEL and WIJSMAN independently arrived at the statistical analysis and solution of the problem.

Assume that one has an aggregation of particles of different sizes in close contact in a powder mass. The contact area of each particle with its neighbors will, on the average, be in proportion to its surface area. At the first the minute contact points the surfaces may actually be welded together. Regarding the many such points of contact and realizing that when the dust is dispersed these points will be ruptured, it is possible to consider what the chance is that owing to lattice imperfections there are one or more excess electrons on one side or the other side of the boundary. One may define *unit contact* as a region of extremely high probability (i.e., certainty) of transfer of one electron to one side or the other on separation. Thus for each unit of contact one particle or the other will have an excess negative, the other an excess positive charge. Assume also that the number of contacts is large on all particles and that the direction of transfer is completely random in any homogeneous system.

Let the number of unit contacts of a certain particle be N. Thus N is proportional to S, where S is the surface area of the particle.

If the number of electrons gained is K, where $N > K$, then the number of electrons lost at the other side of the contacts is $N - K$ and the net charge will be $n = (N - K) - K = N - 2K$ elementary charges. The probability of such an event is given by

$$P_K N = \frac{N!}{2^N (N - K)! \, K!} = \frac{N!}{2^N \left(\frac{N}{2} + \frac{n}{2}\right)! \left(\frac{N}{2} - \frac{n}{2}\right)!}$$

since N is large, STIRLING's formula applies and

$$P_K^N \simeq \left(\frac{2}{\pi N}\right)^{\frac{1}{2}} \frac{1}{\left[1 + \left(\frac{n}{N}\right)\right]^{(N+n+1)/2} \left[1 - \left(\frac{n}{N}\right)\right]^{(N-n+1)/2}} \, .$$

If N is large, such a binominal distribution curve can be adequately approximated by the Gaussian

$$P(n) = \left(\frac{2}{\pi N}\right)^{\frac{1}{2}} e^{-\frac{n^2}{2N}} \, .$$

The average charge will then depend in a simple fashion on the size as follows from the relation

$$\bar{n} = \left(\frac{2}{\pi N}\right)^{\frac{1}{2}} \int\limits_0^\infty n \, e^{-\frac{n^2}{2N}} = \left(\frac{2N}{\pi}\right)^{\frac{1}{2}} \, .$$

Let $N = \frac{\pi}{2} \alpha^2 S$ in which event $\bar{n} = \alpha (S)^{\frac{1}{2}}$. Here α is a constant factor which assumes that the statistical relation between N and S does not disturb the average.

For spherical particles $\bar{n} = Bd$ where d is the diameter.

This indicates that $\bar{n}$ should not vary as the surface but as the $\sqrt{S}$, or as the diameter d if the particles were spherical. As they are not spherical and the true surface S really increases much more rapidly than that of a sphere because of agglomeration and roughness. N is greater than $\frac{\pi}{2}\alpha^2 S_{\text{geom}}$ and approaches $\frac{\pi}{2}\alpha^2 S_{\text{irreg}}$. Thus $\bar{n}=\alpha^2(S_{\text{irreg}})^{\frac{1}{2}}$ will increase faster than the d computed from $\bar{n}=\frac{\pi}{2}\alpha^2(S_{\text{geom}})^{\frac{1}{2}}$ or $\bar{n}$ varies more rapidly than d as observed.

One point should be noted and that is that the Gaussian distribution derived from the approximation to P_n^N requires that there be *more neutral particles than charged ones*. On the other hand, if only 100 particles are observed and the average charge $\bar{n}$ is 1000, for a given size range the number of uncharged particles observed would not be high on purely statistical grounds.

On the basis of this theory, some estimates may be made concerning the number of unit contacts for particles of a given size considering Table 15. For diameters between 2 and 3 microns the data give $N=\pi/2\bar{n}^2 \simeq 4000$. For particles between 0.5 and 1.0 micron $N=\pi/2\bar{n}^2 \simeq 400$, (the actual value is perhaps slightly higher than 400 since small particles lose charges to neutralizations by air ions in their long descent). Now if the surface area (geometrical) is known very roughly, the maximum average size of unit contact can be estimated.

The data indicates for quartz that at least one electron is transferred at each area of contact of 10^{-10} cm². Similar magnitudes occur for all particles observed. The differences existing are marked by the uncertain knowledge of the surface areas involved. These values are therefore *maximum* estimates. Since only 1/100 of the total surface could be involved in contacts, a much more realistic figure for the average area of unit contact would be 10^{-12} cm². This corresponds to one imperfection in 10^4 atoms instead of 1 in 10^6 using geometrical surfaces. If the fraction of the surfaces in contact are still further decreased, then the imperfections required become too large, e.g., such as one in 1000 atoms.

This indicates the nature of *homogeneous or symmetrical* charging and is a logical consequence of known properties of the solid state. It further indicates that *symmetrical charging* will be present whenever solid surfaces are disrupted or contacts broken, the limiting small size for its appearance depends primarily on the number frequency of imperfections, or excess electron distribution among the atoms or molecules of the substance. There is no limit to the upper average value of $\bar{n}$ as size increases except that produced by back discharge in air, or by conductivity. It yields a charging phenomenon that overlies that by other processes and makes interpretation of charge patterns of the asymmetrical class more complicated.

g. It is finally clear that where surfaces differ, there is a good chance that on intimate contact of surfaces charges will be transferred from one surface to the other. Thus there is a net transfer of negative charge from S to SiO_2, and from Ni and Pt to S, and Pt and Ni to SiO_2. There is no strong charge exchange between SiO_2 and a borosilicate glass. The reason for the charge exchange and the nature of the charges exchanged are not certain. For the exchange of Ni and Pt and powdered and solid Ni, the mechanism is the well understood contact potential mechanism of Chap. II. It might be suspected that substances like S and SiO_2 would accept electrons from the metals Pt and Ni on contact. Beyond this, these results tell little. It is, however, clear that dissimilar substances in intimate contact can exchange charged carriers, electrons, or ions, depending on their structure and properties. While with the dusting experiments, it is not always possible to get highly asymmetric charging relative to the homogeneous charging background, if the particles are light in mass owing to cushioning effects of the gas stream or to electrostatic coating of surfaces by fine dusts of the same material, it is occasionally observed.

h. Highly asymmetric and dangerous charging of dust clouds on dispersion or impact with surfaces leading to segregation of charge and ultimate sparks are possible under the following conditions.

1. Impact of sufficiently large and dense particles on an appropriate surface leading to their effective charging, insulation of the surface impacted, precipitation of the charged dust on an appropriately insulated collecting surface, or removal of the dust to the outside air or ground.

2. Separation of a mixture of two powders of different nature intimately mixed by a gas stream and possessing sufficient differences in size or density that mechanical segregation under gravity or in the air stream by differences in mechanical behaviour results in the asymmetrically charged components of the dust being collected or located in different areas or volumes. In this case, the separation of the two materials requires that they are substances that exchange charges on contact.

Both charge processes presuppose that the humidity is sufficiently low in order that the collecting surfaces do not suffer from surface leakage which disperses the charge faster than it accumulates. Substances must also be such a type that they are not very hygroscopic.

i. Substances that charge fairly readily and highly in contact with other substances as given in past literature many of them summarized and reported by GUEST[100] can be listed as follows:

1. Red lead, Pb_3O_4 powder mixed with sulfur on dispersion has sulfur negative and Pb_3O_4 positive.

Segregation is easy because of mass difference. It is useful for discriminating surfaces charged positively and negatively by color on dusting with the mixed powder.

2. It is asserted that *dry* solid acids striking metals charge them positively, dry solid alkali dusts charge metals negatively.

These could be electrolytic effects.

3. S usually charges negatively with other substances but is positive to a few dry acids and SiO_2.

4. Glass charges positively to most powders, but is negative to a few dry alkalis. The effects 3 and 4 again may be electrolytic.

5. HgS dispersed from a brass or metal tube gave the highest electrical charge per gram of any substance dispersed according to RUDGE[101].

6. Corn flour also was highly charged on dusting from brass according to RUDGE[101].

7. Certain types of charcoal finely divided charge highly on dusting from a metal container.

8. Talc also appears to make a fine and uniformly highly charged dust with small particles—if properly dispersed from a metal container.

9. $CaCO_3$ in dry air forms a highly charged uniform dust from a metal container.

10. Sugar dust with glass tubes becomes negatively charged leaving glass positive. Copper tubes are charged negatively by sugar. By blowing through narrow tubes potentials of 20 kV could be obtained. This is reported by BEYERSDORFER[102].

11. Coal dusts electrify heavily. BLACTIN and ROBINSON[103] observed heavy charging in blowing coal through iron pipes. There was no generation if the relative humidity rose above 65%. WALTHER and FRANKE[104] studied coal dust. The dust travelled through a long length glass tube and then through a section of copper tube connected to an electrometer. The *dust* charge indicated by the Cu tube was negative, presumably the dust charged by the glass.

12. BONING[105] observed that snow blown against a lump of ice became charged. The sign of charge appeared to depend on the size of the snow particles.

13. HOAG[106] observed strong electrification on blowing high pressure CO_2 through a bag to create CO_2 snow.

14. THOMAS[107] indicates a very interesting modification of contact electrification between metal, presumably iron, and coal as a function of its state of surface oxidation. Fresh coal was crushed and passed through a 30 mesh BBS sieve. It could be oxidized in air for various times at 350° C. Samples were passed at constant rate over a rapidly rotating drum of iron in a suitable housing such that the particles made several impacts with the metal before discharge to the air. The charge on the insulated metal system was measured. The resistivity of the coal was 10^9 ohms $\times$ cm and increased with increasing temperature. The metal showed a positive charge of the order of 10^{-8} coulomb gram when unoxidized, but as heating in air and presumably "oxidation" of the coal continued, it changed the charge continuously to lower values and finally to 10^{-8} coulomb per gram negative at the end of some 5 minutes heating at 350° C. This action was ascribed to oxidation of the coal, but could have involved changes in moisture content and other factors as far as control factors reported indicate. However, since it came from a specialist in the coal industry who states that about 1% of O_2 is absorbed in such a process and modifies the physical properties of the dust such as coking on carbonization the result may well have been one of surface chemiadsorbtion of O by the coal. In this event, the carbon changes from an electron acceptor with metal to negative O^- ion donor on "oxidation".

These examples beyond those cited may prove of interest and use, although the data about them may be quite inadequate.

j. Before leaving the dust studies, two further investigations by DODD[6], should be mentioned. These deal with the charging of Hg droplets from a glass sprayer and of glass spheres of a borosilicate glass.

An ordinary glass atomizer was used on Hg. High air pressures were required to disperse the Hg. The Hg droplets were positively charged while the sprayer acquired a negative charge in keeping with the usual behaviour of metals in contact with many dry solids. At first, rather peculiar results were obtained. The Hg spray projected too far and discharged against the walls. Table 17 indicates the relation between charge and droplet diameter observed when the walls ceased to discharge the droplets. The variation here was nearly linear with droplet diameter in general agreement with KUNKEL's theory. However here conditions are more exactly defined as drops are spherical conductors. The droplets have a common potential, e.g., that of the surface V at the instant of separation. At separation $q = CV$ where C is the drop capacity, or essentially its radius expressed in esu. Thus here, the charging was done by contact and separation of the Hg from the glass surface electrons transferring to the glass. Although Na$^+$ ions entering Hg and taking electrons from Hg amalgamate with Hg it appears that T is too low to permit transfer of Na$^+$ ions from glass to Hg. Glass accepts electrons as indicated*.

The statistical charging of Hg is precluded even though it is a liquid with a high, (10^{22} electrons/cm^3), electron concentration. These render the Hg so conducting that any disparity in charges occurring by statistical variations as drops separate would equalize by conduction on separation with very little if any net charging. The energy imparted to the small droplets is not adequate nor locaized enough to tear off any of the electron cloud of the metal so that spray electrification does not occur. Thus it is clear that sprayed Hg, or, in fact, *any liquid metal from containers that have the ability to absorb electrons, or yield ions* to the metal, will because of their inertia, density, and nature create very active charge generating mechanisms and separate as charged spheres equipotential with the charged liquid Hg surface.

Table 17. *Charge on* Hg *and droplet size for* Hg *sprayed from glass as observed by* DODD

Diameter d microns	Electrons q of Pos. peak
0.75–0.95	36
1.00–1.45	80
1.5 –1.9	125
2.0 –2.8	175
2.9 –3.9	275
4.0 –5.6	430

This is demonstrated by the first "electric light" observed in 1705 by HAWKSBEE when a partially evacuated dry barometer tube containing liquid Hg showed a striking glow in the dark as the Hg rolled along the glass wall of the container. Here the static contact electrification produced potentials leading to a visible glow discharge between receding Hg and glass in the low residual gas pressure present.

To verify the statistical law for blown dusts, DODD used a "dust" composed of soft Na glass beads, (Minnesota, Mining and Manufacturing Co., Superbrite Type 118) dispersed from a borosilicate glass cup for analysis. The beads were quite spherical and smooth presumably as a result of fire polishing. Their diameters lay in the range from 20 to 50 microns. Electron microscope and optical microscopic examination showed them to be quite smooth spheres except for occasional "blow holes" caused by eruption of vapor in fire polishing. The large size and rapid fall rendered the results less accurate than desired. For the 27 to 40.5 micron diameter beads the charge distribution was Gaussian with an average value of −135 electrons and a standard deviation of 625. The average charge was at about the limit of error in reading charges so that it could well not have been real. When earlier measurements had been made with the beads rolling on a borosilicate glass container during the dispersal strong net asymmetric negative charging of the beads was observed. Thus the marginal negative charge

* Studies in progress in the author's laboratory indicate that indeed electrons sprayed onto clean borosilicate glass in vacuum are not liberated by light of 2537 Å, ($\sim$4.9 ev) or longer wavelengths although surface Na atoms in this glass are photoelectrically ionized leading to current $\sim$10^{-12} amp/cm^2 by 2537 Å. This is in agreement with removal of electrons from Ni, $\Phi_e \sim$6 ev by such glass on impact.

observed was perhaps real but of spurious origin, e.g., contact with the borosilicate glass. The quantity of interest here was the standard deviation of 625 electrons. Applying KUNKEL's theory $\bar{n} = S^2 = 625 = 3.9 \times 10^5$ electron transfers per glass bead of $d = 27 - 40.5 \times 10^{-4}$ cm. KUNKEL's data gave $\langle |q| \rangle = 81.2$ electrons for irregular quartz $d = 3 - 4 \times 10^{-4}$ cm. That made $\bar{n} = 10.4 \times 10^3$ electrons per quartz particle of 3 to 4 microns. From KUNKEL's data this would have given $\bar{n} > 10^6$ transfers per quartz particle with its irregular surface and thus more rapid increase than proportional to d. Thus the glass beads with their smooth surfaces transfer about an order of magnitude less charge than the very rough quartz. Stated otherwise the average *surface area* wherein an electron or ion transfer occurs is about an order of magnitude larger for smooth glass than for quartz with its many asperities, to wit 9.10×10^{-11} cm^2 relative to something less than 3.5×10^{-11} cm^2. The difference, however, could be inherent in the frequency of excess ion or electron containing imperfections in glass relative to SiO_2. For example, with the presence of the alkali atoms in the soda glass the number of displaced electrons per 10^4 molecules might be less than for SiO_2. On the other hand, the imperfections in the Na glass may be occasional loss of ions of Na^+ free to 1 in 10^6 and not by electrons, while in quartz, the imperfections could be O^-, or electrons, present in greater numbers.

DODD made a further study of the influence of H_2O vapor on these beads. Clean beads in equilibrium with air, dried over P_2O_5 showed no significant difference in charging relative to beads at 69% relative humidity at 22° C.

E. The contact charging by rolling of inorganic insulator spheres on metal surfaces

The apparatus designed by J. W. PETERSON[7] for measuring the charge acquired by small spheres rolling down a metal incline and the influence of external fields and moisture on such charging has been presented. While making those investigations, PETERSON next investigated the influence of gas pressure on the charging of his glass spheres.

It had been observed that the saturation charge acquired appeared to vary with the speed of rolling of the spheres. The charge increased as the speed increased. This increase was ascribed to the effect of the potential distribution over the sphere. Since at contact with the metal the potential is small, and the capacity large, as the sphere rolls, the local potential increases rapidly from contact up as the effective capacity of the region decreases. It reaches a maximum when the sphere has rolled 180° from contact. This increasing potential creates a field driving the charge back towards the point of contact. Thus the faster the sphere rolls the less time is available for a loss of charge by back flow along the surface caused by high fields.

In consequence, at higher speeds, a higher proportion of the charge remains on the sphere. Using the high speeds, the spheres began to acquire impressively large charges. Fig. 35 shows the charges per sphere in coulombs acquired as a function of path length at *various pressures in air* with high rolling speed. *Within experimental accuracy all curves had the same initial slope indicating the same initial charging rate.* The saturation charge however, was a function of the gas pressure just as observed by DEBEAU. The results are shown in Fig. 36. Here the saturation charge is shown as a function of the pressure, A at low speed before outgassing; B at low speed after outgassing and C at high speed after outgassing. The curve C shows the same sort of phenomenon observed by DE-

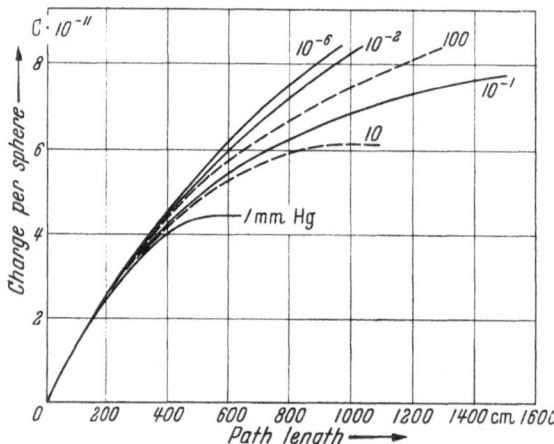

Fig. 35. PETERSON's curves for charge acquired by borosilicate glass spheres rolling on Ni as a function of path length for various pressures of air

BEAU, who also measured saturation charge on fast rolling.

It will be recalled that GILL[96] had suggested that DEBEAU's curve might represent a limit to charging set by the loss of charge by sparking to the surface. PETERSON's observation that *the charging rate was constant but the saturation charge depended on pressure* definitely answered the question in favor of GILL's suggestion. The pressure was *not altering the ability to gain charge*, i.e., the charge transfer mechanism. This was independent of pressure. The

pressure *limited the charge which the spheres could hold and the only way in which this could have occurred was in permitting the spheres to discharge electrically by gaseous discharges.* Under some conditions discharge of spheres by spark to the

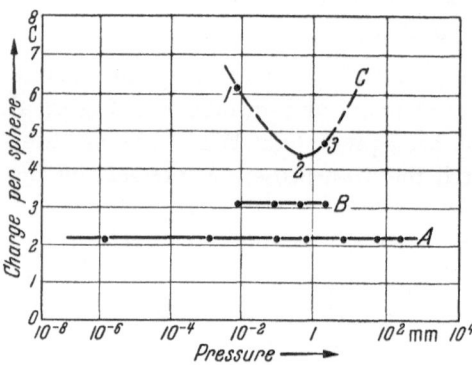

Fig. 36. PETERSON's curves for charge per borosilicate glass sphere rolling on Ni against pressures of air in mm at various speeds before and after outgassing

probe could be seen in the dark. By using soiled and moist surfaces and slow rolling speeds, the *charging rate was reduced* by loss through surface conduction. Using clean spheres and rolling them so rapidly that charging was faster than discharge through surface leakage could prevent the spheres acquired charges such that the total value was limited by spark discharge. At very low and very high gas pressures, higher and higher potentials and charges were required for discharge while at around 1 mm with the geometry of the spheres, the discharge passed so easily, that the quantity could rise to no more than to give any portion of the sphere a potential of some 400 volts.

The work with this system also revealed that the rate of charging was influenced by the condition of the Ni surface. Running a glow discharge on it in air oxidized the Ni, increased its work function and *reduced the rate of charging* for the borosilicate glass spheres.

From these investigations, a number of factors emerged. First, using a group of spheres on an inclined plane was awkward. The spheres had to discharge and

X ray ionization was required to remove charge. This was undesirable as it could alter surface states. It was essential to outgas the system and to control both rolling speed and length of paths. The influence of gas was seen to introduce limitations into the study so that work in vacuum was indicated.

Small spheres on charging stuck to the surface. Rolling contact was preferred to sliding as friction introduces complications.

As a result of these considerations, J. W. PETERSON[108] designed the following equipment for his studies:

a. Design of apparatus. Fig. 37a, b shows schematic views of the apparatus. An accurately ground six-millimeter sphere * of the nonconductor to be investigated rolls freely in the bottom of a nickel cylinder revolving about a horizontal axis. The cylinder is mounted on a stainless steel shaft, supported at each end by glass bearings. The shaft is driven by a synchronous motor and variable gear train, through magnetic coupling. The curvature of the cylinder cross section is high at each end and low in the middle, which confines the sphere to the central portion of the cylinder, while allowing it to roll in a random manner. A probe, which is in effect one-third of a complete Faraday cage, is mounted

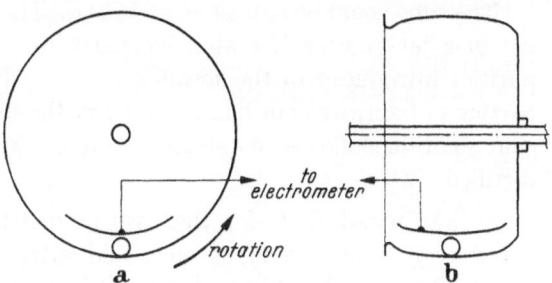

Fig. 37a and b. Schematic diagram of PETERSON's squirrel cage apparatus for measuring charging rate of spheres rolling on an Ni cylinder

ed concentric with the cylinder and spaced eight millimeters from its inner surface. The probe was calibrated by the use of a cylinder with an insulated segment which, together with the probe, comprised an essentially complete Faraday cage. Comparison of the total charge of the sphere with that measured by the probe alone showed that the probe "collected" $33 \pm 4\%$ of the charge of the sphere. The probe can be rotated upward until its shielding screens it completely from the charged sphere, where it can be grounded to provide a reference point. The cylinder is kept at ground potential by means of a *graphite brush* pressing against the shaft.

Low capacity, low leakage circuit leads were made from fine wire sealed into glass tubing, subsequently outgassed and evacuated. Shielding is provided by copper foil wrapped around the tubing. Polyethylene and polystyrene cable were found unsatisfactory because of relatively high spurious charges developed by slight flexing. Such charges leak away slowly.

An inverse feedback type of vacuum tube electrometer was developed, in which the potential difference between the measuring circuit and its shielding is essentially zero, thus minimizing charge leakage while allowing the use of relatively high voltage ranges. The effective grid current of the electrometer tube used was about 10^{-15} ampere, which is small compared with the rates of charging encountered in this work. An adjustable air gap condenser of 40 to 255 mμf

---

* The spheres were ground to a good polish from carefully annealed material, and were round to within half a wavelength of light. The grinding was done by tumbling the blanks in a cylindrical cavity containing abrasive particles, by means of an air jet.

determines the potential difference produced between the measuring grid of the electrometer tube and the feed-back circuit by the charge induced on the probe. A ten-ohm resistor in series with the electrometer circuit supplied a ten-millivolt full-scale potentiometer type recorder, with a full-scale balancing time of two seconds. The chart drive was equipped with a quick change, allowing paper speeds of six or sixty inches per hour.

The pumping system consisted of a mercury diffusion pump, backed by a high-speed mechanical pump. In addition to the standard trap between the mercury pump and the final stopcock, a second liquid nitrogen-cooled trap prevented stopcock grease from entering the experimental tube. Pressures in the lower range were measured by an ionization gauge, which showed that pressures of $3 - 5 \times 10^{-7}$ mm Hg were readily obtained.

The tube could be outgassed at $350°$ C. The nickel cylinder was cleaned by the use of a jet of very fine abrasive particles suspended in water, to remove impurities introduced in the forming process. Buffing was avoided in order that particles of abrasive not be imbedded in the metal. Final cleaning was by detergent solution, followed by steam cleaning and boiling and subsequent rinsing in distilled water.

It was found that, for the case of quartz and borosilicate glass, washing successively in nitric acid, chromic acid, nitric acid, tap water, and distilled water produced uniformly clean surfaces, as shown by uniform charging characteristics. Other methods, such as steam cleaning and rinsing in acetone, followed by distilled water, left surfaces which were not only not uniform but actually had areas which acquired charge of polarity opposite to that of the clean surface. Evidently very great changes in surface properties are produced by even slight contaminations; such dirty surfaces also display rather high conductivity. It is probable that the contamination was by oils and greases, which undoubtedly produce complex and unpredictable local surface structures.

The cylinder was slightly asymmetric, with the result that the sphere was alternately raised and lowered with respect to the probe as the cylinder revolved. This resulted in a total variation of about seven percent in the charge induced on the probe, producing a "wobble" on the recorder trace. Because of its relative freedom of lateral motion, the sphere traveled somewhat farther than the circumference of the cylinder in one revolution of the cylinder. This effect is important only at low speeds and caused a slight overestimate of the charging rate at these speeds.

The cylinder rotated smoothly under the magnetic coupling except at lowest speeds, where bearing friction occasionally produced a jerking motion. Where oscillation of the sphere developed, the results were discarded. Since probe and cylinder were not concentric in cross section, the proportion of charge collected by the probe varied slightly with displacement of the sphere in a direction parallel to the axis; there was negligible variation for large displacements in the direction of rotation. As the charge of the sphere became high its motion became quite erratic, because of the very large image force attracting it to the cylinder causing it to stick. At a charge of about 4×10^{-9} coulomb, measurement became impossible because the sphere adhered to the cylinder and was at times raised considerably above its normal position before it broke loose and rolled back to the bottom of the cylinder.

In order to estimate the surface conductivity of quartz, the probe was grounded in measuring position, then the electrometer switched to a low range of polarity opposite to that of the charge on the sphere. As charge drained from the sphere to the cylinder, the probe appeared to collect charge of the opposite sign. It is assumed that a nearly uniformly distributed charge produces a given potential distribution over the surface of the sphere; it is also assumed that surface conductivity is uniform over the sphere and is approximately independent of electric field strength. Then, without knowing the form of the potential distribution, it can be expected that the potential and therefore the potential gradient is proportional to the total charge on the sphere. Thus, by knowing the total charge, and measuring the leakage current, the surface conductivity can be found in arbitrary units. This procedure is valid as long as the charge distribution has not departed seriously from the initial approximately uniform distribution, and holds for the conditions under which the measurement was made in practice.

Because of the relatively high surface conductivity of borosilicate glass, the measured charge on a borosilicate glass sphere decreased rapidly after it had stopped rolling. This allowed an accurate measurement of conductivity directly from the recorder tracing, on the same voltage scale at which the charging curve was made*. The decrease in measured charge actually represents principally the redistribution of charge over the surface of the sphere, where charge drains from the upper surface of the sphere down to the region near the point of contact. The actual rate of loss of charge from the sphere appeared to be considerably lower than the probe measurement would indicate. If the sphere were revolved a half turn, after standing for a few minutes, bringing the previous area of contact up near the probe, the probe signal became as large, or larger, than it was when the sphere first stopped rolling.

When a gaseous atmosphere was desired, high-purity nitrogen was used, dried by passing through a long trap cooled by liquid nitrogen. At no time did the charging rate or surface conductivity appear to be changed by the repeated introduction of nitrogen.

Discharging the sphere was accomplished in either one of two ways, depending upon its surface conductivity. Spheres having low surface conductivity, such as fused quartz, were effectively discharged by rolling in an atmosphere of nitrogen of about 4 mm Hg pressure, which is near the pressure for minimum saturation charge, as a result of gaseous discharge between the sphere and the cylinder. While the residual charge after this procedure was not zero, it was relatively small and uniformly distributed. Spheres having relatively high surface conductivity, such as borosilicate glass, were discharged by rolling at low speed, which reduced the residual charge to considerably below that resulting from gaseous discharge. In earlier work, discharging was accomplished by X-ray ionization of the gas in the tube. This required near-atmospheric pressure of gas to be effective and with the equipment available required several hours. For these reasons, as well as to

---

* The observations and statement are correct but later study and theory by Wagner indicate it is doubtful whether *accurate* values of the conduction can be derived from the decrease *of apparent* charge of the sample *at rest*.

avoid the unknown effects of X-rays upon the properties of the nonconductors, their use was avoided in the present work.

The rate of charging of the sphere was most conveniently measured by the rate of gaseous discharge to the probe. At about 4 mm Hg pressure, gaseous discharge occurred only to the probe, and not to the cylinder. This method measured the total charge collected by the sphere per unit distance rolled and incidentally provided a check on the probe calibration, when compared with the initial slope of the charging curve in vacuum. The agreement was quite good.

b. Experimental results on fused quartz. Fused quartz acquires a negative charge in contact with nickel. When carefully cleaned, the surface is very uniform, as judged by its charging characteristics. Since the probe was relatively close to the upper surface of the sphere, the charge on the extreme upper part of the sphere was heavily weighted in the induction of charge on the probe. Thus, the probe in effect sampled the surface of the sphere as it rolled underneath it, and if the surface of the sphere were not uniformly charged, a fluctuation was produced in the probe signal. Although uniform charge distribution is the exception rather than the rule, when care was taken to produce such uniformity the signal on the probe varied by one percent or less.

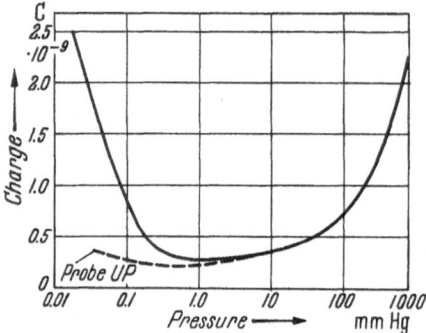

Fig. 38. Peterson's curve for maximum charge of quartz sphere as a function of pressure of N_2 gas in mm

The surface of the quartz sphere became contaminated as a result of outgassing the apparatus, probably by contaminants driven out of the nickel and redeposited on the sphere. Recleaning the quartz invariably returned it to its original charging characteristics. Relatively little contamination appeared to occur as a result of standing in a vacuum of the order of 5×10^{-7} mm Hg. Two quartz spheres from different sources had nearly identical charging characteristics. Therefore, it is probable that results obtained here are generally applicable to fused quartz.

c. Effect of gas pressure. The effect of the pressure of the atmosphere in which charging takes place was investigated in order to determine whether gaseous discharge actually is responsible as indicated by earlier work. The maximum charge of the sphere was measured as a function of pressure of N_2, from 3×10^{-7} to 700 mm Hg, giving the results shown in Fig. 38. These results were reproducible to about ten percent and were in principle independent of rolling speed. At low pressures the measured charge was less if the probe had been swung up before measuring the charge than if it had remained in measuring position. The existence of gaseous discharge from the sphere was established by the observation of individual charge transfers through the gas to the probe. Such charge transfer through the gas to the probe can have only occurred by Townsend discharge or spark. Below 10 mm Hg all discharges went to the probe, and above 40 mm Hg all discharges occurred to the cylinder. At higher pressures the discharges were large, provided the rate of acquiring charge was relatively high at this point. Such large discharges resulted in a very nonuniform charge distribution over the sphere, thus producing a large fluctuation in the charge induced on the probe. Fig. 39 which is a photograph of the recorder tracing, shows the progress of charging at 400 mm Hg, culminating in several large discharges to the cylinder and the resultant "patchy" charge distribution. Where the charging rate was low, the discharge pulses were often so small even at high pressure as to be indiscernible on the recorder tracing. With decreasing pressure the maximum pulse size decreased, probably tending toward a typical Townsend discharge at very low pressures*.

The form of the pressure dependence of maximum charge, the fact that charging rate is independent of pressure, and in particular the fact that gaseous

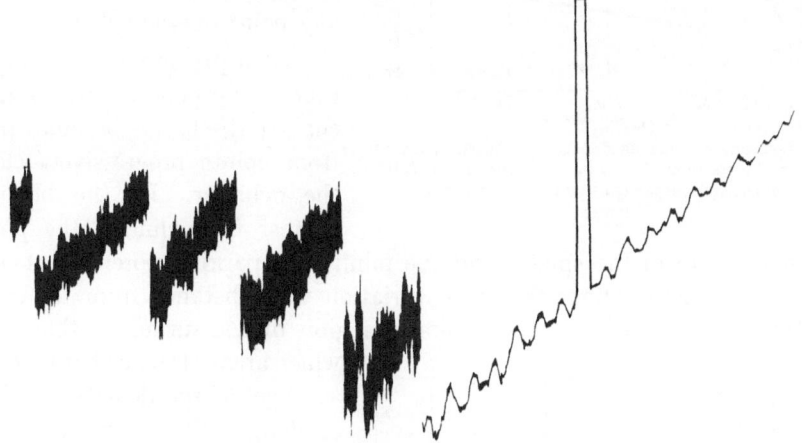

Fig. 39. PETERSON's record for charging curves showing large discharges to cylinder and cylinder wobble. Charge increases downward. At the rectangular pip, the probe was grounded for checking the electrometer

discharges occur to the probe, establish that the pressure dependence of maximum charge is due to gaseous discharge.

The pressure dependence of the maximum attainable charge is very similar to the minimum sparking potential curve for a pair of electrodes, as a function of gas pressure and electrode spacing. To demonstrate the striking similarity, Fig. 40 shows both curves. The usual parameter used in the sparking potential curve is the product pd, where p is the pressure in millimeters of mercury, and d is the gap between the electrodes, in centimeters. It should be observed that, while the minimum sparking potential (V_s) curve is nearly linear in pd above its minimum point, the curve for maximum charge (q_{max}) as a function of pressure falls off gradually, and drops far below the curve for V_s.

While the value of pd for the minimum of the V_s curve is about one in nitrogen, the minimum in the curve for q_{max} is at about two mm Hg. This fact suggests that the discharge occurs from points four or five millimeters from the metal surface. Since the gap between the probe and cylinder is eight millimeters and

---

* It is interesting to note that Townsend discharges occur with glass or quartz cathode and metal anode. This raises the question of γ from such insulators. This has now been quantitatively verified by V. J. ROHATGI in the author's laboratory. There is a photoelectric coefficient of secondary emission γ from borosilicate glass of value 10^{-4} in vacuum, yielding currents of 10^{-12} amp/cm² with 2537 Å radiation from a mercury lamp and 10^{-3} for shorter wavelengths.

both are ground potential, the points of maximum potential on the sphere lie at just about this distance. At pressures lower than about one millimeter the probe inhibits discharges, because it effectively limits the path length available for the course of a discharge. With the probe swung up out the way, however, paths

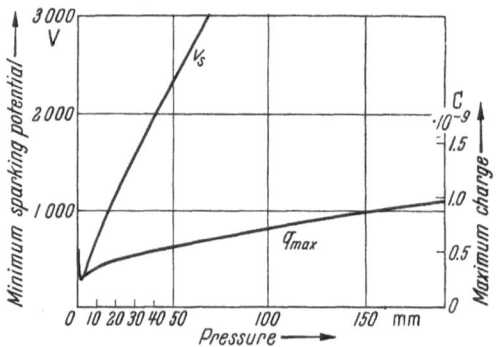

Fig. 40. Comparison of maximum charge on quartz sphere as a function of pressure of N_2 and minimum sparking potential for a pair of electrodes separated 5 mm corresponding to probe distances

of greater length are available, while the potential of the upper part of the sphere becomes greater, facilitating discharge and lowering q_{max} (see Fig. 39). At approximately 10^{-3} mm discharges no longer can take place for lack of sufficient potential at any point on the sphere.

As the pressure is increased above that corresponding to minimum charge, discharge becomes possible from points progressively closer to the cylinder. In the hemisphere nearest the cylinder, the potential at a given point of the sphere and the minimum sparking potential at a given pressure both have about the same variation with distance from the cylinder. For this reason, there is a considerable region of the surface of the sphere in which areas of somewhat higher than average charge density can initiate discharges. With increasing pressure, this region moves downward and increases in area. When a discharge is initiated from a point in this region, it may trigger additional discharges until the whole region is discharged. Thus, with increasing pressure the discharges will become larger and leave an increasingly less uniform charge distribution, if the charging rate at this pressure is relatively high, as shown in Fig. 39. If the charging rate is low, there will be greater uniformity of charge distribution, and the discharges may be localized, and therefore small. At

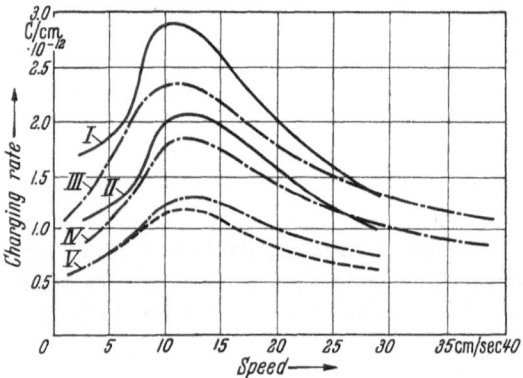

Fig. 41. PETERSON's charging rate of fused quartz as a function of rolling speed under different conditions. (I) Freshly cleaned. (II) Freshly cleaned, at 55°C. (III) Contaminated after outgassing apparatus at 225°C. (IV) After outgassing at 325°C. (V) Further contaminated after standing in vacuum for several weeks; dashed portions shows decreased values resulting from long rolling

pressures below that for minimum q_{max}, a progressively smaller area has sufficient potential to produce a discharge. Therefore, the discharge become smaller, leaving a more uniform charge distribution.

The explanation of the observation that at pressures up to about 10 mm Hg all discharges occur to the probe lies in the probe's greater distance from the sphere. At higher pressures, however, the most favorable distance becomes less than that from the sphere to the probe, and discharges begin occurring to the cylinder.

The largest total charge which can exist on the sphere without producing gaseous discharge will be for the case of uniform charge distribution. After a large discharge has occurred, draining the charge from a portion of the surface, while the total charge of the sphere has been considerably reduced, other areas of the sphere still have sufficient charge to produce a discharge. Thus, succeeding discharges continue to lower the total charge of the sphere, while producing an extremely "patchy" charge distribution. Finally, a roughly steady state is reached, where the total charge is considerably less than that which could exist if it were uniformly distributed. This effect is shown clearly in Fig. 39.

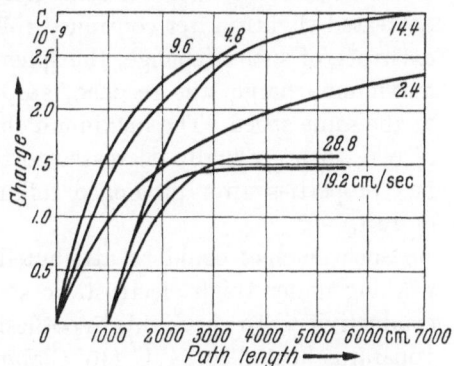

Fig. 42. Charging curves for various rolling speeds for freshly cleaned quartz in vacuum. Curves for speeds greater than 10 cm/sec have been displaced from the origin for clarity

d. The charging rate. The charging rate, (charge acquired per cm of distance rolled), was measured and found to be independent of pressure in the range from 3×10^{-7} mm Hg to 700 mm Hg. Since charging rate is independent of pressure, it was conveniently and accurately determined by measuring the rate of accumulation of charge on the probe transferred from the sphere by gaseous discharge. Curve I, Fig. 41, shows the charging rate of clean quartz as a function of rolling speed. Curves III, IV, and V show the decreased charging rate due to contamination of the sphere resulting from outgassing the apparatus at 225°C, 325° C, and the further contamination of the latter state resulting from standing for two weeks in a vacuum of about 3×10^{-7} mm Hg. These curves all show a maximum charging rate at an intermediate speed of rolling. Prolonged rolling at higher speeds produced a decreased charging rate, such as that shown by the dashed portion of Curve V. Long rolling at low speed was necessary before the charging rate had recovered its initial value.

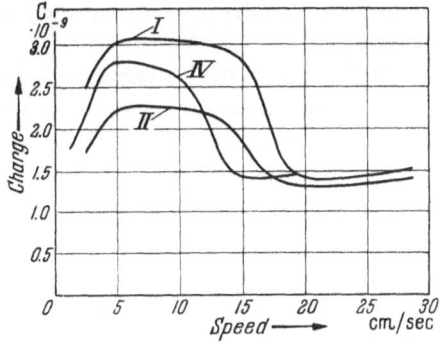

Fig. 43. Maximum charge for fused quartz in vacuum as a function of speed under various conditions. (I) Freshly cleaned, at 55°C. (IV) Contaminated after outgassing apparatus at 325°C

Fig. 42 gives charging curves for newly cleaned quartz at several speeds of rolling in vacuum, showing the total charge on the sphere as a function of the distance rolled. The curves made at a low speed of rolling, (less than 10 cm per second), are shown beginning at the origin and all have the same general form. The curves for high-speed rolling (greater than 10 cm per second) have been displaced from the origin in the figure in order to show clearly that they all have a similar form different from that for low-speed rolling. The initial slopes of these curves correspond to the charging rates of Curve I, Fig. 41. The maximum charge in vacuum increases with speed up to intermediate speeds and decreases sharply at higher speeds, as shown in Fig. 43. The curves representing the same surface condition in the figures are labeled with the same Roman numeral.

The low maximum charge and altered form of the charging curves for high-speed rolling suggested that a change in surface properties occurs during the course of rolling. Further evidence came from the observation that the maximum charge decreased slightly after continued rolling at high speeds. In order to verify the existence of such a change, the sphere was rolled at 15 cm per second until the maximum charge was reached, stopped for twenty minutes and rolled again at the same speed. The maximum charge increased appreciably after the period of rest, then gradually decreased to the original value. Quite evidently the surface properties after prolonged rolling were different from those after a period of rest.

Such changes could be attributed to heating of the surface, by mechanical working under thigh electrostatic stresses, with a resulting increase in surface conductivity. To check this hypothesis, measurements were made with the entire apparatus heated to 55° C (30° C above room temperature) using a clean quartz sphere. The charging curves for this case corresponded in form to but lay below those of Fig. 42. The maximum charges, shown by Curve II, Fig. 43, lie considerably below those for clean quartz at room temperature, shown by Curve I. Curve II of Fig. 46 shows the surface conductivity at 55° C, measured immediately after various speeds of rolling. Because of the difficulty of measuring the low surface conductivity of quartz even at this temperature, the curve is only approximately correct. However, it does show a real increase with speed of rolling. Whether conductivity decreases again above 30 cm per second is doubtful, although the curves for maximum charge all show a slight increase at the high-speed end, as shown by Fig. 43. It is possible that this is caused by minor mechanical instability of motion such as bouncing, though this is not perceptible to the eye.

Surface conductivity can also be measured by the decrease in measured charge of the sphere after it has stood for forty to sixty hours and gives values of conductivity K of about 0.1 at room temperature. Measurements made immediately after the sphere had stopped rolling gave rough values of from 0.4 to 2. Thus surface conductivity appears to increase greatly after rolling, particularly at higher speeds.

e. Borosilicate glass. One reason for choosing borosilicate glass was the desire to investigate the charging properties of a material of relatively high surface conductivity, as compared with quartz. Since the surface conductivity of borosilicate glass in vacuum was found to be three to four orders of magnitude higher than that of fused quartz, this condition was fulfilled. Borosilicate glass charges negatively in contact with nickel, and in general, its charging properties are similar to those of quartz.

Considering the difference in conductivity, it is remarkable that the maximum charges and charging rates were of the same order of magnitude as those of quartz. However, borosilicate glass did not seem to have the high degree of reproducibility which quartz exhibited. In addition to becoming contaminated faster and more seriously than quartz, its charging rate at times seemed to decrease after standing for some time with a high charge. This is not surprising, considering that the high local fields are probably sufficient to cause migration of ions, with a resulting change in structure. Migration of ions in such glass has been observed. Such action does not occur with the more stable and strongly bound quartz structures.

Subsequent prolonged rolling appeared to cause a recovery of the charging rate reduced by standing with high charge.

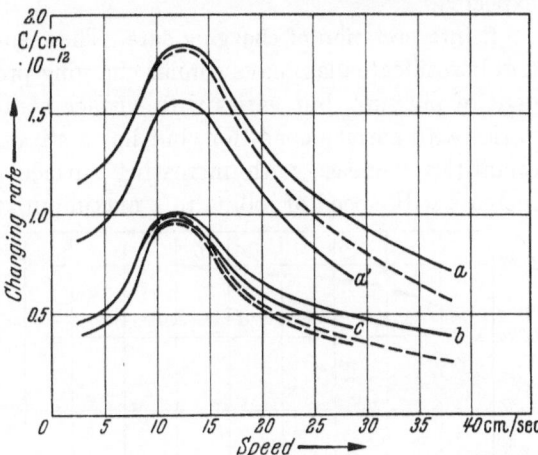

The newly cleaned sphere became somewhat contaminated after standing for a day or two in a vacuum of about 5×10^{-7} mm Hg. The surface appeared to stabilize after a few days and remained more or less constant until after the first outgassing of chamber and Ni cylinder at 225° C, after which the contamination had evidently increased, as in the case of quartz. Outgassing at 325° C still further increased the amount of contamination.

Fig. 44. Charging rate—borosilicate glass as a function of rolling speed. (a) Freshly cleaned. (a¹) Contaminated after standing in vacuum for a few days. (b) Further contaminated after outgassing apparatus at 225°C. (c) After outgassing at 325°C. Dashed curves show decreased values resulting from prolonged rolling

Within experimental error the variation of maximum charge with pressure was identical to that of quartz, although higher rolling speeds were required to initiate discharge at the extremes of the pressure range. At the pressure used for charging rate measurements, the discharge pulses to the probe were larger and less frequent than in the case of quartz.

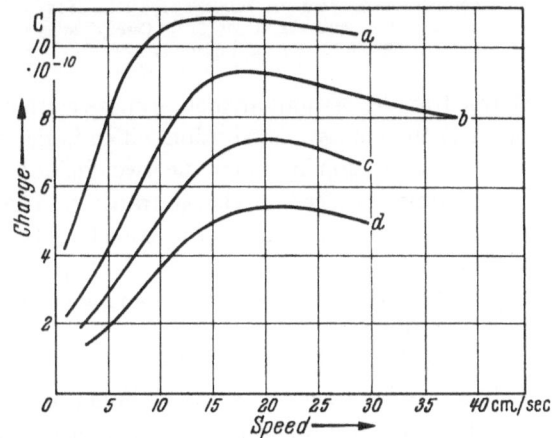

Charging rates are shown in Fig. 44. While having lower values, the curves are quite similar in shape, and the maximum rates of charging occur at about the same speed as for quartz. As in the case of quartz there was a pronounced decrease in charging rates after prolonged rolling at higher speeds, as shown by the dashed curves. Prolonged slow speed rolling returned the charging rates to their original values.

Fig. 45. Maximum charge of borosilicate glass in vacuum, as a function of rolling speed. (a) Freshly cleaned. (b) Contaminated after outgassing apparatus at 225°C. (c) After outgassing at 325°C. (d) Case (c) at 55°C

Fig. 45 shows maximum charge in vacuum. The curves do not show the sharp decrease at high speeds, characteristic of quartz, indicating that conductivity increases less with speed of rolling than in the case of quartz. This is borne out by Fig. 46, showing the conductivity of both as a function of speed. While the values for quartz are only qualitative, they are accurate enough to establish that the relative increase with speed is much greater for quartz than for borosilicate glass. A comparison of Figs. 45 and 46 shows a definite decrease of maximum charge with increasing conductivity. The speed corresponding to the peak of the maximum charge curve increases with conductivity, as might be expected.

f. Interpretation of charging data. The results just described show that quartz and borosilicate glass have similar charging properties. Charging rate is independent of pressure, but varies with surface contamination. Maximum charge also varies with surface condition, but has a strong dependence upon pressure. Both quantities decrease with increasing surface conductivity, and both show an increase with speed of rolling to a maximum at an intermediate speed. Charging rate decreases sharply for both materials, following this maximum. The maximum charge of borosilicate glass decreases only slightly at higher speeds, while that of quartz decreases sharply and then remains nearly constant. The sign of charge is independent of all these variables, if the sphere has been properly cleaned initially.

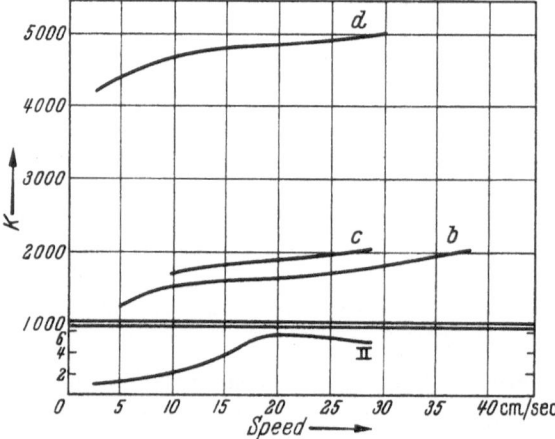

Fig. 46. Surface conductivity as a function of rolling speed.
(b) Borosilicate contaminated after outgassing apparatus at 225°C.
(c) Borosilicate after outgassing at 325°C. (d) Case (c) at 55°C.
(II) Freshly cleaned fused quartz at 55°C

Consideration of these results leads to the conclusion that the processes involved in the charging of an insulator in contact with a metal fall rather naturally into two classes: first, that associated with the direction and amount of charge transfer per unit area of intimate contact; and second, those processes which produce the charge distribution which is measured. It seems important to draw a clear distinction between the two, in order to avoid errors in interpretation.

g. Primary process of charge transfer. The primary process of charge transfer is considered to be due to the transfer of electrons, analogous to the electron transfer which produces the contact potential difference between metals of different work function, where the metal of higher work function acquires electrons from the other.

Because the charge transfer is a surface phenomenon, it is impossible to distinguish between a true work function and the existence of surface states. If it is a true work function—that is, a property of the bulk of the nonconductor—it should not be expected to be the same as that measured by other means. For example, a photoelectric work function would of necessity be correlated with filled levels, whereas in the case of a negatively charging nonconductor (as is usually the case) the so-called work function must be correlated with unfilled levels, which can accept electrons. Furthermore, the levels, if they exist, must be localized, or the electrons would be free to move back toward the point of contact. Since the nonconductors investigated have aperiodic structures, it is at present impossible to attempt a calculation of their energy level schemes, and it is perhaps presumptuous even to use the concept of well-defined energy levels or bands. Nevertheless, the reproducibility and the uniformity of charging, under favorable conditions, suggests that there is at least a distribution of levels or states whose average value is well-defined.

There is as yet little direct experimental evidence to support this hypothesis. There is, however, qualitative support from the observation that borosilicate glass charges much less in contact with lightly oxidized nickel than in contact with relatively clean nickel. The electric double layer produced by electronegative oxygen increases the work function of the nickel, thus decreasing the work function difference between it and the glass (which charges negatively in both cases and, therefore, presumably has a higher effective work function).

h. Secondary processes governing redistribution of charge. There appear to be two secondary processes governing the amount and distribution of charge over the surface of the nonconductor: conduction over the surface of the sphere and the gaseous discharge from the sphere where the charge is sufficiently great, which has already been discussed.

i. Surface conduction. Assuming an equilibrium contact potential difference to exist between the nonconductor and metal in contact, electrons will be transferred to a region of the nonconductor which is effectively in contact with the metal. The area of the nonconducting sphere in physical contact with the metal can be defined as that area for which the separation between the two surfaces is small enough for electron transfer between the two. This area depends upon the shape and smoothness of the two surfaces, and upon pressure at the point of contact, which can be greatly increased by strong image forces, when the sphere is highly charged.

As the sphere rolls, the charged area is separated from the metal surface and its potential with respect to the metal increases. Thus a potential gradient is established along the surface of the sphere, producing a flow of electrons back toward the point of contact. The magnitude of this discharging current depends upon the potential gradient and surface conductivity. For a given current, the charge removed from an element of area will depend upon the length of time which it spends in a given high field region. Therefore, the rate of *loss* of *charge* by *conduction* back toward the point of contact would be expected to *decrease* with increased speed of rolling. The amount of charge removed should not be expected to be inversely proportional to speed of rolling, since the *increased remaining* charge at higher speeds increases the potential gradient and thus the discharging current. In addition, all previously charged areas of the sphere produce a flow of electrons toward the point of contact particularly when the revolution of the sphere brings them close to the metal. This loss again depends on conductivity and time. Thus, increased speed of rolling tends to decrease the rate of loss of charge.

If it is assumed that, for a given speed of rolling, the rate of acquiring charge is proportional to the uncharged area remaining on the surface of the sphere, and the rate of discharge is proportional to the total charge on the sphere, the *net rate of charging*, (net charge acquired per unit distance of rolling), is given by

$$\frac{dq}{ds} = a\left(1 - \frac{q}{q_s}\right) - k q \tag{4.1}$$

and

$$q = \frac{a}{B}(1 - e^{-Bs}), \quad \text{where} \quad B = \frac{a + k q_s}{q_s} \tag{4.2}$$

$$\left(\text{maximum charge } q_m = \frac{a}{B}\right).$$

q_s represents the total saturation charge of the sphere in the absence of discharging. The quantity $1 - q/q_s$ in the early state of charging represents the fraction of the surface of the sphere which has not yet been charged and therefore can accept electrons at the rate a per unit distance rolled. After the sphere has acquired appreciable charge, this quantity also takes account of the fact that previously charged areas have lost some of their charge by leakage, and therefore can accept an additional charge. k is a constant of proportionality for the rate of electron flow from the previously charged areas toward the point of contact and is assumed to be directly proportional to the surface conductivity. Surface conductivity as indicated earlier can be measured. It is represented by the symbol K, which is expressed in arbitrary units, that are equal to the apparent positive current to the probe per volt of probe signal, multiplied by 10^{15}.

By substituting a/q_m for B into (4.2) yields,

$$q = q_m (1 - e^{-a s/q_m}) \tag{4.3}$$

$$\frac{dq}{ds} = a (e^{-a s/q_m}). \tag{4.4}$$

All the quantities in these equations are measurable.

Experimentally, as described earlier, it is very difficult to remove all charge from the sphere. It is simple, however, to remove most of the charge, leaving a uniform distribution of relatively low charge density. Therefore, this procedure was used, and the charging curves were made with such an initial uniform charge on the sphere. In this case

$$dq/ds = a (q_s - q_i - q) - k (q_i + q),$$

where q is the increase in charge over the initial value q_i, and

$$q = (q_m - q_i) (1 - e^{-a s/q B}) = q_m' (1 - e^{-a s/q_m}) \tag{4.5}$$

$$\frac{dq}{ds} = \frac{(q_m - q_i)}{q_m} (e^{-a s/q_m}) = \frac{q_m'}{q_m} (e^{-a s/q_m}). \tag{4.6}$$

Thus, the predicted charging curve for the case of a uniform initial distribution of charge is the same as that for zero initial charge, except for a multiplying factor. The charging rate is smaller by the factor q_m'/q_m. Charging rates measured under these conditions have been corrected to the value for zero charge, in order that they may be compared for different conditions.

To check experimental agreement with Eq. (4.6) the charging curves for quartz at 55° C, similar to those of Fig. 42, have been replotted on semilogarithmic paper, plotting $(q_m' - q)$ against distance rolled. Eq. (4.5) predicts a straight line for such a plot, and Fig. 47 shows good agreement for the low-speed curves. The similarity of low-speed charging curves for quartz, demonstrated in Fig. 42, is thus explained by the fact that they are all of the predicted form, while the curves for high-speed rolling depart rather seriously. In addition, the maximum charge is much lower at high speeds in the case of quartz.

This effect could be accounted for by either a decrease in the rate of initial electron transfer, or an increase in the rate of discharge, as a result of continued high-speed rolling. The first possibility seems improbable when it is noted that the net rate of charging is always higher for high-speed rolling than for the lowest speeds, while the low speeds have final charges greatly in excess of those for high-speed rolling. These considerations led to the conclusion that conductivity increases after high-speed rolling, which was subsequently confirmed by the curves of Fig. 46, showing surface conductivity after prolonged rolling at various speeds.

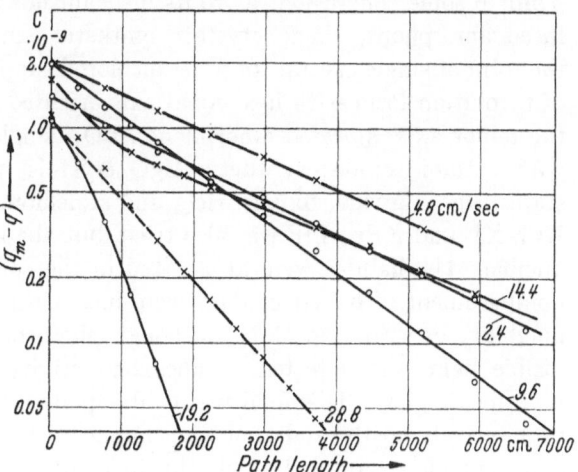

The increased surface conductivity of the sphere was attributed to surface heating at high rolling speeds, which seems to be confirmed by the large increase in conductivity measured at 55° C. These results indicate that an increase in surface temperature of about 5° C for borosilicate glass and 15° C in the case of quartz are sufficient to account for the increased conductivity at high speeds. It appears that a relatively shallow surface layer becomes heated, and that it is cooled by conduction to

Fig. 47. Semi-log plot of the charging curves for fused quartz at 55°C. Equation (4.5) predicts a straight line for such plots

the bulk of the sphere when it is not rolling. This surface cooling appears to take place more rapidly than the uniformly heated sphere itself would cool by conduction, as shown by the rapid recovery of the surface during a period of rest. Because of its much lower thermal conductivity, the quartz surface would be expected to cool less while rolling, and therefore show greater surface heating than borosilicate glass.

It is not surprising that there should be surface heating at high speeds under the rapid application and removal of strong image forces to an element of area, as it approaches and departs from close contact with the metal surface. That relatively high surface charges are required to produce such heating is evident from the comparatively slight decreases in charging rates, after prolonged rolling at 4 mm Hg pressure where gaseous discharge limited the surface charge to low values.

The decrease in the charging rate of quartz due to contamination probably reflects a change in effective work function, as well as an increase in conductivity. This appears to be so, since the rate of charging and maximum charge are less for the contaminated case than for clean quartz at 55° C, although the latter case had the higher conductivity. It is probable that contamination similarly alters the work function of borosilicate glass, while increasing its surface conductivity, as shown in Fig. 46.

F. Extension of PETERSON's investigations by WAGNER

1. Experimental techniques

WAGNER[109] has recently concluded a study in which PETERSON's techniques were extended and improved primarily in order to ascertain the nature of the charge transfer mechanism. The work was extended to include as many other substances as would yield reproducible results. Control of conditions was improved to the point where *modern outgassing* techniques could be applied to both surfaces. The charging was investigated relative to the work functions of several metals. Studies of surface conductivity were made.

The data throw considerable light on certain aspects of contact charging and point to some conclusions as to its mechanism. The substances tried were SiO_2, fused amorphous, single crystals synthetic, single crystals natural; Al_2O_3, in the form of single crystals of pure uncolored sapphire, in spheres and in ellipsoids of revolution form with hexagonal axis oriented along the major axis and along the minor axis; sintered amorphous Al_2O_3, single crystal of ruby, Al_2O_3 colored with chromic oxide; synthetic MgO crystals, periclase, in various "colored" states, e.g., Mg rich, oxygen rich, and stoichiometric. The alkali halides, NaCl, KCl, KI, and KBr. LiF was also tried, but abandoned because it was not reproducible. The halides were also tested in the colored form obtained by electron bombardment. The latter data were inconclusive because of the lack of reproducibility generally speaking, although photoconductivity of colored samples in visible light was detected. Some conductivity studies were carried out with borosilicate glass. Diamond was contemplated but the grinding of a sphere was prohibitively costly and time consuming. Initially, grinding and polishing was done outside. Later, through the ingenuity of Mr. C. GRANT, of the Physics Department shop, a convenient and rapid technique for grinding and polishing samples was developed and the MgO and halide spheres as well as quartz spheres were prepared right in the laboratory. Measuring and recording was done through the use of a vacuum tube electrometer circuit designed by Dr. N. SEATON and later by use of a vibrating plate electrometer designed and kindly built for him by F. A. MULLER and H. DEN HARTOG of the University of Amsterdam. One series of measurements was made using surfaces of known work function measured by the Kelvin method for reduced and oxidized Ni such as used in the rotating cylinder. The glass envelope collapsed through some flaw in the glass after one exemplary run was made. Later a tube was designed with 3 rotating cylinders, one of Cu, one of Pt, and one of Ni, on the same shaft. The same sphere could rolled in each. In each, the contact potential relative to a standard outgassed W filament could be measured by the method of ZISMAN. In this apparatus, the sphere of SiO_2 could be separately heated and outgassed away from the metals. ALPERT vacuum techniques were used, pressures remaining around 10^{-9} mm. With this, the variation of saturation charge with work function of the metal was studied.

Probably the greatest problem was found to be that of proper cleaning techniques. Both SiO_2 and Al_2O_3 were sufficiently stable so that they could be cleaned with HNO_3, chromic acid, HNO_3 and rinsed in distilled water without being altered. The SiO_2 crystalline surface was examined by electron diffraction after grinding, polishing, cleaning and running in the cylinder. It was found to be undamaged in the processes. This observation is in agreement with the findings of FINCH[110], on quartz and Al_2O_3. LEISE[111] found the same for polished LiF.

After outgassing the samples gave charging curves reproducible to within 10%. Al_2O_3 was also reproducible but owing to preferential charging at the certain positions relative to the hexagonal or optic axes, the asymmetrical charging caused a wobble that made measurement of average charge rate less certain. For MgO the problem was more difficult. It is apparently soluble in acids and alkalies. Here cleaning was achieved by washing, degreasing in chemically pure CCl_4, C_6H_6, ethyl alcohol and distilled water. Here the effective agency after grease removal was the action of distilled water which appears to have dissolved off a layer possibly 100 Å thick without spoiling the polish at 450 fold magnification. Evidence for this comes from the behaviour. Cleaning with acids etched the sample in a pattern of cubic symmetry.

Least reproducible were the alkali halides. Here the surface polish could not be preserved after cleaning. Rinsing in quasi-saturated solutions of their own species rendered clean but altogether too rough surfaces. In the end, polished spheres were wiped with a water moistened *glass* cloth that had in turn, been cleaned by the rigorous techniques as had the SiO_2. In outgassing the apparatus by heating above $350°$ C, the quartz became contaminated by impurities that came off the metal. The spheres had to be removed during the baking out of the apparatus.

2. Basic theory

In order to present and to discuss the results, PETERSON's theoretical equation for the process must be recalled. He assumed (a) an "effective contact potential difference" characteristic of the two surfaces in contact governing sign, initial charging rate and saturation charge density in vacuum if it were non conducting; (b) a surface or bulk conductivity. Here PETERSON in his paper considered 2 effects. (1) An *increase* of *charging* owing to *seepage* of charge *over* a *volume*, or surface of the regions adjacent to the contact area by conductivity. The suggestion was made *in the event that contact charging involved absorbtion of ions or electrons into states in the body of the lattice.* So far, the charge exchange appears only to involve exchange of ions on the surfaces of the materials. It is therefore unlikely that this factor actually occurs in charging situations*. In consequence it was omitted in reproducing PETERSON's paper in Sec. E.i. of this chapter. There is no evidence that it appeared in any studies.

(2) A small but finite conductivity tending to discharge the sample. As the sample rolls the charge separated *at* the *contact*, (which does not return by field emission back discharge), rises to higher potential causing a leakage current *around the surface*, or *through* the *insulator*. In the case of borosilicate glass it appears to be a surface leakage. (c) Gas discharge. This form of loss was precluded in all this work by using pressures *below* 10^{-5} mm, except where gas was needed to discharge the sphere under which conditions 4 mm pressure pure N_2 was intro-

---

* Very recent evidence from the work of ROHATGI, and work reported by R. G. FOWLER to the author points strongly to bound surface electronic states of high surface mobility in contrast to body states in the case of borosilicate glasses.

duced. N_2 does not affect charging if pumped off as soon as used. The threshold potential for this discharge in N_2 could be estimated from PETERSON's data neglecting the small effect of the dielectric constant of the quartz. The potential of any position on the uniformly charged sphere is caused by a point image charge $-q$ in the metal, located one radius of the sphere below the metal surface, with q the total charge on the sphere. The potential difference between the top of the sphere and the metal plane is then

$$V = -q\left(\frac{1}{3r} - \frac{1}{r}\right) = \frac{2}{3}\frac{q}{r} \text{ in esu.} \tag{4.7}$$

Wit $q = 2 \times 10^{-10}$ coulomb and $r = 0.3$ cm, $V \sim 400$ volts in good agreement with the minimum sparking potential of N_2.

PETERSON set q the charge on the sample at a given rolling speed as

$$\frac{dq}{dt} = \alpha(q_s - q) - \beta q \tag{4.8}$$

so that

$$q = q_s \frac{1}{1 + \beta/\alpha}\left[1 - e^{-(\alpha+\beta)t}\right]. \tag{4.9}$$

When

$$t = \infty, \quad q = q_m = \frac{q_s}{1 + \beta/\alpha} \tag{4.10}$$

the apparent saturation. Actually, PETERSON's Eq. (4.3) was written in terms of dq/dx the charge per unit distance rolled.

Here βq represents the discharging current due to surface leakage (assumed by PETERSON), and q_s is the *assumed "saturation" charge* in the absence of surface conductivity, while q_m is the *maximum observed*, or *apparent saturation charge*.

The quantity α represents a proportionality constant which gives the fractional charging rate at small charge. For quartz β is so small that it can be neglected and PETERSON verified the exponential nature of the relation for quartz rolling at speeds below 10 cm/sec.

When conductivity is present, saturation is achieved when $-\beta q$ equals the charging rate $\alpha'(q_s - q)$. The α' is the value of α in the event that is altered by greater effective area of contact produced by the *electrostatic forces*, as electrification proceeds, of which more later. The observed saturation charge q_m becomes $q_m = \frac{q_s}{1 + \beta/\alpha'}$ and the charging rate $\alpha' + \beta$.

In this theory the relative motion merely supplies new area. That is, it is *not* assumed to *yield friction* at local hot spots on the metal causing thermionic or other charge transfer. HARPER's results with metals show that charging is produced by motion normal to the surface, e.g., contact and separation. It was in line with this concept that PETERSON and WAGNER also *insist on the use* of the term *contact electrification*, in contrast to the old expressions of tribo or frictional electrification which should be reserved *for a few special cases* involving friction as a mechanism in its own right. As will later be seen, in most cases, the *frictional effect is not important, charge transfer requiring chiefly intimate contact*. True tribo electrification can result for two identical substances on asymmetric rubbing but is a rare occurrence.

3. Experimental observations

a. Saturation charge. WAGNER made the observations which follow: Fig. 48 shows the initial charging rate, fractional rate and maximum charge for quartz against Pt, as a function of rolling speed taken under improved conditions. Fig. 49 shows $q_m - q$ plotted against time for 6.3 cm/sec rolling speed. The

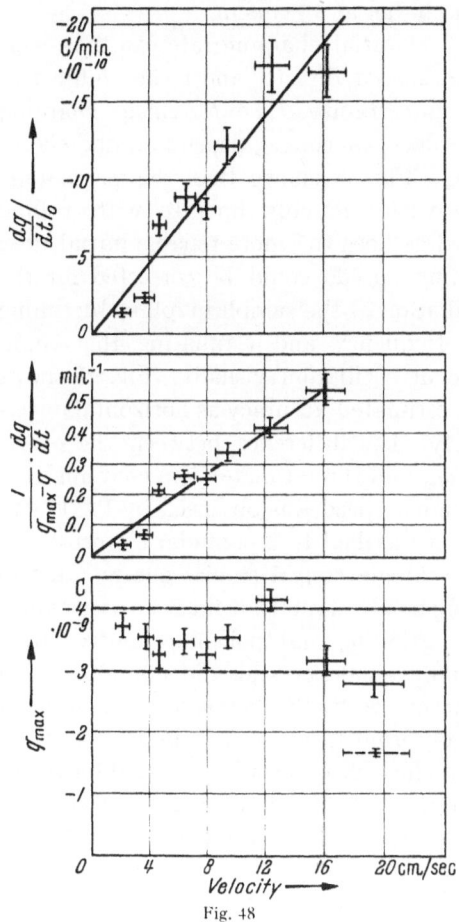

equation predicts a straight line. For no rolling speed, (except 19 cm/sec), was the deviation of the points from a linear behaviour greater than shown in Fig. 48. One difficulty in testing the equation is that it becomes difficult accurately to *evaluate maximum charge.* Except at 19 cm/sec, the rolling becomes so erratic before q_m is reached that the sphere sticks to the back of the barrel. Thus all values of q_m were obtained by fitting experimental points to a straight line on semi-logarithmic paper, using q_m as an adjustable parameter. The sensitivity of this method

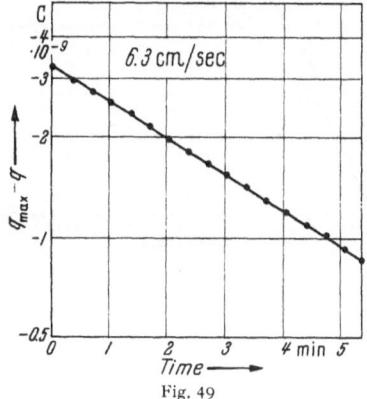

Fig. 48

Fig. 49

Fig. 48. Initial charging rate, fractional charging rate, and extrapolated maximum charge of quartz on Pt, as a function of rolling speed after WAGNER. Lightly dashed lines indicate 30% spread in extrapolated maximum charge. Lower value at 19 cm/sec is observed maximum charge at this speed. At all other speeds maximum charge was too large to observe directly

Fig. 49. WAGNER's data for $\log(q_m - q)$ plotted against time for 6.3 cm/sec rolling speed

is illustrated by the vertical uncertainties shown in Fig. 48. With the exception of the 12 and 19 cm/sec curves where 63% of q_m was reached experimentally only $1/e$ of q_m could be observed. The 19 cm/sec curve is shown in Fig. 50.

The lower curve using the observed maximum charge is not exponential. That is the value of q_m *observed* is too low to follow PETERSON's law. The upper curve plotted so as to fit the exponential yields a q_m which is higher. Since it was impossible to saturate at any other rolling speeds, it might be that PETERSON's law does not hold and that no curve is exponential. This drastic conclusion is

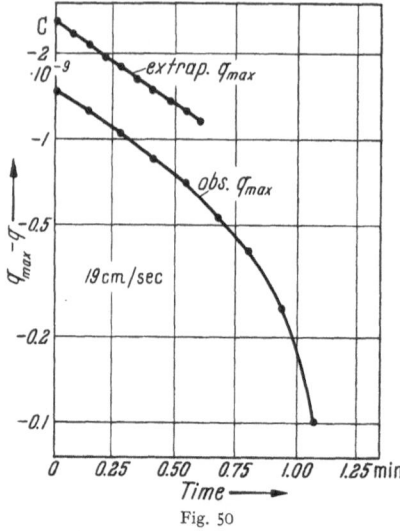

Fig. 50

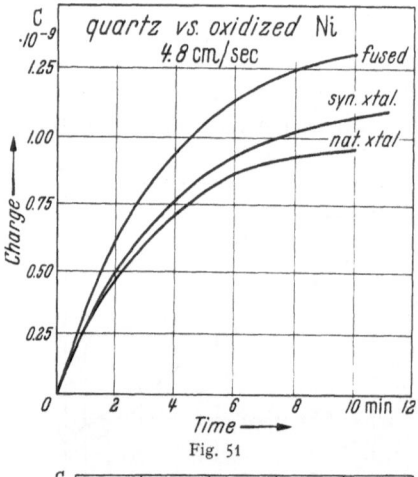

Fig. 51

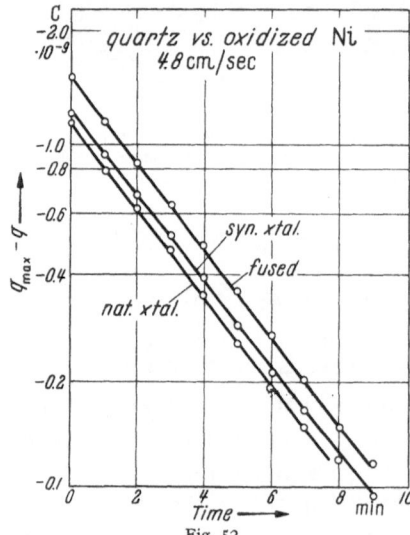

Fig. 52

negated, however, by the fact that observations of the curves indicate that at lower rolling speeds, the saturation is *not as low* as at 19 cm/sec. That is, the saturation charges observed experimentally at lower rolling speed were greater than the observed saturation charge at 19 cm/cm.

The initial charging rate can, however, be measured directly and is not subject to this uncertainty. The *percentage* charging rate does, of course, depend on the choice of q_m. Fig. 48 shows that the percentage rate varies roughly linearly with rolling speed as does the more reliable initial rate. Rolling speeds could be corrected for the oscillation of the sample in rolling by timing the frequency and estimating the amplitude of oscillation visually. Fig. 48 shows the estimated accuracy as horizontal uncertainty. The difference between the observed q_m and the estimate from charging rate at higher speed was observed by PETERSON and is ascribed to a secondary process.

b. Single crystal studies quartz. In studying single crystals, it must be ascertained that grinding and polishing spheres from initially single crystals did not alter their structure. Actually *electron* diffraction shows the outermost, say 50 Å to be monocrystalline. The type of disorder reported for X-ray diffraction studies shows disorder of microns depth, but only consisting of crystallites tilted up to about 4°.

Fig. 50 shows charging curves at 4.8 cm/sec rolling speed on three freshly cleaned quartz specimens, a fused sample, a synthetic single crystal, and a natural single crystal. Saturation charges were respectively -1.4×10^{-9}, -1.2×10^{-9}, and -1.1×10^{-9} coulombs. Fig. 52 shows semi-logarithmic

Fig. 50. Semilogarithmic plot of charging curve, freshly cleaned fused quartz against Pt at 19 cm/sec. Lower curve uses observed maximum charge; upper curve uses extrapolated maximum charge

Fig. 51. WAGNER's charging curves at 4.8 cm/sec for fused, synthetic, and natural single crystals of quartz against oxidized Ni

Fig. 52. $\log(q_m - q)$ against time for fused, single pure crystal and single natural crystals of quartz

plots of all three, yielding the same percentage rate 0.28 min⁻¹. Initial charging rates were 4.0, 3.2 and 3×10^{-10} coul/min reflecting the differences in saturation charges. There was *no* indication of *axial asymmetry* of charge in any of the samples—even at the slowest rolling speed as indicated in Fig. 53 when the periodic oscillation is due to a slight axial asymmetry of the cylinder. The failure to observe any anisotropy in crystalline quartz appears to be in disagreement with observations by HARPER[112], who reported large anisotropy effects in quartz-quartz charging. However, uniformity, cleanliness and outgassing here were far different from those used by HARPER. Such cleaning and treatment as WAGNER and PETERSON have used wipes out anisotropies caused by inadequate cleaning techniques, e.g., patches. The charging behaviour of the three samples was almost identical to the accuracy of measurement, but the *conductivities* were not. Fused and synthetic quartz initially discharged less than 0.02% per minute when rolling stopped. Natural quartz discharged between 0.6 and 0.3% per minute depending on sample orientation. Thus natural quartz had a conductivity two or more orders of magnitude higher than the others and one which was anisotropic. This could readily be expected in a natural crystal.

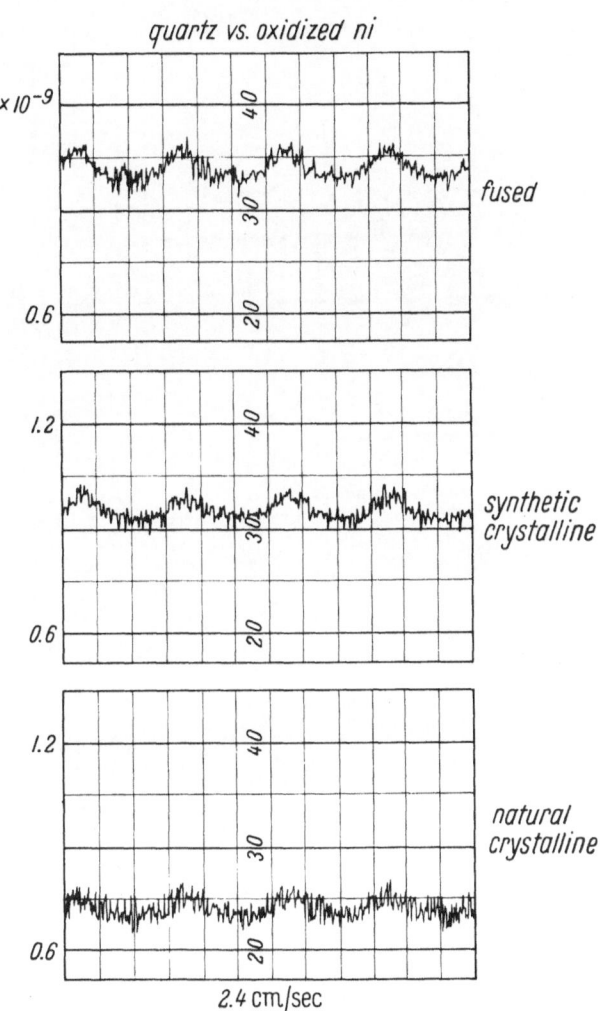

Fig. 53. Sections of saturated part of charging curves for three types of quartz. Long periodic oscillation is intrumental. Assymetry is shown by short period irregularities

c. **Aluminum oxide, synthetic single crystals, (white sapphires).** Using a *sphere* made of monocrystalline pure white sapphire yielded a small negative charge which was distributed *very* non-uniformly over the surface. While the effect could have been caused by uneven distribution of impurity left over from cleaning, the nature of the phenomenon after repeated cleaning indicated that this must be caused by anisotropic charging relative to the crystalline axes. Al_2O_3 crystallizes in the hexagonal system. Thus two ellipsoids of revolution were ground—one with the hexagonal axis parallel to the major axis; the other with it parallel to the minor axis of the ellipsoid. The orientation of the two samples could readily be checked as sapphire is birefringant and crossed polaroids at once indicate the axes by the ring and cross pattern centered about the optic or C axis. Fig. 54 shows the anisotropy produced in the sample with the optic axis along the minor elliptic axis exposing the ends of the crystal to the probe as it rolled, the pattern with the optic axis along the major axis, the pattern with the sphere, together with that for the *normal asymmetry* observed with quartz for comparison when all samples rolled at 3.2 cm/sec. A sintered Al_2O_3 sample proved of no use as it visibly had imbedded in it a contaminating layer probably some of the polishing compound that further grinding and polishing with Al_2O_3 failed to remove. The fluctuation of the charging curve (wobble), produced once each revolution by the slight asymmetry in the mounting of the revolving cylinder clearly seen for quartz is masked entirely by the asymmetrical charges on the Al_2O_3.

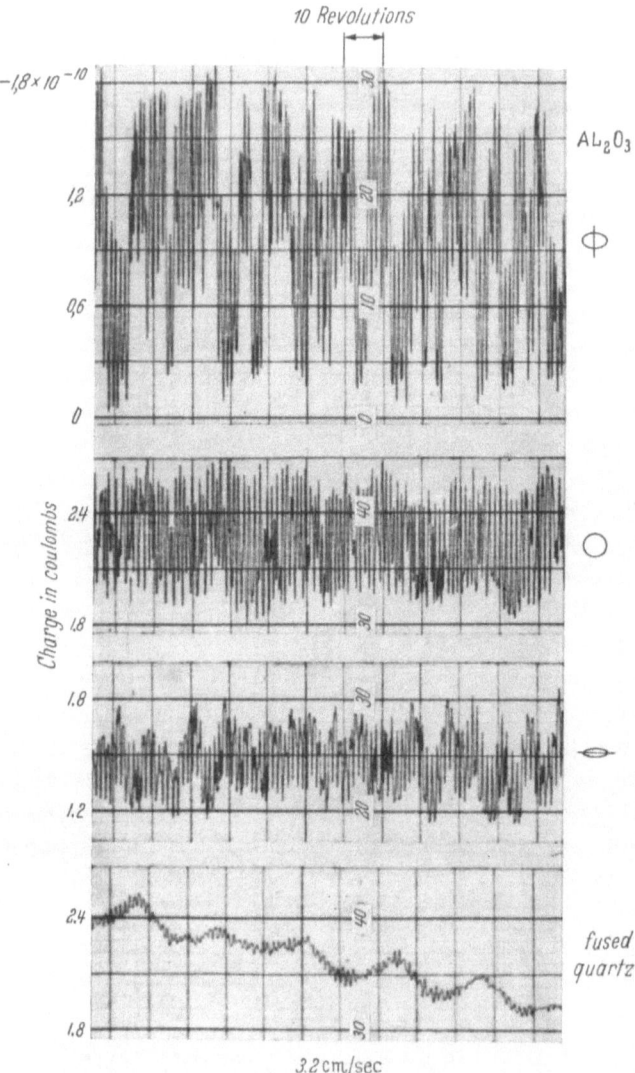

Fig. 54. WAGNER's curves for ellipsoids and spheres of single sapphire crystal. Top *C* axis along minor ellipse axis; next sphere and third *C* axis along major ellipse axis. Bottom for fused quartz illustrating crystallographic anistropy of charge on sapphire single crystals

It was impossible to resolve the number of probe fluctuations per revolution which was of the order of one or two. This would have indicated whether one or both ends of the optic axis, or the normal to it, yielded high charging rates. While the average charge appeared to be somewhat more negative for the high symmetry ellipsoid, this was not established reproducibly. No estimate can be made of the true charge nonuniformity because of the averaging effect of the probe and the marginal response time of the recorder. The values in Fig. 55 are lower limits. With data taken at a later date at still slower rolling speed, there appeared to be *more nearly two, rather than* one *fluctuation per revolution* of the low symmetry sample indicating that both ends of the axis, or both ends of parallels to the axis, accepted the same amount of presumably negative charge. That is, with the method at hand, all that was noted was a difference in charge between the ends of the C axis and the lines parallel to the axis. It was *impossible* to *say* which *regions accepted a maximum or a minimum*. The asymmetry was greatest when the optic axis and its normals rotated under the probe. It did appear, however, that *both ends, or parallels*, were charged to the *same* sign.

The average charge on sapphire did not increase exponentially to a maximum as did the other substances tested. It reached maximum charge and then slowly decreased to a lower saturation charge. If rolling stopped for a long time and then resumed again, the specimen recharged, but to a lower intermediate maximum and then declined to about the same steady value. Such effects could well be associated with the high asymmetry of charge. There can well be charge dissipative actions invoked by the extremely highly localized charging in this case which appear on prolonged rolling.

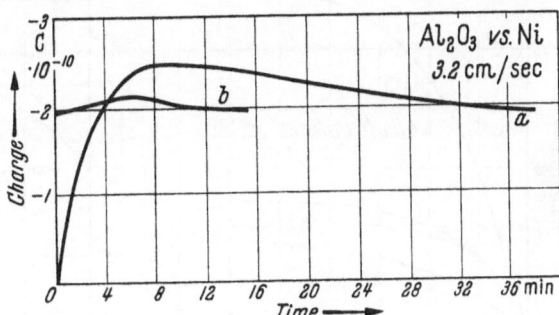

Fig. 55. Lower limits of charge as a function of time for Al_2O_3 single crystal. Note the charge for maximum and decline in curve (a) and the later repetition after standing for some time in curve (b)

The Al_2O_3 was spectroscopically designated by the manufacturer, Linde Air Products Co., to be 99.98% pure PbO being the major impurity. Also available was a single crystal sphere of synthetic ruby with 99.23% Al_2O_3 and 0.75% chromic oxide. Ruby and sapphire charged in the same fashion as far as observations could determine. Thus this particular type of body impurity did not affect charging.

Quartz was reproducible in its behaviour to within 10—20% as between runs and different samples. With Al_2O_3 the charging was an order of magnitude less than quartz and the reproducibility was within a factor of two. If the abolute uncertainty remained of the same order in such measurements rather than the *percentage uncertainty*, this action would be expected. Constancy of the absolute uncertainty of charge implies the reproducibility of the circumstances involving the state of the Ni cylinder, the degree of cleanliness of the surfaces, chance contamination, etc., etc. If these allow of only a certain reproducibility of the charging measurements as between runs and samples, then the observations are plausible. Since cleaning techniques, surface conductivity, etc., were similar in these two cases, this conclusions is certainly logical.

d. Alkali halides. Here reproducibility was not better than by a factor of 3 which can, in considerable measure, be led back to the methods of cleaning and polishing as well as to the rather feeble charging. Consistent with the view that the cleanest sample charges most vigorously and uniformly is the observed fact that the highest charge measured on a particular sample was also the most uniform. Fig. 56 shows curves considered the most reliable ones for the halides because of high charge and surface uniformity. Here the cubic symmetry precluded axial unisotropy. NaCl was considered the most reliable and then in order

KCl, KI, and KBr. LiF charged positively to the same order of magnitude as the others, but was not reproducible. All these substances charge positively against Ni as was observed for NaCl by DEBEAU and has also been reported by others. Their charging rate is five to ten times as small as quartz and yields a fractional charging rate of the order of 0.5 min^{-1} at 3.2 cm/sec rolling speed.

e. Magnesium oxide crystals, (periclase). When properly cleaned, single crystals of MgO exhibited a *positive* saturation charge of the order of that observed on alkali halides*. Reproducibility was within a factor of two. This substance in the form of a clear monocrystalline lattice of the cubic system, called Periclase in the mineral world, is of interest in that it can be additively coloured by heating

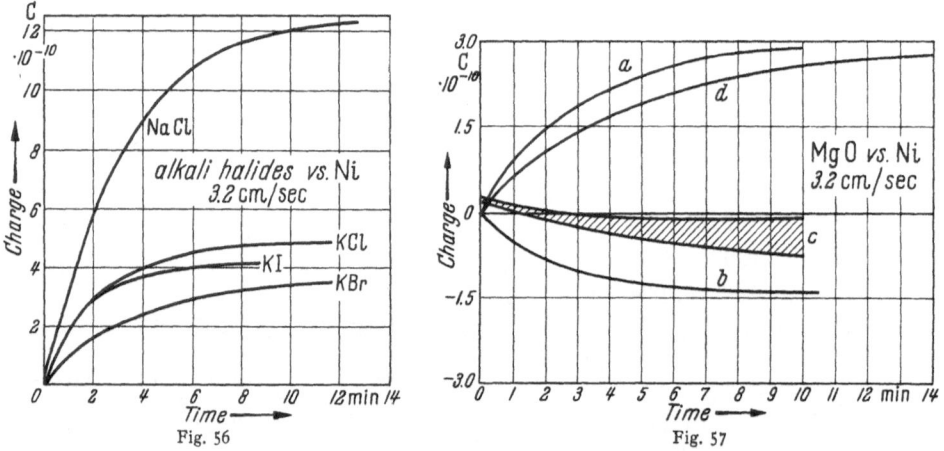

Fig. 56. Charging curves for alkali halides against slightly oxidized Ni according to WAGNER

Fig. 57. Charging curves for freshly cleaned MgO: (a) oxygen rich sample; (b) same sample after overnight vacuum firing at 1200°C; (c) same sample after 15 min polish with Al$_2$O$_3$; (d) same sample after 1 hour polish

in an atmosphere of Mg vapor or of O$_2$ gas. Alkali halides were also colored by electron bombardment in these studies but the damage and lack of reproducibility produced in surface cleansing covered up any differences that might have been present. In a sense the above statement could also be applied to most of the "coloring" studies with MgO for the cleaning by distilled water may well eradicate the centers that are near enough to the surface to produce effects. This gloomy prediction was borne out in studies of the samples of MgO bulk enriched with Mg or with O$_2$, for in the heating the surfaces of the polished spheres were sufficiently disturbed so that cleaning had to be drastic. Fortunately, it happened that cleaned manufacturers samples heated to 1200° C over night at 10^{-6} mm of Hg pressure, a treatment known to remove excess oxygen present in the crystal as it came from the manufacturer**, did not injure the surface and produced a stoichiometric crystal showing changes. The first effect was to change the sign of the charge. Fig. 57 shows curves of a freshly cleaned manufacturers sample (a) the sample after vacuum firing (b) (c) the same sample after a 15 min polish with Al$_2$O$_3$ and (d) the sample after an hour's polishing. Curve (c) is cross hatched

---

* GANDY engaged in solid state studies on MgO reports that in surface studies in order to achieve reproducible results he resorted to the same chaning techniques as those used here.

** A qualitative check on the 2850 Å band characteristic of excess O$_2$ in MgO indicated that this band present in an unheated sphere was completely bleached by the vacuum firing.

to show the large non-uniformity in a *partially polished* sample. In the other curves, the non uniformity lay within a few percent. The sample was cleaned identically between runs.

Another effect unique to vacuum fired MgO was the hyperexponential discharge of a charged sample to what appeared to be a lower steady state value. This is shown in Fig. 58. The initial charge was -1.3×10^{-10} coulomb and the final charge was -0.6×10^{-10} coulomb. The initial time constant was of the order of four hours. The original charge was restored on subsequent rolling. No discharge could be detected in 12 hours for the untreated samples including the treated sample after it had been repolished. It was clear that the vacuum firing had

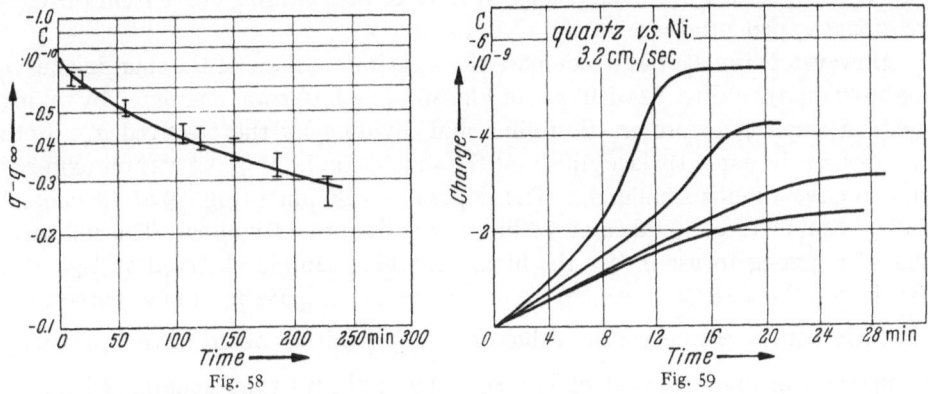

Fig. 58. Apparent discharging curve for stoichiometric MgO. Curve drawn through experimental points has been taken from a Pyrex discharging curve, to show the similarity

Fig. 59. Charging curves, freshly cleaned fused quartz: bottom, sample ground 2 min with 30—60 μ SiC; next to bottom, ground 5 min with 20—40 μ SiC; next to top, polished 10 min with 0.3 μ Al$_2$O$_3$; top, polished 45 min with 0.3 μ Al$_2$O$_3$

changed the MgO from a probable O⁻ ion donor to Ni to either an acceptor of electrons or possibly of O⁻ ions from the slightly oxidized Ni. Polishing locally heated the surface in the presence of air and restored the surface to its oxygen rich state. The partial discharge of the vacuum fired (stoichiometric) sample could have been caused by a surface conductivity owing to one type of mobile carrier to yield the leakage current partially discharging it.

f. Effect of surface finish. With fused quartz and MgO a study of the variation of charging with surface finish was made. Both substances showed the same behaviour except that the data on quartz were much more reproducible. Fig. 59 shows that charging curves for the same sample of freshly cleaned quartz with varying degrees of surface finish. *The saturation charge is the higher the smoother the surface* as would be expected on contact charging. This was qualitatively reproducible regardless of the order in which surface finish was varied, e.g., whether the sample became gradually smoother or gradually less smooth. These curves were carried to saturation in spite of the danger of losing the sample during the most erratic stages of rolling *. Polishing beyond the top state showed no change in charging within the overall reproducibility of the data.

g. High saturation charges. The very high saturation charge and the peculiar change in shape of the charging curve in the case of these very clean quartz

---

* This occurred 3 times out of 4 requiring breaking of the seals and reprocessing. The curves obtained in this fashion proved so valuable that they justified the time.

specimens on clean Ni above some 2×10^{-9} coulomb of charge presented some very puzzling features. Between $2-3 \times 10^{-9}$ coulomb of charge the spheres begin to roll erratically. As this occurred, the spheres became highly charged and stuck to the probe, the back of the cylinder or up $30°$ from the usual rolling position at the bottom of the cylinder. The visual erratic rolling was accompanied by considerable irregularity in charge distribution the amplitude of oscillation on the probe signal going from the usual $2-3\%$ at 2×10^{-9} coulomb to as high as 15%. As saturation of charge occurred, this irregularity declined to about 1%. By estimating the increased amplitude of motion, rough correction for rolling speed could be made in some cases. This correction where justified by its accuracy may account for the observed change in slope of the charging curve from the conventional exponential rise to an accelerated rise.

However, it is essential to compare the logarithmic slope of the charging curve for fused quartz of Fig. 52 with that of Fig. 50. The latter was computed by taking dq/dt at small q, before erratic rolling and dividing by the observed q_m. Both curves had the same surface finish on the same sample. Fig. 52 was on oxidized Ni and gave a much smaller q_m. The slopes are 0.28 min^{-1} (Fig. 59 at 4.8 cm/sec) and 0.07 min^{-1} (Fig. 59 taken at 3.2 but normallized to 4.8 cm/sec). This indicates that it is wrong to use q_m for the highly charging sample observed at *high q* to divide into the dq/dt at *low q*. Were it possible to compute q_m at low charge from the early part of the curve, the value of $\left(\frac{1}{q_m} \frac{dg}{dt} \right)$ initial would have been more normal and in line with that of Fig. 50. This indicates that something happens at high q to produce an apparent change in charging rate leading to a high q_m and/or q_s. Since the erratic rolling and sticking indicate electrostatic forces on the sphere comparable with its mass, it appears likely that contact areas increase so that rate goes from α to α' and thus both *dq/dt and q_m* or q_s increase. This will later be seen to be the case.

h. Charging and work function. PETERSON had observed a decrease in charge on the borosilicate glass when the Ni surfaces were slightly oxidized. Further, to test this, the whole system was baked in O_2 or H_2 gas at some $250°$ C. A Kelvin contact potential measuring device was connected to the charging tube by means of a side arm. It had a sample of the Ni stock used for the cylinder opposite an Au electrode so that the work function of the Ni could be measured under conditions for which charging curves were run on oxidation and reduction. Preliminary results indicated that after bakeout in O_2, the Ni had increased its work function relative to the Au by 1.1 volts. In this instance, the charge on freshly cleaned quartz decreased by a factor of two. Since the work function of the Au may have changed, (this is unlikely), not too much credence was placed on this result, even though gold is assumed to remain unchanged. A misfortune to the apparatus resulting from poorly tempered glass, destroyed the device after one run. To further study the problem the Ni surface was altered by subjecting it to ion bombardment in H_2 and O_2 gases by means of a glow discharge. After each bombardment a charging characteristic was run for the particular sphere under test. The sample could not be left in the tube during the discharge and used without cleaning because of the various materials sputtered onto it. Charging characteristics were invariably poisoned more by O_2 than by H_2. This was because with O_2 there was more sputtering of the metal than with H_2.

The cleaning of the sphere necessitated opening up the chamber after bombardment which subjected the metal surface to room air. However, in view of the major alteration produced by bombardment, it was not believed that the surface of the metal was seriously altered by a brief exposure to air*, when cold. The air was rapidly removed after the sphere was inserted. Fig. 60 shows several charging curves for freshly cleaned fused quartz after the treatment indicated. The discharge conditions amounted roughly to 0.1 ma/cm² current for 10 min time with the Ni rotor as cathode in the normal glow discharge of the gas indicated. In curves (a), (c) and (e) it was impossible to saturate the sample. It is clear that this treatment in all cases changed the saturation charge by more than a factor of 2.

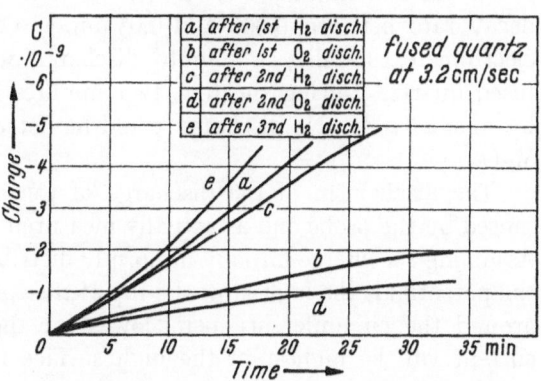

Fig. 60. Effect of oxidation and reduction of Ni by glow discharges in O₂ and H₂ on charging curves for quartz. Here the work function of Ni is reduced possible a volt by H₂ relative to O₂ treatment. Note the increase in charging rate as work function decreases

Attempts to observe similar changes in charging with positive electrification of MgO and of KCl proved futile. The acceptance of a negative charge by the Ni surface in contact with KCl or MgO acting as negative ion donors was not altered by oxide bombardment or reduction with H₂. On the other hand Al₂O₃, an assumed electron acceptor, gave results in conformity with quartz. Here with sapphire just after a 30 min discharge on the Ni surface in O₂, the Al₂O₃ developed an average charge of $+0.3 \times 10^{-10}$ coulomb while subsequent to H₂ bombardment produced a charge of -4×10^{-10} coulomb. In both cases, the non-uniformity was about 1×10^{-10} coulomb after 10 min rolling at 3.2 cm/sec.

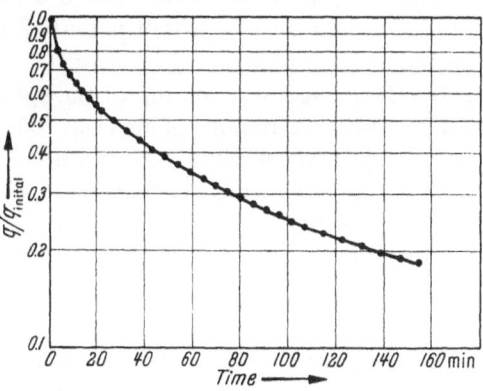

Fig. 61. Logarithmic plot q/q_s against time for charged borosilicate sphere as sensed by the probe owing to surface leakage. Note the hyper exponential decline owing to charge redistribution by surface leakage

i. **Leakage currents.** In general, surface and/or bulk conductivities of all samples used were so small that long time measurements on the discharge of the charged sample were impossible owing to competing drift and leakage currents registered by the electrometer. Leakage currents in borosilicate glass were of such a magnitude as to be amenable to observation. Fig. 61 shows the discharge curve plotted to a semi-logarithmic scale. The borosilicate glass curve is clearly hyper-exponential. The extremely fast initial early discharge rate of the glass makes it *impossible* to give any well

---

* Alteration of Ni treated by bombardment by air at room temperature has been observed to be pretty slow by HUBER in her studies on secondary electron emission.

defined decay time as a measure of conductivity though towards the end there appears to be an approach to an exponential decay. With the other substances, the discharge, when it was measurable at all, was too slow to permit measurement of the overall shape of the curve and only the initial decay could be observed. Typical rates for alkali halides, the discharge of which appeared linear for times of the order of minutes were $2-4\times10^{-2}$/min. Natural quartz yielded an initial decay rate between 0.006 and 0.03 min^{-1}. On one exceptionally good day for electrometer stability a minute discharge was observed from highly charged fused quartz. During the first two minutes the charge decreased very rapidly by about 0.02%. No discharge could be detected for sapphire or for untreated MgO.

The mechanism of the discharge of the sphere by surface conductivity as sensed by the probe and as actually measured in the case of the glass is complex. Assuming the charge initially uniformly distributed over the surface of the sphere the potential is the highest at the top of the sphere below the probe and decreases around the circumference being lowest at the point of contact. The leakage current will be highest in the high surface fields causing the charge to drain rapidly from the top and accumulate towards the equatorial belt of the sphere. That this actually occurred was clearly demonstrated with borosilicate glass. A sphere was charged and allowed to "discharge" along the hyper exponential curve for some time. Then the sphere was slowly rotated through 90° to expose its equatorial belt to the probe. The apparent charge was found to increase to a value less than but near its initial value. Then the leakage current to the metal begins to increase until the potential becomes too low to cause much flow.

This clearly indicates the charge decline as sensed by the probe. Were the sphere a conductor of constant potential, the decline in charge would be exponential in time and plots of log q_a against t would be linear in t. Actually, the probe senses the charge near the top of the sphere with a uniform charge on the sphere and initially maximum potential at the top. In view of the redistribution of charge before much actual loss can occur, the probe senses a decline when there is little real loss. Thus the initial slope of the log q_a curve gives the *apparent leakage current*; this will be hyper-exponential. The true leakage is given when the sphere is more nearly an equi-potential. Thus under no circumstances should the initial slope of the discharge current be used as a measure of surface conductivity.

j. Charging under ultra high vacuum with measured work function. The second charging tube with three metal barrels and devices for real outgassing and contact potential measurements was to eliminate the effect of adsorbed layers on quartz as a result of macroscopic cleaning or as a consequence of adsorbtion of residual gases from the metal, and to compare the charging characteristics of *clean* quartz against a given metal with its work function actually measured under good modern vacuum techniques.

PETERSON reported that vacuum bake out with the sphere of quartz or borosilicate glass in place in general *decreased* charging, (the saturation charge of the quartz decreasing by as much as a factor of 2), and increasing the surface conductivity of the glass by a measurable factor. He concluded that heating drove more polluants into the sphere than it removed. Repetition of the measurement confirmed PETERSON'S findings and indicated the charge on the quartz after bakeout to be extremely uniform indicating uniform contamination.

In the new tube, the sphere was kept in a side tube at room temperature while the rest of the system was baked. There was no straight line path between sphere and any hot surface. After the first bakeout, the sphere developed a small non-uniform *positive* charge against all metals. On recleaning the sphere charged negatively as it had before bakeout. Copper deposits were observed after bakeout even in parts of the side tube not in direct line of flight from the hot portions. The pressure after bakeout was 10^{-8} mm even with the quartz side tube unbaked. The quartz side tube was then placed inside the bakeout oven. Here the quartz sample was only exposed directly to heated surfaces of quartz or glass and not to metal. With the side tube prebaked for some minutes in room air before inserting the cleaned sphere, on subsequent baking of the apparatus when well evacuated *there was no reversal of sign* with bakeout. Bakeout *lowered* saturation charge of quartz against Cu by 25% at 6.3 cm/sec rolling speed, and against Ni by more than 10%, The maximum charge against Ni had been too large to measure before bakeout. These small changes could be ascribed to either quartz, or metals, on bakeout. The sample was then baked in the side tube at about 700° C for an hour. With the side oven hot the initial pressure in the system was 2×10^{-9} mm. Introducing the sphere raised the pressure to a maximum 100 times higher after which pressure gradually decreased. Baking was stopped when the pressure with sample heat was 6×10^{-9} mm. Removing the sphere from the oven reduced this pressure by a factor of 3. No change in charging characteristics was observed after this treatment.

Even when the sample was *not carefully cleaned*, as evidenced by extremely small non-uniform charging before bakeout, baking the system and subsequently baking the sphere at red heat for one hour while the system cooled (at a pressure of 10^{-8} mm) gave a charge only 30% lower than in the previous thorough outgassing. The charge *before* bakeout was but *10% of that observed after bakeout.*

The work function differences of the 3 metals Cu, Pt, and Ni as measured after bakeout were discouragingly small. The saturation charges could be measured with quartz for the 3 surfaces. In ultra high vacuum after repeated bakeout rolling the sphere against Ni after the sample had been saturated against Cu lowered q_s by 27% while the work function of Ni was 0.5 volts higher than that of Cu. No correlation could be established between work function and charge differences on sapphire. While there were pronounced characteristic charging differences from metal to metal, there was no sensible correlation with work function. Against Pt it charged to $+1.25 \times 10^{-10}$ coulomb, against slightly oxidized Ni, the average charge was zero and against slightly oxidized Cu the average charge was zero or slightly positive. Charging characteristics varied in an unpredictable fashion depending on the histroy of the sapphire sample. The charge against Pt was slightly higher when the sample had rolled previously on Ni than when it charged to saturation after it had been directly placed in the Pt track. The Pt work unction was observed to be 0.2 volts lower than that of the slightly oxidized Ni or the Cu which were equal within experimental accuracy. Here it is clear that the Al_2O_3 was on the border line between some sort of a negative ion donor or an electron acceptor so that its previous history could have had some significance.

4. Interpretation of results

a. Surface and bulk conductivity. As indicated, the rate of discharge of nearly all samples studied, except borosilicate glass were too small for accurate evaluation of surface conductivity.

Thus while an "apparent" conductivity could be derived from the initial slopes of the logarithm of the quantity of charge against time for a very few of the materials studied, it was only with borosilicate glass that an approximation of the true surface conductivity could be made. In order to consider the effect of conduction on the evaluation of the saturation charge from PETERSON's equation, the only measurable data in the samples used come from the evaluation of the initial slope of the hyper-exponential charge decay. This "apparent" conductivity sets a high value β_a for the upper limit of β.

Calling the "apparent" initial discharging rate β_a one can establish limits on the hypothetical saturation charge q_s by observations of β_a and q_m. For convenience it is possible set

$$q_s = q_m\left(1 + \frac{\beta}{\alpha}\right) \cong q_m\left(1 + \frac{\beta}{\alpha + \beta}\right) \tag{4.10a}$$

to the first order in $\frac{\beta}{\alpha}$. Call $\gamma = \frac{1}{q_m - q}\frac{dq}{dt}$, the observed fractional charging rate. Then $q_s \cong q_m\left(1 + \frac{\beta}{\gamma}\right)$ since $\gamma = (\alpha + \beta)$. Since $\beta_a > \beta$ then

$$q_m < q_s < \left(1 + \frac{\beta_a}{\gamma}\right)q_m$$

to the first order in β/α.

For alkali halides using observed β_a and γ values the conductivity lowers q_s by less than 10 to 20% at 3.2 cm/sec. For Al_2O_3, MgO and fused synthetic crystalline quartz the lowering is neglegible. The observed q_m for natural quartz is 20—25% lower than for synthetic pure crystalline or amorphous quartz though γ is the same.

This difference in q_m is barely beyond the limit of experimental uncertainty. Thus conductivity can be assumed to be the cause of difference in these curves. The value of β is, however, so small compared to α that the observed values of γ are the same within 10% even with observed differences in q_m. The synthetic and natural quartz crystals came from the Brush laboratories. They reported that spectroscopic analysis indicated the impurities present in the natrual quartz to be Rb, Cs, Cu, Ba, Cr, Mn, Ti, Zr, and Ag, while both samples contained traces of Li, Na, K, Mg, Al, and Fe and unspecified oxides though the oxide content was 0.04 weight percent in natural and 0.01 weight percent in synthetic quartz. In consequence, the conductivity β changes q_m in no case by more than 20—25% and in most cases by much less. In what follows, therefore, β will be neglected and q_m will be set as q_s.

b. Dependence of rate of charging on surface exposed. Having disposed of β in these studies as negligible relative to α the values of α may be safely taken as closely given by $\frac{1}{q_m - q}\frac{dq}{dt}$ and q_m may be set equal to q_s. It is now of interest to ascertain whether the α derived on the assumption that the rate determining process in contact charging is simply that of exposing uncharged area to contact is in agreement with observation. To this end, one utilizes the theory of rolling contact developed by BOWDEN and TABOR[52]. The area of contact between a six mm diameter quartz sphere and the Ni surface is $\sim 3 \times 10^{-8}$ cm^2 using 10^5 g/mm^2 for the yield pressure of cold worked Ni and assuming full plastic flow of the metal. This high value for Ni is justified by the treatment in manufacturing the Ni cylinder. The asperities of the quartz of the dimensions of the finest abrasive used and with optical finish of the order of 10^{-5} cm in linear dimension plastic flow would occur even if the Ni were smoother than the quartz. Consider an atomically smooth Ni plane on which the quartz sphere rests. Assume the sphere to be covered with conical asperities of half angle 45° and height 10^{-5} cm.

It must be noted that this shape chosen for the asperities is arbitrary but facilitates computation. The asperities could as well be hemispherical bosses or other shapes. They will slightly alter the change of area with load and could alter values of charge rate by factors of 2, but not orders of magnitude. If the total plastic flow area supporting the load is 3×10^{-8} cm^2 and the apparent, elastic contact area of a sphere on the yielding plane is $\sim 5 \times 10^{-7}$ cm^2 as computed from HERTZ's equations for 6 mm spheres on a Ni plane, about 1500 cones support the load. Each cone sinks one fourth its height into the Ni giving an *effective area for charge transfer of value* $\sim 3 \times 10^{-11}$ cm^2 per cone. Since the area of the base of the cones is 3×10^{-10} cm^2 the ratio of integrated contact area (after every cone has touched to the apparent area of the spherical surface, $(4\pi r^2)$, is about 0.1. The individual contact areas are separated by about $10d$ where d is the radius of the base of the individual contact area cones and is about 2×10^{-5} cm. In one second a strip of width containing $\sqrt{1500}$ rows of cones, each row containing $v/2 \times 10^5$ cones at a rolling speed v is covered. Since the area of contact per cone is 3×10^{-11} cm^2 the rate of exposure of area to contact is $\dfrac{dA}{dt} = 3 \times 10^{-11} \dfrac{\sqrt{1500}\, v}{2 \times 10^{-5}}$. Using $v = 3$ cm/sec, $\dfrac{dA}{dt} \cong 0.01$ cm^2/min. Since the integrated contact area is $A \cong 0.1\,(4\pi r^2) \cong 0.1$ cm^2, the fractional rate of exposure of new area in the early stages of rolling is $\dfrac{1}{A}\dfrac{dA}{dt} \cong 0.1$ min^{-1}. The experimental value for $\alpha = \dfrac{1}{q_s}\dfrac{dq}{dt}$ is 0.06 min^{-1} at 3.2 cm/sec. The remarkably good order of magnitude agreement between these two values strongly supports the notion that the charging rate is governed only by rate of exposure of new uncharged contact area. Surface roughness of the Ni should perhaps reduce $\dfrac{1}{A}\dfrac{dA}{dt}$ but not alter its order of magnitude. The reduction is due to the fact that the apparent area will be lowered, lowering the number of cones indenting the Ni at any time and thus lowering dA/dt while A will not change materially. If the Ni were covered with rough hemispheres of 10^{-3} cm raduis α becomes ~ 0.05 min^{-1}.

The evaluation of the charging rate has been summarized by WAGNER in equation form as follows:

The initial rate of exposure of area to molecular contact con be shown to be

$$\left.\frac{dA}{dt}\right|_{t=0} \cong \left(\frac{2E}{r}\right)^{\frac{1}{3}} \frac{v}{\sqrt{2p}}\,(m\,g)^{\frac{2}{3}} \tag{4.11}$$

and the area on the sphere which ultimately undergoes molecular contact is

$$A_s \cong \sqrt{2} \, \frac{(16 \, E \, r^2)^{\frac{2}{3}}}{p} \, (m \, g)^{\frac{1}{3}}. \tag{4.12}$$

Here E is the effective YOUNG's modulus given by

$$\frac{1}{E} = \frac{1}{E_Q} + \frac{1}{E_{Ni}} \tag{4.13}$$

where E_Q and E_{Ni} refer respectively to quartz and Ni. The yield pressure of Ni is p, the rolling speed is v, r is the radius and mg is the weight of the sphere.

If the charge density separated is σ since

$$\frac{dq}{dt}\bigg|_{t=0} = \sigma \left(\frac{dA}{dt}\right)\bigg|_{t=0} \quad \text{and} \quad q_s = \sigma \, A_s \tag{4.14}$$

$$\alpha = \frac{1}{q_s} \frac{dq}{dt}\bigg|_{t=0} = \frac{1}{A_s} \frac{dA}{dt}\bigg|_{t=0} \cong \left(\frac{m \, g}{1024 \, E \, r^5}\right)^{\frac{1}{3}} v$$

or

$$\alpha \cong \left(\frac{m \, g}{E \, r^5}\right)^{\frac{1}{3}} \frac{v}{10} \, \sec^{-1} \tag{4.15}$$

and

$$q_s \cong \sqrt{2} \, \frac{(16 \, E \, r^2)^{\frac{2}{3}}}{p} \cdot (m \, g)^{\frac{1}{3}} \, \sigma. \tag{4.16}$$

c. **Maximum or saturation charge.** α. *Surface roughness.* It is next of interest to consider the maximum charge and the factors determining it. The experimental data indicate that the maximum charge $q_m = \frac{q_s}{1 + \beta/\alpha}$ (4.10) appears to approximate the true saturation charge q_s quite closely for the substances studied. This followed because β in these substances was very small, certainly less than 0.1 of α. This would not be so for borosilicate glass and many other substances, e.g., for ice. It is thus q_s that is the important quantity which will throw a light on the nature of the charge transfer process since α has been shown only to represent the rate of exposure of contact area.

Concerning the value of q_s it will be noted that this decreased as surface finish became increasingly rougher. This is a direct consequence of the areas available to contact and charge transfer. If the pits on the surface of the sphere, relative to the asperities are greater in depth than the distance of penetration of the asperities into the surface, the area occupied by these pits can never make contact, or charge up. Now the penetration of the assumed conical contact points of the sphere were of the order of 0.25×10^{-5} cm. Any pits deeper than 2×10^{-5} cm would never be exposed to contact. Observations of the roughened surfaces using a microscope showed the pits to have depths 10^{-4} cm and to be of the same length in lateral extent on the surface. As revealed by Fig. 59 the area of the quartz sphere which can make contact with the Ni changes by a factor of about 2 when the surface finish goes from that characteristic of 5×10^{-3} cm diameter SiC abrasive to an optical polish. The variation of saturation charge at lower rolling speeds might actually be used quantitatively to estimate the quality of surface finish using the techniques above.

β. *Effect of electrostatic forces.* The measurements with quartz on clean Ni indicated high saturation charging q_s. Where despite erratic rolling and sticking, the charging curves were followed to saturation observations indicated a deviation from the exponential increase for normal charging with an accelerated charging rate. It was indicated at that point that this probably could be ascribed to an increase in charging rate α to an α' caused by deeper penetration of the surface cones responsible for dA/dt into the surface. Greater surface also implies a larger saturation charge. It is now a question whether in fact such an increase will lead to the observed shape of the curves of Fig. 50 viz. an initial charging curve of essentially normal type with low saturation charge which reverses slope through a point of inflection when charges increase byeond $\sim 3 \times 10^{-9}$ coulomb and finally reaches a saturation q_s. WAGNER has carried out such a computation.

Consider that the normal force on the sphere,

$$F_n = m g + \frac{q^2}{4 r^2} \tag{4.17}$$

where r is the sphere radius and q is the instantaneous charge. The second term is a first approximation to the image force term. Now initially at $t=0$ the charging rate is given by the quantity $\left.\dfrac{dq}{dt}\right|_{t=0} = \sigma \left.\dfrac{dA}{dt}\right|_{t=0}$ (4.14), where σ is the charge density separated at contact. $\left.\dfrac{dA}{dt}\right|_{t=0}$ depends on the normal force, the rolling speed v, the effective YOUNG'S modulus E and the radius r of the sphere according to

$$\left.\frac{dA}{dt}\right|_{t=0} = \frac{v E^{\frac{1}{2}}}{2^{\frac{2}{3}} r^{\frac{1}{2}}} F_n^{\frac{2}{3}} = K F_n^{\frac{2}{3}} \quad \text{and} \quad \left.\frac{dq}{dt}\right|_{t=0} \cong \sqrt{2}\sigma \left(\frac{E}{4 r}\right)^{\frac{1}{3}} \frac{v}{p} F_n^{\frac{2}{3}}. \tag{4.18}$$

Now the probability of acquiring more charge as forces increase is $P(q)$. This is equal to

$$P(q) = 1 - \frac{\sigma A_T}{\sigma A_u} = 1 - \frac{q}{\sigma A(q)}. \tag{4.19}$$

Here σA_T is the charge separated by areas already touched which equals q and σA_u is the charge to be separated by areas A_u that would *ultimately* touch if F_n held constant at the value characteristic of the instantaneous value of q. Now the normal force per asperity is F_n/n. Here n is the number of asperities that support the sphere at any instant. The overall area which could ultimately touch at a given F_n/n is given by $A_u = A(q)$ and this is proportional to $F_n^{\frac{1}{3}}$ that is $A_u = A(q) = M F_n^{\frac{1}{3}}$ where M is constant. Thus

$$P(q) = 1 - \frac{q}{\sigma M F_n^{\frac{1}{3}}} = 1 - \frac{q}{L F_n^{\frac{1}{3}}} \tag{4.20}$$

where L is a constant

$$L = \sigma M.$$

The value of

$$\frac{dq}{dt} \text{ is } \frac{dq}{dt} = \sigma \frac{dA}{dt} = \left.\frac{dA}{dt}\right|_{t=0} P(q) = K F_n^{\frac{2}{3}}\left(1 - \frac{q}{L F_n^{\frac{1}{3}}}\right) \tag{4.21}$$

with

$$K = \sigma \left(\frac{E}{4 r}\right)^{\frac{1}{3}} \frac{v}{p} \sqrt{2} \quad \text{and} \quad L = \frac{(16 E r^2)^{\frac{1}{3}} \sqrt{2}}{p} \sigma. \tag{4.22}$$

Here p is the yield pressure of Ni.

True saturation occurs when

$$\frac{dq}{dt} = 0 \quad \text{i.e. when} \quad q_s = L F_n^{\frac{1}{3}} \tag{4.23}$$

or in terms of q_{s_0} when $q^2/4r^2$ can be neglected,

$$q_s = q_{s_0}\left(1 + \frac{q_s^2}{4r^2 m g}\right)^{\frac{1}{3}} = L(m g)^{\frac{1}{3}}\left(1 + \frac{q_s^2}{4r^2 m g}\right)^{\frac{1}{3}}. \tag{4.24}$$

Direct integration of the expression for dq/dt appears extremely difficult. In any event, it can be seen from the expression that the charge will increase more rapidly in time than the previous exponential with the same initial slope because the area available to charge deposition is, itself, an increasing function of q until all surface available is brought into contact. Under appropriate conditions this action could cause $\frac{d^2 q}{dt^2} > 0$. Using data available, examination of $\frac{d^2 q}{dt^2}$ has shown that regeneration cannot occur unless the theoretical saturation charge of q_s is at least 3.5 times the maximum charge of 5.4×10^{-9} coulomb observed experimentally. Numerical integration has shown that if the actual positive curvature observed is caused by increase in contact area *only*, the saturation charge must be between five and ten times that observed.

Admittedly the theory is quite rough, especially as it involves the shape of the asperities. If the theory is valid to within a factor of three, as it appears to be, at lower charge the positive curvature cannot alone be ascribed to the change in F_n i.e. to $m g + \frac{q}{4r^2}$ by charge attraction. Thus *some* of the *curvature* is to be *ascribed* to *increased rolling speed* in *consequence of erratic rolling* which cannot be corrected for. Both effects become important only when q exceeds 2×10^{-9} coulomb.

It is now of interest to show that the charges q in excess of some 4×10^{-9} coulombs could account for the erratic rolling and sticking and that $F_n \sim 2 m g$. For the sphere sticking to the probe all that is required is that the force of gravity on a 0.3 gram mass be overcome by the electrostatic forces. Neglecting the polarization forces on the sphere, which are relatively unimportant, complicate calculation and in the end merely increase the forces calculated by some tens of per cent, the electrostatic force is that given for the uniformly charged sphere by its image force in the metal probe. This force is $q^2/4r^2$. The charge q at 5×10^{-9} coulomb is 15 esu and $r = 0.3$ cm. This leads to an electrostatic image force of 0.6 grams which is more than adequate to support the sphere. With polarization forces added the sphere will be capable of sticking at even lower charge but will be observed to stick only when forces and erratic motion are such as to bring it close to the probe.

The instances where the sphere is observed to stick and ride up the rotating cylinder through an angle of 30° before it rolled back, appears to present a little more difficult problem. In this instance, the sphere suffers a restoring torque of $m g r \sin 30° \sim 50$ dynes. In consequence of indentation of the Ni, while the normal force is large, this force acts through an *obvious* lever arm, of the order of 10^{-3} cm. However, when this sticking occurs, the rolling is erratic and the probe senses non-uniformity in surface charge of the order of 15% at a maximum.

The unequal charge distribution presumably comes from the erratic rolling. If this occurs as the sphere having some more heavily charged patch rolls uphill along with the cylinder from contact, the attraction between the patch and its image charge in the cylinder causes a counter torque. For example, if there is only 5% excess over the average charge of 2×10^{-9} coulomb concentrated at a patch with its center at $0.2\,r$ uphill from the contact point of the sphere, the resisting torque is of the order of 10^2 dynes $\times$ cm which is in excess of the 50 dynes $\times$ cm resulting from the mass of the sphere at 30° inclination. Thus sticking is readily accounted for when it is observed.

d. The true charge separation and q_s. It is next essential to discuss the magnitude of the quantities determining q_s on the basis of the picture which is here being developed for the behaviour of quartz. In this picture, it is assumed that quartz, in which electrons can presumably occupy vacant surface states when in contact with a metal such as Ni which has free electrons, will accept them and charge up. This is a surface charge exchange *not involving the interior*. It is this process which must now be considered in detail. Using the integrated contact area A of the sphere of the order of 0.1 cm² just derived the "apparent" "contact potential" between quartz and Ni driving electrons from Ni to quartz may be calculated. Assume A to be a plane at an average separation d of say 5 Å from the quartz. The value of d is judged to be the order of magnitude of the *intimacy of contact at the cones over which electrons may transfer*. It is not a critical quantity as will be seen. The capacity of the condenser system so formed is charged to q_m or in the case quartz to $q_s \sim 5 \times 10^{-9}$ coulomb. The capacity C is $\dfrac{A}{4\pi d} \sim \dfrac{0.1}{4\pi \, 5 \times 10^{-8}} \sim 1.6 \times 10^5$ cm $= 1.75 \times 10^{-7}$ farad. It is assumed that the contact potential V_c charged the capacity so that $V_c = q_s/C = 2.9 \times 10^{-2}$, or 26 *millivolts*. This *apparent*, or observed, contact potential V_s is surprisingly low relative to data on the change of q_s by change in work function of Ni. It would, in fact, be expected that contact potentials would be more in the order of say 5 volts so that this result indicates some sort of a difficulty. Now the distance over which charge transfer could occur is certainly not greater than 20 Å and the other quantities are all fixed in order of magnitude so that the calculation fails for some reason not considered. In consequence, one must look at the problem from another angle.

Consider what happens when two uncharged plates of area A one of Ni and one of quartz with the electron affinity of quartz say 5 volts higher than the work function of Ni approach each other. As indicated pictorially in Fig. 62 the free electrons in the Ni see a narrowing barrier for transfer onto the quartz. As the barrier narrows, (and its height diminishes), electrons begin to penetrate the barrier and enter the quartz until at the distance of closest approach $\sim$5 Å electrons transfer freely in both directions, with an equilibrium charge $-q$ established on the quartz and its image in the metal. At this point, using the capacity computed above and $V_c = 5$ volts $q \sim 8 \times 10^{-7}$ coulomb. If now the plates are *separated at constant charge* the barrier seen by electrons on the quartz is the sum of the usual surface barrier less the applied field set up by the separated charge of value $q_c = 2.4 \times 10^3$ esu and the field X becomes $X = 4\pi \, q_c/A = 9 \times 10^7$ volts/cm. Such a field permits field emission through the barrier, (essentially back leakage of previously separated charge). This back discharge or leakage will continue until the distance of separation x_0 is reached at which field emission ceases. The field emission lowers q, thickens the barrier and emission ceases. If back leakage occurred, it could readily reduce the charge to the 5×10^{-9} coulomb observed for q_s. If the flux remains approximately normal on separation the field caused by residual charge density observed is $4\pi q_s/A$ which in this case is $4\pi \times 15/0.1$ esu

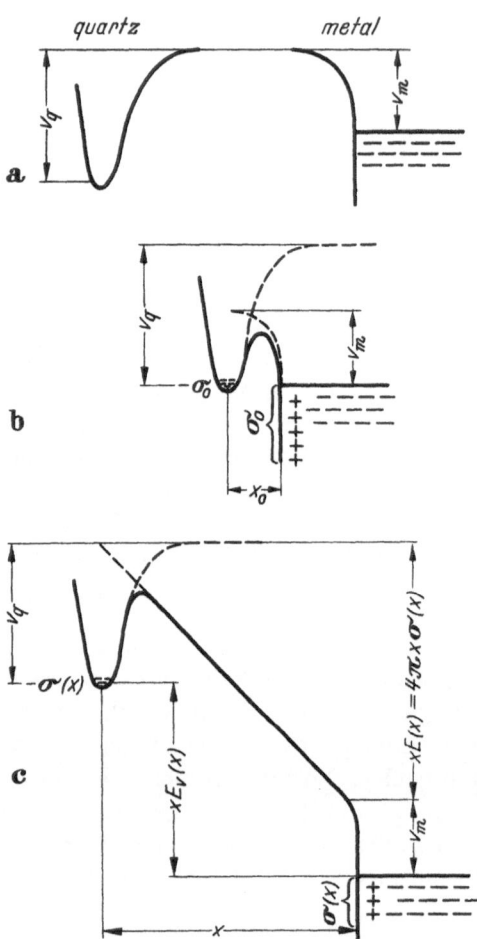

a

b

c

Fig. 62 a—c. Pictorial sketch showing WAGNER's concept of electron exchange by quartz and Ni on contact. Diagram illustrating possible role of field emission: (a) as uncharged surfaces approach; (b) at distances of closest approach, with equilibrium charge density σ_0 separated; (c) after separation to a distance x, with charge σ left on surfaces. $E(x)$ is the field which affects the barrier for field emission; xE_v (x) is voltage that would be read by a meter

per cm $= 5.6 \times 10^5$ volts/cm. This value is of interest since it is known that the surface of the quartz is not that of a plane. It has on it assumed cones of some 10^{-5} cm high and an equal radius.

As the planes separate to 10^{-4} cm, or more, then the fields causing emission are no longer fields between the plane surfaces of indenting cones but the fields at the quartz cone tips. Practical observations on field emission breakdown indicate that this ceases to yield measurable currents between relatively smooth surfaces (e.g., such as those of the SiO_2 and Ni when the field strength falls below some 10^5 volts/cm.

It is thus clear that initial charging is determined by V_c and leads to a charge q_i which on separation of surfaces reverts by back diffusion to the observed charge of q_s, an "apparent" contact potential V_s and determined by the point at which back discharge ceases with a field $X_s \sim 10^5$ volts/cm.

The action of a field emission conditioned back discharge was first mentioned by HARPER[53] in his study of contact electrification between metals in order to account for the same phenomenon, to wit, higher known values of V_c for the metals than the observed charge would account for.

One very important problem that this back discharge to a value q_s and X_s for saturation charge raises is that of *how* the *increase* of *effective contact potential* V_c can *cause an increase in* q_s since the value of the initial charge at q_i is determined by V_c, while q_s is determined by the field X_s at which field emission ceases. Actually this charge q_s while related to the point at which field emission ceases is also determined by the ability of the condenser to discharge as a result of field emission currents before separation reduces the field to a non-emitting value.

Regarding the nature of the field emitting equation and the nature of the separation process HARPER has concluded for metals that q_s depends on V_c which implies that the ratio $\dfrac{q_i}{q_s} = \dfrac{V_c}{V_s}$ remains constant and thus that q_s varies as V_s. This then would make the value of q proportional to the contact potential difference as was observed by HARPER for metals where V_c is known and for quartz against a metal as appears to be the case here. How valid HARPER's reasoning is and whether it applies to quartz-metal systems is an open question.

e. Influence of rolling speed on q_s. One of the outstanding puzzles in this study and that of PETERSON was the apparent decrease of q_s with rolling speed on prolonged rolling. At slow speeds the curves were regular and extrapolated to the same saturation value q_s. Above 10 cm/sec for quartz, the rolling speed increase appeared to decrease q_s. At 19 cm/sec the saturation charge q_s could be measured and was nearly 50% lower than the value calculated at the beginning of the rolling. WAGNER also observed this for clean quartz on Pt of high work function and low contact potential V_c.

It was observed by PETERSON and in WAGNER's work that the higher charges created erratic motion, sticking etc., because of the high charge on the sphere as has been indicated above.

Within the uncertainties of measurement with erratic rolling no great change in α could be observed and the increased indentation and consequent contact-surface increase by electrostatic forces was not observed in these studies with oxidized Ni or clean Pt. That is α remained virtually constant.

PETERSON studied the effect and concluded that q_m in quartz at high speeds was low because surface conductivity β was increased by an increase in temperature. At any rate, he definitely experimentally associated an increase in temperature of the quartz with the decreased value of q_m. He though that the heat increased β and some measurements seemed to confirm the appearance of an increase in the *initial decline* of charge with a heating to 60° C. This increase in *apparent* conductivity may not have been adequate.

It is not impossible that in PETERSON's work the heating caused increased contamination and an enhanced surface conductivity of the quartz to the extent observed. This was not the case in WAGNER's work where the metal and quartz were *really* clean.

The present work using *very* pure and clean quartz has clearly indicated that β for clean *quartz* cannot be the cause and that $q_m = q_s$. It also confirms the same general decline in q_m with speed observed by PETERSON.

It is perhaps unsafe to speculate much on this phenomenon. However at the high rolling speeds as the quartz sphere charges, rolls erratically and sticks, it is *certain that surface forces cause heating of the quartz*. Since this heating concentrates on WAGNER's hypothetical cone points and since these are excellent heat insulators, it is not improbable from observations of BOWDEN and TABOR that the asperities on the quartz become *very* hot. It is therefore perhaps not unexpected that where the heating takes place so that the whole surface heats to 60° or more, the conelets are raised to temperatures where thermionic emission or volume conduction is possible. Such currents and local conductivity will *increase* and *prolong back discharge beyond the value* of the limiting fields of 5×10^5

volts/cm for field emission. Thus q_s will be lower. At the start before the heating becomes appreciable, the extrapolated q_m or q_s is not much below that for slower rolling speeds. The general surface heating eventually reaches such proportions that that back discharge by the vaıious agencies drains charge off faster than it accumulates. Thus q remains at a lower level and saturates at a q_s below that indicated by the initial q_s before heating progresses. That is why saturation can be observed at 19 cm/sec rolling speed on Ni or on Pt and why the charging rate is low. Whether the thermionic back discharge associated with heating is caused by electron emission or O^- transfer cannot be answered at this point.

f. **The nature of charge exchange mechanisms.** The close agreement between saturation charges for monocrystalline and amorphous fused quartz indicated that there was no correlation of electron surface sites with long range crystal structure. This was also indicated by the absence of any anisotropic charging as indicated by the very sensitive probe.

It can thus be concluded that the basic mechanism as regards SiO_2 is one of open surface states available to electrons and the availability of electrons from the metal surface, there being an effective "work function" or better "contact potential" for transfer of electrons from metals to quartz. This has a value probably in the order of volts. The saturation charge is governed by this potential but is diminished by field emission back discharge and at hot spots at high speeds by some sort of thermionic and increased conductivity emission. Charging is strictly contact charging and obviously a surface effect with little diffusion of the electrons bound on the surface into the interior as indicated by the conductivity.

In the case of sapphire Al_2O_3, it is very clear that some sort of charge transfer occurs on contact. However, the transfer is crystallographically very anisotropic. It is essential to analyze the significance of the curves of Fig. 54. Looking at the ellipsoid with hexagonal, optic, or C axis, parallel to the minor axis of the ellipse, it is clear within the resolving power of the probe that there is a maximum of charge sensed when either *end of the axis, or the parallels to the axis*, are nearest the probe. The probe, however, is not very discriminating since it does sense some of the charge at 90° from the maximum charge direction. The amplitude of the major oscillations should represent the influence of the 0° and 90° orientations of the optic, or hexagonal axis, relative to the probe. If the probe were very discriminating it could give zero charge at the 90° and 270° or 0 and 180° positions and the maximum charge at the 0° and 180° *or* the 90° and 270° positions. However, as the probe integrates the charges which it sees the amplitudes observed are not as extreme as may exist. Furthermore, the ellipse wobbles and what is worse, moves irregularly from the front to the back of the cylinder causing the irregularities noted in Fig. 54. There are thus areas of charge exposed that are off axis with less dense charges.

The first question to resolve is whether the opposite ends of the axes charge oppositely. The difficulty in controlling the rolling made it uncertain whether the peaks were registered at the 0° and 180° positions or at the 90° and 270° position or only at the 0° or 90° positions. However, it was observed on slow rolling that there were on the average of 1.5 peaks per complete revolution which, with wobble and low time resolution indicate *charging of both the ends of the axes* with the same sign. The amplitudes of the oscillations observed, therefore, represent *not* the maximum 0—180° or the 90—270° charges and the minimum 90° and 270° or 0—180° charges but some lower charges at the top and higher charges at the bottom owing to the response time of the system and probe averaging effect.

It should next be noticed that the median line represents some sort of an *average* charging curve if such an expression has any significance. In the case of the sphere, the areas of the optic axial surface are weighted less heavily than the equatorial areas as sensed by probe. Thus the mean of the spherical charging curve oscillation shown in Fig. 54 represents more nearly the average charge on the equatorial areas.

The whole significance of these charging curves is obscure for two reasons: First, it is not clear whether the highly charged portions represent the surfaces at the ends of the optical axes or whether they represent the sides parallel to these.

Secondly, there is no way of being certain whether if the ends are the selectively charged faces, whether there is any charge separation on faces parallel to the optic axis. For using spherical or elliptically shaped forms, the probe always senses both end charges as well as lateral charges in virtue of rocking and oscillations. Only by contact of very carefully ground faces, cylinders about the optic axis or planes normal to it, can an answer to these questions be found.

Thus the use of median curves to represent charging characteristics must be regarded with caution. It would, however, be natural perhaps to assume the charge transfer to occur at the ends of the axes with relatively little on the surfaces parallel to that axis.

It is probable that the proper charging to consider is given by the envelope of the peaks ignoring the minima and realizing that the peaks even then represent a minimum of *possible* charge. This asymmetry can only be interpreted as indicating that if charging is caused by electron transfer from the metals the localized surface states accepting electrons are only at sites having a definite relation relative to crystallographic axes. In these localities, the charge densities must be very high.

The next significant circumstance is that there is not close correlation between the relative work functions of the metals and the saturation charge, (such as is estimated by the median and/or upper envelope of the charging curve), which differs materially from the behaviour of SiO_2. There was one observation in which while oxidation of the Ni yielded $+3 \times 10^{-10}$ coulomb on the Al_2O_3, Ni reduced in H_2 yielded -4×10^{-10} coulomb on Al_2O_3. This made it appear as if Al_2O_3 picked up electrons from clean Ni. Finally, the indications with the very clean metals that the charging varies with the past history of the specimen as well as the tendency of the sapphire to acquire a positive charge when rolled on fairly well oxidized Ni, all point to the possibility that in this case, the charging depends on ionic transfer of loosely bound ionic states on the crystal lattice. Whether the negative charging corrseponds to transfer of AlO^+ or other positive ions from the lattice to the Ni or Cu, or the removal of loosely bound O^- ions from the slightly oxidized metal surfaces while the positive charging is caused by a loss of O^- ions to the more heavily oxidized Ni surface, is sheer conjecture. In any event, the charging here is obviously not the simple electron transfer depending on metal work function observed with SiO_2.

Again it must be recalled that Al_2O_3 on prolonged rolling on all metals showed an increase of charge to a maximum and then a decline even when rolled at 3.2 cm/sec. Here the very slight electrification precluded heating as in quartz. If allowed to sit for some hours after rolling, the charge on rolling again rose slightly to a maximum and then declined. Here again a small temperature rise at asperities could cause an alteration of the charge exchange between the two surfaces. It must be recognized that while sapphire is a very hard and apparently stable form of Al_2O_3, that Al_2O_3 intrinsically is an amphoteric electrolyte and Al_2O_3 is slightly soluble in H_2O relative to SiO_2. It is not impossible that the rigorous acid cleaning treatments have left the weaker hexagonal axis ends of the crystals in a much more reactive chemical state than was the case for SiO_2.

The behaviour of the alkali halides was uniform, consistent, and in keeping with their known properties. Electrons are very strongly bound to the Cl^- ions and it takes on the order of nearly tens of volts in ultra violet illumination of alkali halides to produce free electrons within them. They are ionic conductors as electrolysis experiments show. They are not good electronic acceptors as they have a relatively small electron affinity, and do not appear to have vacant electronic surface states although impurities are said to cause these. The uniformity of behaviour of all halides of different origins render a common impurity unlikely. It is thus hardly likely that they would accept electrons from metal lattices even if these had low work functions. On the other hand, they have Cl^- ions and positive alkali ions which can dislocate on the surface or inside and wander in fields leading to their volume conductivity. Since most metals have positive metal ion lattices, they are, in general, quite capable of binding negative ions such as O^- or Cl^-. The binding forces may not be particularly strong as metal surfaces contaminated with Cl^- will evaporate them at temperatures of the order of $800°$ C, or perhaps less, depending on the surface. Thus it is not strange that all of these salts show definite positive charge on intimate contact with relatively clean metal surfaces such as Ni. The readiness with which they will charge a metal depends not on the work function of the metal but it does depend on the electro-chemical affinity for Cl^-, B^-, or I^- ions, and on the degree to which the halogen ions are loosely bound or dislocated on the crystal surface lattice. The difficulty of achieving clean surfaces of halogens without a relatively high degree of surface roughness renders any quantitative evaluation difficult and causes irreproducibility. With surfaces roughened to an indeterminable degree, the calculations for α and for q_s etc., are virtually impossible. The roughness of the surface no doubt also introduces a much higher percentage of dislocated and separable surface states for it than for smoother surfaces. Again since the halides melt at relatively low temperature $\sim 1000°$ C and ionic dislocation increases exceedingly rapidly with temperature another uncontrollable variable appears with local heating. Thus it was even difficult to arrange the halides in a reliable so called contact electric series, possibly related to strengths of ionic binding and frequency of appearance of imperfections. The low values of saturation charge despite a relatively small conductivity must be carried back to the surface roughness. Were clean optical surfaces practicable, charging might be nearly as high with SiO_2. The sign of charge and behaviour thus point to a definite transfer of loosely bound surface states of Cl^- ions to the metal.

With MgO despite the failure to obtain clean cut and different results in the rolling measurements for colored samples containing excess Mg or excess O *in bulk*, because of the alteration of surface by cleaning techniques, data were straight forward and relatively easy to interpret where the *surface* condition was known. MgO cleaned with organic solvents and *then distilled H_2O*, which probably dissolves off as much as 100 Å of surface, charged *positively* against Ni with q_s about an order of magnitude less than quartz. It crystallizes in the cubic system and charges isotropically and symmetrically. Its surface leakage is low. Such MgO heated in air is oxygen rich. When such a cleaned specimen is heated in vacuum at 1200° C for some hours it is known to lose O and become stoichiometric MgO. Such MgO on rolling on Ni develops a *negative* charge. That is, it accepts either electrons or perhaps O^- ions from the slightly oxidized Ni surface.

If the vacuum fired stoichiometric MgO is then polished again, it reverts in its behaviour to that of the normal O rich MgO. This could be ascribed to the chance that the thickness of the layer of MgO which was cleaned of O by heating was so thin that it was removed by polishing. This seems unlikely for the diffusion constant of O_2 in MgO $\sim 2 \times 10^{-8}$ cm²/sec at 1200° C would leave no excess O in the crystal after 10 hours heating. It is more likely caused by the local heating of the MgO surface in air in the process of polishing. For surface temperatures at points during the polishing process reached at least 1200° C, sufficient to cause readsorbtion of the oxygen. Thus with MgO it is clear that if it is oxygen rich it will yield O^- ions to the metal and acquire a positive charge. The process is a contact process with definite α. The value of q_s was less than for SiO_2 and saturation could be observed. Whether there was any loss of the initial charge transferred as a result of back discharge and high potentials as with quartz is doubtful. The surface of MgO was not as smooth as that of quartz after water treatment since it is slightly soluble. The amount of charge transferred per unit contact area probably depends on the surface concentration of loosely bound O^- ions on the MgO and the area of the Ni surface in contact that is free from O and capable of accepting O^- ions. With MgO deprived of its excess O by heating MgO is in a state certainly to accept any loosely bound O^- ions on slightly oxidized metal, in this case Ni surfaces, or it might in lieu of O^- accept electrons.

That both are possibly acceptable leading to the negative saturation charge of O free MgO is indicated by the peculiar conduction behaviour of such charged MgO. Here there was a hyperexponential decay of surface charge of MgO to a constant value of half its initial charge over some four hours after rolling O free MgO. If the electrons accepted from the Ni can contribute to a surface conductivity of the MgO this would cause the initial hyper-exponential discharge. The O^- ions however which can diffuse much more slowly over the O poor MgO at room temperatures thus might represent the residual surface charge left on the MgO.

The insensitivity of O deficient MgO to changes in work function of the Ni might indicate that electron transfer to the O deficient MgO did not occur. However, the two stage conductive loss of charge of the charged O deficient MgO sample indicates clearly the presence of two negative carriers coming from the Ni which was partially oxidized. However, the very strong binding forces of the O deficient MgO for O$^-$ ions and/or electrons might mask the effect of contact potential changes on q_s or q_m for electrons transfer. Beyond this, speculation is futile and the conclusions above are suggestively explanatory.

It is thus proper to summarize the conclusions derived on the behaviour of q_s for various substances in contact with metals by stating that in one instance there was a clean cut example of electron transfer related to the "contact potential" of the metal in causing strong reproducible charging. In two other cases there was strong indication of ion transfer for positive charging of the spheres and some ion transfer and probable electron transfer not related to work function for negative charging. In the case of one other crystalline oxide, Al$_2$O$_3$, that of an element showing strongly amphoteric characteristics in less rigidly bound form, strongly asymmetrical charging associated with its crystalline or optical axis indicated probable ion exchange with metal surfaces unrelated to electronic work function. This study of pure clearly defined substances against the surfaces of metals has thus led us well along the path towards understanding the nature of a *few instances* of solid-solid contact charging, indicating that even where metals are concerned, the *nature* of the *process* is *not always the same* and depends on such matters as adsorbtion energies and electrochemical affinities.

G. Contact charging of ice on ice

From the studies on the electrification of dusts and the preceding considerations on the contact charging of insulators and metals enough has been learned concerning the nature of the charging processes to make one inclined to assert that the *contact of one clean surface on the clean surface of the same material* should yield *only symmetrical* but *no asymmetrical charging*. In fact, the evidence in support of this is so strong and logical that when asymmetrical charging is reported on contact of quartz on quartz, or of crushed amorphous quartz on crystal quartz or other amorphous quartz surfaces, the author would unequivocally assert that the two contacting surfaces were *not in* the *same state, or else not clean* especially where cleanliness is so very difficult to achieve. Where such charging was observed by J. W. Peterson [97] in the author's laboratory in 1949 the author can assert that the conditions of the measurement even though superior to the general run of such studies were *not such as to insure equal purity*, or cleanliness, of both surfaces and the recent work of Wagner on *really clean* quartz-metal contact confirms this conjecture. Likewise, the asymmetric charging observed for quartz on quartz by W. R. Harper [112] can only imply inadaquacy of cleaning techniques to the extent that it was asymmetrical. Again the exacting criteria established for surface cleanliness of quartz by Wagner make this conjecture plausible.

On the basis of such experience, the author committed himself to stating in 1950 [113], that there was no reason a priori to expect *friction of ice on ice* in *thunderclouds to lead to electrification*. To his mortification in 1954, E. J. Workman and S. E. Reynolds [29] reported that strong static electrification could be observed in *friction* of ice on ice *provided there was a temperature difference between ice surfaces* and that certain contaminants were not present below given concentrations. Earlier investigations, for example, Simpson and Scrase [114] in 1937, suggested that friction of ice on ice might produce thunderstorm cloud charging on the basis of the heavy charging accompanying the blowing of snow along drifts. CHAPMAN]₃ reported electrification on the shattering of snow flakes in 1950. PEARCE and CURRIE[115] in 1949, CHALMERS[116] in 1952, and NORINDER and SISKNA[117] in 1954 reported studies of electrification from ice-ice friction. These observations were not, however, made under the carefully controlled conditions applying to the investigations of WORKMANN and REYNOLDS. Unambiguous results required that temperature and contamination differences be taken into account in such studies. The studies revealed that in *frictional* contact *the ice formation which is the colder acquires positive charge* unless it contains NaCl contamination in concentrations in excess of 10^{-5} molar. At first this action was most puzzling, except that in this case there was imposed the condition *that there must be a temperature difference in excess of about 2° C* before electrification could be observed. This indicated that at least there was *a difference, (that in temperature)*, required for the phenomenon to occur. Extensive study eventually led to the conclusion that involved in the *"frictional" process* leading to charge separation was the pressure or frictional melting of the ice followed by regelation. The *supposed pressure melting of ice* had long been known and the mechanics of ice skating and skiing was ascribed to this principle by J. THOMPSON[118].

Later studies of BOWDEN and TABOR[52] indicated that for really cold ice surface the *melting* was *not due* to *pressure but to friction.* Nevertheless pressure produces friction at contact points causing melting followed by regelation so that the essential facts remain the same except that the *melting is a frictional* rather than a presure phenomenon. Thus in terms of the studies of BOWDEN and TABOR the *frictional* process of this sort even between impacting cold pure water crystallites against warmer larger graupel pallets could achieve this mechanism.

With the suggestion that *melting and regelation* were involved in the electrical separation, it at once appeared natural to inquire whether the striking *freezing potentials* observed by WORKMAN and REYNOLDS might not be responsible for the charge separation. For if on impact there is melting the colder surface will grow by the unilateral freezing of the aqueous film from the cold side outward. But such a freezing process in the presence of traces of impurity lead to potentials between the water phase and ice during freezing as seen in Chap. I. If then, mechanical separation occurs before regelation is complete, the warmer particle separates with the aqueous film and the colder particle with the newly formed ice carrying a charge.

At first consideration, this process ran into apparent contradictions in sign of charge until further studies of electrification accompanying the freezing revealed that the *polarity of the electrification* was determined *not only by the kind and amount of substance* in solution but also by the *nature of the substrate upon which the solution is frozen*, particularly when the substrate is ice. The polarity which arises during the freezing of the substrate ice tends to be transmitted to the solution frozen on the substrate, *irrespective of the constitution of the later solution.* That is, once a substrate has frozen in such away as to give a certain sign of charge to the freezing surface, it will continue to freeze with that charge even though it later encounters a solution of salt which might initially have yielded a different sign of charge. This effect of the substrate supports the argument that charge is separated during the frictional contact as a result of the formation and re-solidification of a liquid layer at an ice-ice contact. The varied and complicated programs of experimental investigations of the charging phenomenon establishing this rule are detailed in the *published reports** of the work and find no place in this book. Enough has been reported in Chap. I concerning the freezing potentials and is stated here to account for the mechanism of charge. This mechanism is basically a physical-electrochemical process involving segregation and separation of ions at the advancing interface of ice and liquid water. The exact nature of the processes is perhaps not well known or understood. However, the action is not far removed from the action of different ions on surface tension at liquid-gas interfaces, now carried over to a different phase change—ice-water, involving perhaps compatability of certain crystalline forms with the ice structure.

The important point to be noted here is that the ice-ice electrification *requires a physical difference* between the ice particles impacting e.g., a temperature difference in addition to some basic *frozen in structural feature which differentiates the two impacting forms*. Thus *electrification on contact does not occur by asymmetrical charge transfer between structurally really identical surfaces.*

The particular virtue in the findings of the action of the substrate lies in the relation of this condition to charge segregation in thunderclouds. The charging is produced by impact of the very cold small crystallites of ice that thave grown by sublimation. These are the colder and consist of a substrate of very pure water. They encounter the larger and warmer graupel pellets of glaze ice which may contain impurities. On impact, melting and regelation the smaller colder crystallites take on the positive characteristic of their substrate, pure water and escape on separation with a positive charge while the locally moist graupel pellet carries off negative charge. This leads to an accumulation of positive charge on the dome-like top of the thunderclouds through the light crystallites carried on by updrafts and concentrates the negative charge in the lower portion of the cloud cell on the growing graupel pellets. These will eventually carry negative charge down in the hail or rain unless the charge is drained off to ground by a lightning stroke.

H. Other possible charge transfer mechanisms

1. Asymmetric rubbing—True tribo electrification

It had been observed by WORKMAN and REYNOLDS that in rubbing two ice samples at the same temperature in the form of rods such that one rod was stationary while the other rod was run back and forth at right angles to the first, as one would bow a violin string, the rod that did the bowing usually charged positively to the stationary rod. This they soon associated with the heating and melting of the bowed rod by friction at one spot while the bowing rod remained relatively cold. The electrification was in that case, however, the consequence of regelation. Apparently quite independently P. S. H. HENRY[119] described a series of experiments in which *two rods of the same material could be charged by asymmetric rubbing*. In this study merely two insulator rods, cut from the same piece of

---

* To date no papers have appeared on this aspect in scientific journals.

material, were rubbed at right angles to each other the moving bow stroking the other piece at right angles at always the same place.

The chance of contamination playing a role in this instance is ruled out both by nature of the rods and reversal of roles in bowing. Thus the effect is a real one not caused by differences in cleanliness. Therefore, in this case, a *true frictional electrification was achieved*, one rod charging positively, the other negatively. By reversing the roles of rubber and rubbed, the signs of charge on the two rods reversed. The measurements were not carried out under controlled conditions with cleaned surfaces, or surfaces of known structure or composition, or in the absence of air, etc. Nor were the substances of different nature studied so that the specific behaviour of different substances could be noted and the relative charging effects compared. The problem is fundamental and merits further study for a wide range of substances some of preferably known nature under controlled conditions. If this is done more light will be thrown on the actions present.

The *explanation on such meagre data given by* HENRY *is undoubtedly in the right direction*. The confined surface that is stroked is unquestionably heated by the stroking. Heating is on a twofold scale, that is a *general background heating* of the *whole rubbed surface* and a *localized higher heating* of *asperities* as indicated by BOWDEN and TABOR. Both these effects will vary widely with the nature of the substance and perhaps surface finish. The question is then how a difference in temperature can cause charge transfer. Obviously for hard substances where asperities are raised to high temperatures, thermionic emission of electrons from these is possible and the heated surface should be positive. However, this extreme process may and need not take place. If there are loosely bound ionic carriers of one sign or the other on both surfaces since, free carrier densities are given by the Boltzmann equation $N = N_0\, e^{-A/kT}$, where A is the binding energy, the hot surface will have an excess of carriers of higher diffusion coefficient over those of the cold surface. Thus more ions will move from hot to cold than in the reverse direction leading to charge transfer.

The asperities on both surfaces will initially be heated equally by the friction. These have small volumes, small heat capacities and perhaps low heat conductivities. But as the rubbed overall surface temperature generally rises, the asperities will achieve higher temperatures on that side. Such temperatures can readily reach values such that one way charge transfer will be assured even if A is relatively high. The heating hypothesis was checked by SHAW and JEX[122]. In an earlier study they stated that a warmed glass rod rubbed on a cold one produces a separation of charge which reverses on rubbing when the other rod is warmed and the first is allowed to cool. Much care must be used in such measurements. Thus if heating were achieved by warming in a flame differences of a chemical nature could be produced. Heating in a furnace, even one outgassed can still contaminate the heated surface. Heating may remove an aqueous film from one surface and leave it on the unheated surface. Thus such heating experiments require exceptional precautions which were *not* taken by SHAW and JEX or other earlier workers.

However, since glass is an ionic conductor, where dissociation increases rapidly with rise in temperature with alkali ions constituting the mobile units, temperature charging of the type indicated, is not unlikely. Until, however, really carefully controlled studies are made on such systems, the mechanism can be assumed to be likely but has in no sense conclusively been established. It can however be stated that unless very great temperature differences are produced between the two surfaces that do not equalize on contact, the *potentials* created by *thermal migrations of ions will be very small* and saturation charges will not be great. It is important that studies under controlled conditions be undertaken. One of the most interesting problems involved in all ion transfer studies lies in knowing something more of the distance of approach of surfaces for ion transfer. It is not likely that penetration of the barrier as depicted by WAGNER for electrons will occur for ions. Thus ion transfer may require somewhat closer approach of surfaces and hence reduce effective areas of contacts.

2. The Henry model for contact charging of solids having a common ion

In his discussion of contact electrification, HENRY[121] considers other conditions applying to contact of dissimilar bodies leading to transfer of charges beside that of the phenomenon of thermal diffusion across a barrier with a temperature difference alone. Consider for example soda glass and a borosilicate glass or better, borosilicate glass and quartz. The soda glass is rich in Na while borosilicate glass is poor, ($\sim$4—6%) and quartz has very few as impurity. Of the Na atoms only very small fractions $N = N_0 e^{-E_i/kT}$ are in the dissociated state as free ions where E_i is the ionization energy of Na atoms *in the glass*. The ions are mobile for they conduct the current. If now two surfaces are in contact and contact points are heated equally the law of mass action alone will insure that more ions cross under diffusion from soda to borosilicate glass, or from that glass to quartz, than vice versa so that the soda glass should become negative to borosilicate glass and the latter charges negative to quartz as observed by WAGNER. The charge transfer per unit area, provided contact time at intimate contact permits equilibrium to be established, will then be in proportion to the difference in common ion concentration in the two glasses, or until diffusion has carried so many ions over that the field created across the contact boundary prevents further diffusion. If there is a potential barrier to be overcome for transfer of Na$^+$ ions to the borosilicate glass, then only that portion of the excess ions that have the energy to surmount the barrier will cross. Transfer again will continue until the *opposing field created by diffusion plus the potential barrier* permit no more transfer.

HENRY has derived a rough relation for contact charging between substances with common ions based on this principle which may be broadened to include the effects of difference in temperature and imposed external fields. Such a relation is useful since it states in relative quantitative terms the factors influencing a case of charge transfer. Its illustrative value though actually applicable to relatively few instances of observed charging, is of importance. He starts by considering the potential energy of the same ions at the dissimilar surfaces for the two substances to be placed in contact. Fig. 63 shows that at surface 1 the image forces cause the potential energy to fall from 0 far away to a minimum V_1 very close to the surface. As the ion approaches the surface more closely repulsive forces of electron shells cause the potential to rise again. The same applies on the right for surface 2. As surface 2 binds the ions more strongly V_2 is lower than V_1 and $|V_2| > |V_1|$. In between the two surfaces separated by the minimum distance of approach d_0 the potentials rise to a maximum V_m.

The concentration of the ions per unit area of surface 1 is N_1 and that for surface 2 is N_2. The temperatures of surfaces 1 and 2 are T_1 and T_2. The rate of loss of ions from 1 to 2 by diffusion will be proportional to $N_1 T_1^a e^{-\frac{V_m - V_1}{kT_1}}$ and the rate of loss of ions from 2 to 1 will be proportional to $N_2 T_2^a e^{-\frac{V_m - V_2}{kT_2}}$. The exponent a of the temperatures is akin to the exponent $1/2$ or 2 evaluated on different assumptions in deriving a similar equation for thermionic emission of electrons. As it is a relatively unimportant term the exponent will be set as indefinite and equal to a for ion transfer. In view of the difference in the values for transfer from 1 to 2 and vice versa a net flow of charges will occur tending to equalize the levels to which ions are held by building up a counter field. Let the electrostatic potentials built up at the electrodes

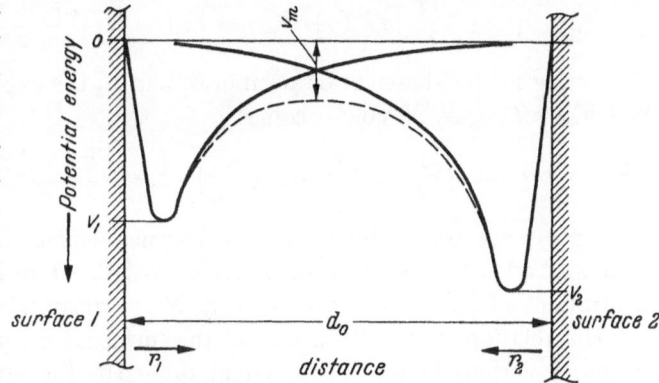

Fig. 63. HENRY's schematic sketch of surface potentials at two surfaces having common ions but in different concentrations and binding energies

because of transfer be E_1 and E_2 with the maximum of the potential between equal to E_m. As can be shown in general at the close approach needed for ion transfer the potential energies eE_1, and eE_2 will be small compared to the potential energy troughs. Thus eE_m will alter the maximum total energy V_m slightly from that in the absence of any diffusion potentials. This difference will be ignored for simplicity. Putting these quantities into the expressions for the rates of diffusion and equating rates at equilibrium there results

$$N_1 T_1^a e^{-[V_m - V_1 + e(E_m - E_1)]/kT_1} = N_2 T_2^a e^{-[V_m - V_2 + e(E_m - E_2)]/kT_2} \qquad (4.25)$$

Actually, if the ions are created by dissociation of the solid by thermal action, N_1 and N_2 the ratio of N_1/N_2 will be decreased by the ratio $e^{-\frac{A_1}{kT_1}} \big/ e^{-\frac{A_2}{kT_2}}$ or by

$$e^{-(A_1 T_2 - A_2 T_1)/kT_1 T_2} \quad \text{and if} \quad A_1 = A_2 \text{ by } e^{-(T_1 - T_2)A/kT_1 T_2} \simeq e^{-A\Delta T/kT^2} \qquad (4.26)$$

which must also be included.

If an external electrostatic field X is added in addition to that caused by the net charge densities $\pm\sigma$ resulting from charge transfer

$$E_1 - E_2 = e D (4\pi\sigma + X) \qquad (4.27)$$

with D the dielectric constant.

Let $V_1 - V_2 = \Delta W$ the work to transfer ions from surface 1 to surface 2, e.g. ΔW is the equivalent of the difference in ionic work function between 1 and 2.

Taking logarithms of the equation above there results

$$\left.\begin{array}{l} \log N_1 + \log T_1^a - [V_m - V_1 + e\,(E_m - E_1)]/k\,T_1 \\ = \log N_2 + \log T_2^a - [V_m - V_2 + e\,(E_m - E_2)]/k\,T_2. \end{array}\right\} \quad (4.28)$$

The inclusion of temperature differences which is both realistic and useful produces complications in the reduction to simple terms. If $T_2 - T_1 = \Delta T$ is not too large compared to $\dfrac{T_1 + T_2}{2}$ it is convenient algebraically to manipulate the equations in order to arrive at HENRY's relation.

$$\left.\begin{array}{l} 4\pi\,e\,D\,\sigma - \Delta W + k\,T\log\dfrac{N_2}{N_1} - e\,D\,X + \\[2mm] + \dfrac{\Delta T}{T}\left[a\,k\,T + (V_m + e\,E_m) - \dfrac{V_1 + e\,E_1 + V_2 - e\,E_2}{2}\right]. \end{array}\right\} \quad (4.29)$$

Correcting for the dissociation term in N_1 and N_2 the expression for the ratio term which is $k\,T\log N_2/N_1$ now becomes

$$k\,T\left[-\frac{A_1 T_2 - A_2 T_1}{k\,T_1 T_2} + \log\frac{N_{02}}{N_{01}}\right] = -\frac{A_1 T_2 - A_2 T_1}{T} + k\,T\log\frac{N_{02}}{N_{01}}. \quad (4.30)$$

This in essence adds a temperature difference dependent term to the term ΔW, that is it adds the heats of dissociation to ΔW. Here N_{02} and N_{01} are the concentrations of ionizable substance e. g. Na in glasses.

This relation of HENRY then gives the surface density of charge σ separated in terms of various factors. It is equal to $-\Delta W$ the effective contact potential difference between the surfaces in the absence of differences in concentration. In the event that $N_2 \neq N_1$ the second term gives the mass action factor which can occur with the other factors small, or with a finite ΔW in the more usual cases. The $-e\,D\,X$ term is the linear influence of the electrical field postulated by GILL and ALFREY in consequence of their findings and is in agreement with PETERSON's and MEDLEY's work*. It will only occur with transferable ions but in all measurements the presence of fields X must be carefully excluded as a factor in the event such transfer is possible. The last term depending on $\Delta T/T$ represents the influence of temperature difference between surfaces. It depends in a small measure on the external temperature factor a in the emission relation and represents the thermal diffusion term considered earlier. The more important terms are the differences between $V_m + e\,E_m$ and $\dfrac{V_1 + V_2 + E_1 + E_2}{2}$ which are not negligible.

They measure the influence of the potential energy hump between the surfaces which *must* be *surmounted by ions*, but can be *penetrated* by *electrons*. As the surfaces approach $-V_m$ increases, or relative to V_1 and V_2 *the hump decreases*. On separation the situation reverses. As the surfaces recede the whole term increases but if it gets too large compared to $a\,k\,T$ the transfer of ions becomes too slow relative to the motion of separation for the important assumption of steady state to apply and the theory is invalid.

If ΔW is a function of T it would appear as an added term in the last expression. Since $\Delta T/T$ is small this term would be negligible.

---

* In the case of these substances the ions are loosely bound and fields needed to transfer them across a boundary need not be high.

The value of an equation of this sort, which is approximate and somewhat idealized is that it clearly delineates the roles of the various important factors active in contact and in *tribo* or *frictional* electrification where this occurs. It clearly indicates the importance of $-\Delta W$, of the effective work function, of concentration, effects of external fields and of thermal diffusion. It, however, neglects back leakage or discharge.

Its probable worst defect for most charging processes, even that of contact charging as WAGNER and PETERSON's work show, and certainly where frictional or tribo effects enter in as indicated by the work of BOWDEN and TABOR is that it relegates temperature effects to small ratios of $\Delta T/T$. In actuality local temperatures and possibly temperature differences at real contact are high. Thus the more exact rather than the approximate equation must be used.

The real difficulty is that depending on temperature effects. For in any frictional or contact processes where asperities are raised to high temperatures, the processes are quasi adiabatic. Again what are the durations of contact and what are the temperature differences under contact? Finally can equilibrium exchange occur under these conditions? Aside from these vital points, about which little can be said without a great deal of analysis for particular cases, as *initially stated contact electrification between surfaces where common ion exchange can occur is relatively infrequent* in processes actually observed. The analysis is of considerable illustrative but little practical value.

In the more practical realm, it is clear from modern knowledge of surfaces and surface contamination that no metal surfaces are ever clean nor can they remain clean over minutes of time unless the metals are flashed to near their melting point in ultra high vacuum and unless the pressure remains below 10^{-9} mm in an initially outgassed vessel. Thus most surfaces are usually a composite of bare and occupied sites. These hold atoms or molecules bound with energies extending from 0.1 volt or less up to serveal volts. The same applies for relatively clean crystal lattices as WAGNER's work shows. At close contact at asperities some of these molecules are displaced and transfer of molecules and charges on ions is not the ideally simple exchange exhibited by the equation for a single common ion species but a complicated reciprocal exchange of different ions of different species at different points of the same surfaces. It is this sort of condition that makes precise analysis so very difficult.

3. Charging of solids by rupture of surface dipoles

Among other processes invoked by HENRY[121] as possible mechanisms active in static electrification is one involving the existence of a polarized, or oriented, double layer at a solid surface *. That is a layer with ions of one sign at the surface bound to ions of the opposite sign below. Such double layers could either be inherent in the structure of the surface or be created in certain crystals by stressing the crystal as in the piezo electric effect of quartz. If such double layers exist

---

* This action forms the basis of a generalized theory of P. BÖNING derived from basic colloidal chemical theory and set forth in book form in 1938 and in alter papers which have only recently come to the authors attention owing to publication largely in specialized journals, see preface.

HENRY assumes that in the frictional process it might be possible to "rub off" one sign of charge on the dipole leaving the other sign behind.

In principle this partakes of the nature of the mechanism involved in spray electrification. It was also considered possible that it might apply to the electron cloud at the surface of liquid metal surface. In those cases, it appeared as shown in Chap. III, that it was *impossible by mechanical effect*, e.g., in this case by gas blast, to concentrate enough energy at local points to disrupt the dipoles of the liquid. In the case of Hg, the electrons bound by image forces with energies of binding in the order of volts the difficulty becomes obvious. In the case of water there was evidence of the breaking of hydrogen bond linkages on rupture leading to excess OH^- or H^+ ions *on small droplets*. It is doubtful whether single OH^- or H^+ ions were ever separated from the double layer by bubbling, spraying, or impact. Droplets of tens or more molecules could be torn off rupturing a hydrogen bond and yielding small droplets of water with excess H^+ or OH^- ions that evaporate down to the size of bi or tri-molecular complexes in dry air. The majority of the charging in water however comes from the deeper lying adsorbed and "uncovered" impurity anions and excess cations on larger droplets.

In the case of surfaces such as metals or other substances having oriented superficial dipoles, the magnitude of the asperities are such that in most cases it is extremely unlikely that thin enough layers can be ripped off to remove the excess of the outer ion layer of the dipole in such structures as say Na^+Cl^-. This is shown nicely in BOWDEN and TABOR's studies of friction. Layers much thicker than monolayers are disrupted and torn in the slip-stick process.

In other words, the degree of frictional or true tribo-electrical charging of this character is probably so exceedingly small as to be negligible since asperity peaks in contact do not approximate rakes with prongs of single atomic dimensions.

Where however, there is an acceptor surface having the power to bind ions of the donor dipole lattice, this exchange can occur *provided contact is intimate* and *either thermal dissociation by local heating*, or *acceptance of loosely bound dislocated*, or *excess surface states is possible*. In the event that the affinity of the approaching surfaces for the halogen ions is greater than their binding energy much breaking of dipole bonds at the surface and charge transfer of the favored ion can take place even with mild contact heating. This process is of the type of contact charging discussed above and does not apply here.

If by double layers is meant the loosely bound chemi-adsorbed layer of O^- on surfaces of Ni, binding ergy ~ 0.1 to 1 ev then a clean surface of higher binding energy or even commensurate binding energy on *contact*, aided by local heating will separate with considerable charge. Such again is the sort of action observed by WAGNER and discussed earlier.

It is however, doubtful whether more of such layers are taken off by friction than by contact. If on the other hand, the polar layer is such a complex one as an adsorbed ion layer on the outside of a relatively thick layer of metal oxide, or other relatively loosely bound insulator, on a metal surface, frictional processes as well as contact processes could be effective in removing considerable charge.

In the event that a polarized layer is created on a surface by such a process as piezo electrification of quartz or Rochelle salts, the question as to whether static electrification can be achieved by *rubbing*, or *frictional* contact, has been in part answered by a measurement made in the author's laboratory by J. W. PETERSON[122].

In this case, in a *fairly clean chamber relatively clean crushed amorphous quartz particles* were allowed to slide down an incline made of *two* pieces of optically polished and cleaned quartz X cut surfaces. Placing these quartz specimens under pressure along the Y axis created one surface with a negative piezo-electric charge upward and next it a surface with the positive piezo-electric surface charge upward. The two quartz crystals originally came from the same piece. The quartz sand was caught in a Faraday cage bucket and its charge measured after sliding down the positive, or the negative, quartz plate and also down the same plates when uncharged at zero stress. The external fields of the quartz were *not such as to induce much polarization charge* since the particles were dry. The quartz sand which fell onto the plates from an amorphous quartz funnel was always negatively charged as it reached the Faraday cage. On the average with no stress on the crystal, the charge was small 8×10^{-11} coulomb. There were 1.2×10^4 grains averaging 0.4 mm in diameter, with 4×10^4 electron charge per particle. Maximum stressing produced a surface charge *on the crystal* of 2.5×10^{-10} coulomb. Of this at most 1.5×10^{-11} coulomb was transferred to the sand in addition to the charge with no stress. Since the sand in its descent made contact with only half the surface area of the crystal *a maximum of about 12% of the available surface charge* of the crystal was transferred to the sand. The *amount of charge transferred* was the *same for both signs of charge*. The air was dried over liquid N_2. Wall gases gradually contaminated the quartz reducing the zero stress charging. This could be restored to its previous value by pumping several hours at 1×10^5 mm Hg pressure. The charging of the quartz particles came from the contact with the amorphous quartz receiving buckett.

PETERSON concluded that since the piezo electric charging amounted to less than 20% of its total charge when the crystal was strained by a force of 5×10^6 dynes it appeared that the piezo electric effect was certainly not a great factor in tribo-electrification.

The author desires to call attention to the fact that while PETERSON achieved a piezo electric charge effect of 20% with quartz sand on stressed quartz crystals that the *author is not certain that these results can be ascribed to the tearing off of oriented surface dipoles*. While PETERSON's cleanliness was good by past standards of investigation, well cleaned surfaces, *dry* air, etc., the pressure of 10^{-5} mm was *not* good, the gradual contamination on standing with change in charge, etc., pointed to the possibility of surface layers. The surface fields of the quartz were not negligible and the *observed 20% charging could have been caused by impurity ion transfer though probably not by $e D X$*. The surface charge was relatively low about 0.3×10^{-11} coulomb per cm² compared to charges of the order of $\sim 10^{-9}$ coulomb per cm² in contact between quartz and Ni. Thus in the light of present day experience, the study will have to be repeated under really clean conditions, a relatively difficult task in view of mechanical requirements.

In the same class with piezo-electric polarization and tribo-electrification is that caused by pyro-electrification. In this case, certain asymmetric crystals on being heated develop an electric polarization analogous to the polarization created in stressing certain crystals. Other crystals, like quartz, that are heated asymmetrically, or have sharp temperature gradients, will be mechanically stressed producing a piezo-electrification which is sometimes confused with pure pyro-electrification. It exists only while there are temperature gradients and not at equilibrium at a fixed temperature as for true pyro-electrification. Irrespective of origin, pyro or piezo orientation presents the same problem in regard to tribo-electrification. The effect will be small if it is even real.

Pyro or pyro-piezo electrified substances, e.g. sands in that state, show a remarkably high apparent viscosity. Mr. STROMBERG, engineer for Del Monte Properties sand plant at Monterey in 1925, called the author's attention to this phenomenon with a very pure blown quartz sand having some little feldspar and mica admixed. It was heated in a drier and placed on a leather conveyer belt where it was subjected to very unequal cooling as well as considerable electrification of the grains by contact with belt and between dissimilar mineral constituents. The heavy temperature gradient created piezo-electrification of the grains. This, together with the many charged particles, made the sand in this condition stiffer than cold molasses when it was stirred with a paddle. When cold the sand was quite loose and gave little resistance to the paddle. The phenomenon was investigated by C. A. RINDE[123], in the author's laboratory, who confirmed the explanation in his MASTER's thesis.

4. Luminosity in contact charging processes

In studies involving the crushing of substances like sugar which show strong contact charging and in certain tribo-electric studies luminosity is seen in the dark. The question often arises as to whether this is caused by contact, or static, electrification. Each instance of such luminous flashes must be studied as a separate phenomenon relative to the circumstances. Thus Hg moving up and down a barometer tube with 10^{-1} mm of air or other gas pressure creates very high contact charging of the glass resulting in a rather bright glow that follows the retreating Hg surface as it moves. Undoubtedly the spheres and sands used by PETERSON or WAGNER at an appropriate gas pressure could well have given fairly distinct diffuse glows as they discharged. Usually in the presence of contact charging at higher pressures, the sparks occur over such short distances as to be pretty well masked, so that the spark manifestations only occur when the surface or the mass builds up to a sufficiently high potential to initiate a common filamentary spark. Usually, some metal or conducting surface is needed to collect the charge and so to yield a large enough quantity to cause a spark to materialize.

On the other hand, there is no a priori reason why the flashes produced when crushing sugar crystals should be of an electrostatic origin. There are a number of phenomena divorced from electrification yielding luminosity. Thus in clean water cavitation produced by a clean propeller blade produces a faint luminosity. The same sort of luminosity is produced in cavitation resulting from strong ultra-sonic excitation. In these cases, the actual kinetic energies produced in these adiabatic processes are sufficient to yield shock waves capable of causing excitation and ionization of the water vapor or gas molecules concerned. With modern techniques shock waves can be produced in gases which travel at 10^5 cm/sec, reach temperatures of 30000° K completely ionizing gases like Argon. The light from such shock fronts is dazzlingly brilliant. It is not impossible that in crushing sugar crystals kinetic energies imparted to the microvolumes in the form of shock waves produced on rupture can excite and ionize the crystal face or the adjacent gas to luminescence without any intervention of electrification and discharge. Actually whatever discharging occurs in the crystal mass, (not at the mortar or pestle surface), will be symmetrical charging and it is unlikely that these will build up potentials on separation leading to a diffuse gaseous glow such as observed unless at low gas pressure.

Finally, in any true tribo-electric generation involving friction the interaction of the asperities, if one substance is harder than the other, will heat these. If the speeds are high with much pressure whole small chunks of one or both surfaces are thrown off frequently heated to incandescence. If they reach their combustion temperature in the air, they will burn as bright sparks. This is the time old principle of flint and steel. It is the principle involved in the misch metal flint used in cigarette lighters. Here the steel rips off small asperities in the misch metal and raises them to combustion temperature in the air. Misch metal is used since it slowly oxidizes in room air but burns brightly in air above about 300° C. It is therefore easy to ignite. Even friction of steel on steel is sufficient to throw off sparks as can be noted at the brake shoes of a railroad train when the brakes are applied.

Another very early example of combustion on adiabatic compression is the "Fire Syringe" that used to be shown in demonstration lectures. In this a stout glass tube of some 2 cm diameter with thick walls is sealed at one end and a bit of cotton saturated with CS_2 is placed at the bottom. If now a fairly loosely fitting plunger is sharply pushed to the end of the cylinder compressing air and CS_2 vapor adiabatically, there is a bright flash of flame as the CS_2 catches fire.

With all these other manifestations of luminosity produced by processes which *might* also be producing static electrification great care must be used in identifying any particular luminosity observed with static sparks or discharges.

V. The generation of static charges by processes involving ionization of gases

A. Introduction

Gases are usually very slightly electrical conductors. Normal gases such as the atmosphere are subjectet to a continuous bombardment of ionizing radiations such as cosmic rays from outside the earth, γ radiations trom the earth's crust, α, β and γ rays from radioactive emanations and dusts exuded or carried aloft from the earth, nowadays man made radioactive materials from nuclear devices, and in confined spaces, with metal walls from the minute radio active content of the walls. The various sources share about equally in creating the 20 new ion-electron pairs per cm³ per second usually measured. These figures are subject to considerable variation depending on local factors. Loss of ions in the atmosphere is largely by recombination. This process is slowed down somewhat by the attachment of carriers to larger aerosols present in the atmosphere. Thus generally speaking there are about 1000 to 10000 ions positive and negative per cm³ of air under normal conditions. These ions lead to very weak currents for applied fields and serve only to yield initiating carriers when fields reach proportions leading to gaseous breakdown. On the larger particles in suspension in the atmosphere—such as the Langevin ions of some 10^6 molecules—there may from statistical accumulation of charge by encounters with ions be from one to several charges[76, 77]. For particles of 1 micron, 10^{-4} cm diameter with 10^{10} molecules charges up to 10 electrons can be observed, the median being around 3 as KUNKEL[77] has shown, with the greatest number in a fixed charge interval at zero charge.

It is seen that in consequence of the natural charges, the air is so slightly conducting that objects statically charged will lose their charge more slowly through the atmosphere than they will over insulators. This was COULOMB's early discovery when he suspended charged spheres by silken threads and observed that doubling the number of sliken suspension threads increased the loss of charge by a factor of the order of half the loss by atmosphere and one thread. On the other hand, substances normally considered insulating in terms of currents of common experience such as thoroughly seasoned dry wood stick 1 m long and 1 cm diameter at 60% relative humidity at 20° C will, for example, discharge a clock type of electrostatic voltmeter of 500 cm capacity charged to 5000 volts with a time constant in the tens of seconds.

Obviously the natural conductivity of the air will be augmented by ionization produced by flames and various types of exhausts from chemical and industrial processes. It appears even likely that evaporated ocean spray particles may carry aloft charges created by various mechanisms such as bubbling[93]. Such charges, however, are rapidly dispersed to low concentrations. Thus normally carriers of this type play little more than a discharging role and do not lead to serious static accumulations, acting generally as feeble leakage agencies.

Again there are numerous agencies that do render gases conducting. Among these agencies there are several that by their nature can lead to a ready separation of charge and heavy accumulations of static charge.

The conditions for separable charge generation are simple. The charge must be generated such that one sign predominates in the gas phase and that the other sign resides on some object or surface of sufficiently different inertia that it remains relatively fixed while gas currents carry the other charge off by convection, or that the charges once segregated, are separated by adequate mechanical forces to build up the required fields on insulators. This means that the ionizing agency must act to liberate charge from a fixed surface, or a solid surface of such size that mass segregation of the solid from the gas stream allows of ready mechanical separation. A more remote case is that where electrification of opposite signs are on carriers of such different mechanical properties that under existing conditions the charges will physically *separate* to different locations in sufficient measure to build up high potentials. Agencies rendering gases conducting are:

a. X-rays or ultraviolet light liberating electrons of some energy from a solid surface, e.g., photoelectric electron liberation.

b. Thermionic effects by which very hot bodies, solids or particles emit electrons, including herein processes in flames. Heated bodies may sometimes emit positive or negative ions, but usually the electron process predominates as the more important factor.

c. Highly electrically stressed asymmetrical conductors of positive or negative sign causing localized corona breakdown, sending appreciable currents into the air of sign corresponding to their polarity. This is one of the most prolific and effective source of unipolar charge emission in gases.

d. Other discharges such as sparks, arcs, or glows in a large measure create carriers in equal quantities. Under rare circumstances diffusion of fast electrons to surfaces could cause some charge separation, e.g., surfaces in contact with plasmas usually carry a slight excess of negative charge. However, charge accumulation is limited by ambipolar diffusion and amounts to very little.

e. Independent of gaseous breakdown, the polarization of liquid or solid conductors in a high field with separation while in the field and mechanical segregation of separated particles yields effective static electrification, but is properly not ionization of gases, though it can effectively charge a gas with a fine mist of charged particles.

B. The various gaseous electronic mechanisms

1. Photoelectric and thermionic emission of carriers

The photoelectric effect liberates its electrons according to the Einstein equation

$$\tfrac{1}{2} m v^2 = h v - \varphi e.$$

here $h v$ is the energy of the liberating photon, v being the frequency of the light, h the Planck constant and φe the work of escape of the electron against the image force in the metal, of from its binding deeper in the atom for X rays. While hv ranges into the millions of volts for γ rays and kilovolts for the X rays these agencies will rarely figure in common static charging problems until the nuclear power sources enter industry and probably not then. If the photoelectric liberation figures at all, it will be in the near ultraviolet, say not much more than the 4.9 volt, 2537 Å radiation from the mercury arc. Under such liberation, while φe may be somewhat less, electrons are emitted with average energies of the order of a fraction, (0.6), of an electron volt. Thus $\tfrac{1}{2} m v^2$ is very slight. In the presence of a gas, the photo efficiency of emission is much reduced, e.g., by a factor of the order of 100 at 760 mm in low extracting fields. Much more effective as far as emitted currents of electrons per cm² of surface go is the *emission of electricity from hot bodies*. Here if bodies are heated up to incandescence quite large current densities of electrons can be expected. These carriers are liberated at relatively low velocities corresponding to perhaps 0.1 ev or less. The densities of emission current are such as to yield space charge limited currents, which very rarely happens for photo emission. Here again, the action of the gas in causing a loss due to back diffusion is manifest.

If now such electrons are emitted from a grounded electrode and there is somewhere in the gas another insulated electrode, electrons or ions created by attachment of electrons to molecules, will diffuse across the gap and charge the

opposing electrode negatively until they set up an apposing field just counteracting the energy of emission or of thermal agitation in the gas. Thus electrostatic
charging from such sources occurs but the potentials acquired are very low indeed.

If now the gas be set in motion, the gas stream can carry the electrons or ions
from the grounded emitting surface up to the isolated collecting electrode down
stream from the emitter. This situation is a common one to all hot furnaces or
sources where the flue or exhaust gases carry excess electrons from the highly
emitting wall as well as some positive and negative ions from the combustion and
deposit them on the walls of the smokestacks, flue, or isolated surface. Here then
the charge acquired by the electrically isolated surface, metal or insulator, will
be greater than that owing to thermal energy of emission. However, if only
electrons and negative ions are carried by the gas, the potential reached by the
electrode relative to the emitting surface will only be such that the potential V
achieved by metal collector or stack will create a field X such that the drift
velocity v of carriers in X is equal to the velocity of the gas stream u.

To carry out this analysis assume the isolated surface to be a metal sphere
of radius r inserted into the gas flow of velocity u with some of the space charge
cloud emitted by the hot surface across wich the gas blows.

The capacity of the sphere will be $C = r$ cm in esu. The surface field created
by the charge is $X = 4\pi\sigma$ with σ the charge density. The total charge is $q = 4\pi r^2\sigma$ and the potential is $V = q/r$. Thus $\sigma = X/4\pi$, $q = X r^2$, $V = X r$ and $v = X k$,
with k the carrier mobility. The sphere will charge until $v = u$. Thus the potential
of the isolated spheres relative to the earthed carrier source is $V = \dfrac{u\,r}{k}$.

The units here are electrostatic. If $r = 10$ cm, $u = 10$ cm/sec, $k = 500$ cm^2/esu
sec as for normal gaseous ions at NTP then $V = 0.16$ esu or 48 volts. For electrons
with drift velocities much higher, e.g., of the order of 6×10^5 or more, V would be
small indeed. Thus static charging would not appear to be a serious problem.

Assume now that the gas has solid particles of ash, say of the order of 0.2 micron diameter, 2×10^{-5} cm, a common size for such particles yielding a bluish haze
in stack gases. These will pick up electrons or ions and will charge by diffusion*.
If the particles are spheres *of a single charge* each, such particles by STOKES' law
have a value of $k = 1.4 \times 10^{-2}$ cm^2/esu sec $= 4.7 \times 10^{-5}$ cm^2/volt esc. In this case
$V = 100/0.014 = 7.2 \times 10^3$ esu $= 2160000$ volts which is a dangerous potential*.

Larger particles will have lower velocities and the charges can be higher still.
The velocity u of the flue gases entering into V will also influence charge accumulation.

It is thus quite clear that any source of unipolar carriers liberated from a surface can, in the presence of even relatively slow gas currents, serve to charge
isolated surfaces in the gas stream to dangerously high potentials if there are solid
particles that can convey the charge in a mechanical fashion so that the motion
of the gas stream exceeds the drift. Since the flue gases are somewhat conducting
by having positive and negative ions present as well, the isolated body will tend
to discharge by gaseous conduction. The charge acquired will depend on the

---

* Actually the charging of dust particles and spheres of all sizes in the presence of ions of
one sign in fields the reader is referred to R. LADENBURG: Ann. der Physik **4**, 863 (1930)
and to M. M. PAUTHENIER and M. MOREAU-HANOT: Jour. de Phys. et de Rad. **3**, 591 (1932).

relative rate of acquisition of charge by the sphere from charged dust particles and loss by conductivity.

Since the ions have mobilities greater than the negative dust particles if these ions are plentiful unipolar charging will not occur. Thus in the rather hot exhaust or flue gases unless excess electron emission and charging of dust is very effective relative to ion density in the gas stream the charging will be small. However, as the flue gases proceed some distance and cool the negatively charged dust particles will persist while the faster positive ions will recombine with negative ions or diffuse to the stack walls. Thus the cooled stack or flue can and does create notable charge concentrations.

Even if there are equal numbers of ions and electrons created in a combustion process, without excess thermionic emission from the surfaces and walls, the dust particles in general will have a negative charge on the average owing to the higher diffusion velocity of electrons. Thus if these particles are removed rapidly enough, before they recombine with positive ions from the gas they can impart a negative charge. For effective charging, however, preponderance of electrons from thermionic emission from solid surfaces is required plus sufficient cooling time for the normal negative and positive ions from the flame to recombine.

It may be remarked in passing that accumulation of charge from such gases is facilitated by the fact that the gases are warm and dry aiding insulation and isolation of surfaces.

It may also be noted that creation of excess carriers of one sign need not be confined to thermionic *electron* emission. For certain heated salts emit negative or positive ions at temperatures usually below that for thermionic emission. Other substances emit copious positive ions such as the pure Fe and Ni alloys with a few percent of alkali atoms such as Na or K added. These are the Kunsman catalysts and emit really good current densities of Na^+ and K^+ ions at 800 to 1100° C.

Considerable study of such flue gases have been made in relation to electrostatic precipitation of dusts by the industries engaged in electrical precipitation on both in America and Europe.

Among those groups that have engaged in extensive research in this area are the Western Precipitation Company and the Research Laboratory of the Research Corporation at Bound Brook, New Jesery. In this field, the dusts to be removed are reduced to the proper electrical conductivity and deliberately charged by negative wire corona discharge and then removed by transverse electrostatic fields.

In general, static charging of exhaust gases from flues and stacks create no great problems of electrostatic hazard as the flue or exhaust gases are non-explosive and in general the possible collecting surfaces exposed to the gases are normally grounded. On the other hand, GUNN[13] reports that airplane exhaust gases can charge the plane at a rate of several microamperes of current under proper combustion conditions. These depend on speed, gas mixture and engine type and apparently depend on creation of solid particles in the exhaust.

As regards the general character of the conductivity in flames little progress has been made in recent years, except in the studies of specific chemical reactions.

A summary of data as of 1946 was given by the author[124]. In the flame itself as distinct from any wall ionization results from thermal processes and through chemical reactions. Depending on flame temperatures, ionization is in part equilibrium, in part owing to special processes, and in part thermionic emission from solid particles. The plasmas are fairly dense so that probing electrodes become at once surrounded with plasma sheath. Drift velocities cannot be measured and used to identify carriers since the fields across the gas are nullified by sheath formation. Electrons are present and may remain free indefinitely. Certainly there will be some attachment to form negative ions in the cooler portions. Diffusion will dpend on electron and ion density. If this is less than 10^8 per cm³ then free electron diffusion can be expected and surfaces touched by the flame and not rendered thermionically emitting by remaining cool will collect negative charge. The gas loses electrons fairly freely and the gas stream will acquire a significant positive charge. If densities approach 10^{10} per cm³, the diffusion is ambipolar and the walls in contact with the flame will achieve a low negative charge depending on positive ion mobility. Here potentials in general will be low and the gas plasma carried along by the flame velocity will remain essentially neutral.

2. The unipolar corona discharges

A highly electrically stressed wire or metal point opposite a plane or cylinder will readily cause a local corona type of breakdown at that conductor. In these cases, the fields adequate for breakdown occur very near the wire or point. Thus the carriers of sign opposite to the charge on the point, or wire, are quickly drawn into the point, or wire, projecting charges of the same sign as the point or wire into the surrounding gas. Owing to the momentum given the gas ions near the point by the fields and to the lower outer field regions, the ions created in a gas, or air, by the wire or point will proceed far from the discharge region. Under these conditions a gas stream or insulated belt moving transversely to the gas stream can receive the excess charge and mechanically transport it to a collector. In fact, in corona discharges, at atmospheric pressures, there is an appreciable electric wind conveying the carriers away from the point with it. The energy available in conveyor belt or material particles in the transverse gas stream receiving the ions can then raise an appropriate insulated receiver to any desired potential. In fact, potential will only be limited by the breakdown potential of the insulation provided. The corona sprayer acting on a plastic belt moving normal to its field constitutes the charging mechanism of the van de Graaf static generator which has been operated up to four million volts in a pressurized tank.

The basic physics of corona discharges in air have extensively been studied by the author and his school for over 20 years[125]. The corona current varies with potential and geometry over a wide range of values. For the small point to plane corona in air at NTP, e.g., a point of 0.10 cm diameter corona sets in for potentials of the order of 8 Kv at about 0.1 to 1 microampere and goes up to some 100 microamperes before breakdown. For some 5 cm length of central electrode wire of diameter 0.015 cm coaxial with a cylinder of 2.9 cm diameter corona sets in at NTP in air around 5 Kv with 1 microampere of current for the positive wire and at 5.5 Kv for the negative wire with 2 microamperes. Currents extend on up to 100 microamperes before sparking. At threshold the positive corona is noisy and intermittent, goes over to a steady glow at somewhat higher potential and becomes very irregular and electrically noisy when pre-breakdown streamers set in before sparking. The negative corona in air is regularly interrupted shortly above threshold. Pulses last some 10^{-7} to 10^{-8} sec. Repeat rate increases with potential from ~ 1000 near threshold to $\sim 10^6$ before spark breakdown. The pulses are exceedingly electrically noisy. The details of the processes and references will be found in the authors's article in the *Encyclopedia of Physics*, Vol. XXII[126].

The corona discharges constitute one of the most useful as well as one of the most amazing and destructive charging processes known. Their ability to deliver considerable quantities of segregated charge of one sign into the air and plaster it on to surfaces in suspension or solid surfaces makes it most susceptible to exploitation for electrostatic generation and charging.

It is useful in electrostatic charging as in the use of charged plates in the printing and photographic process known as Xerography and in high speed writing or printing of messages. It is the basis of the charging process in the best controlled high potential electrostatic generators. It can, as will be seen, be used in the elimination of undesirable electrostatic charging and is so used.

It occurs around high tension systems especially those coated by dust and salt spray ending in discharge over the insulator strings by creating leakage paths. It is a constant nuisance around all sorts of high tension systems requiring insulating, shielding, and increase in radius of conductors. It presents a very serious hazard in aircraft in flight when charging by precipitation static produces corona discharge from sharply curved surfaces which is so noisy that it masks all radio signals and has resulted in numerous plane losses in the past.

3. The effect of plasmas from arcs and glows in static generation

The glow type of discharge is usually surrounded by a glass envelope so that it cannot contribute ionization to the surroundings. The open air arc consists of a plasma channel which is self contained. The plasma has a very high charge carrier density. Thus ambipolar diffusion constitutes the charge transfer if any, through the sheath. There are rather strong vertical convection currents set up about the arc channel in view of the high temperatures of channel and electrodes. Whether these convey any appreciable charge away from the arc is doubtful. The electrodes vaporizing carbon, or metal, or sputtering the same could transfer charges to solids placed in the convection stream. The high current density carbon arcs also have jet flames emerging from the anode which could carry upward considerable charge segregation on condensed carbon or metal particles as they cool. However, in regions sufficiently close to such arc processes to acquire static charges of any magnitude, the conditions are not conducive to the use of isolated systems that charge. Thus these discharges do not play any role as potential static charging mechanisms. That unipolar charge segregation from such are plasmas is difficult is attested by the extensive researches carried on to use these as ion sources of various sorts, with indifferent success.

4. Induction charging in auxiliary fields

The use of an auxiliary field to charge conducting particles by induction as they separate from a grounded nozzle to accumulate charge at high potential by allowing the charged particles to fall into an insulated collector has long been known. It was first used by Lord KELVIN in his "water dropper" static machine. At a later date, a similar device using ball bearings on an endless conveyor belt to restore them to the dropper was tried by W. F. G. SWANN, in order to achieve high potential. The device worked well, but being more complex than the van de Graaf principle, which developed simultaneously, it was not further developed.

This general charging mechanism can be active unknown to operators in numerous industrial processes as stray fields are always present and can lead to electrostatic charging attributed to other mechanisms. It is effective and powerful and must be kept in mind.

C. Static elimination

1. Introductory

Possibly in closing, a few remarks could be made on static prevention. The Problem is a vast one and one which usually requires a separate or unique solution for each set of circumstances to be met. Some general statements can, however, be made.

In the first instance, large accumulations, with high potentials leading to dangerous sparks require metal accumulators. Further metal forms such a large part in moving machinery that one of the accumulating surfaces if not both will be metallic. Thus all metal surfaces at which generation may be going on, or on which charges may accumulate should be bonded together or properly grounded. In transferring gasoline from truck to tank, or barrel to tank all metal parts exposed to contact with gasoline or its sprays should be grounded. This is point number one. Automobiles, gas trucks and all wheeled vehicles with newer tire fabrics are becoming well insulated from ground. Charging on all but humid days is very high—up to 10^5 volts with large capacity. How successful the dragging of chains on the ground is in discharging the body is not certain. At any rate, it can do no harm and may do good. D. BULGIN[126] indicates the solution to be use of tires rendered conducting by some additive, with resistance $\sim 10^9$ ohms.

The real problems occur, where, while one side of the charge generating system may be metal and grounded the other part of the system is a plastic, a fabric, or a non conductor of some sort. Here the plastic can gain very heavy charges relative to metal or other parts. This can lead to sparks, to undesirable adhesion, repulsion of fibres, accumulation of dusts by electrostatic action, occasional dangerous shocks, possibility of ignition of inflammable vapors. Solutions must be looked for in several directions.

2. Reduction of charging by increasing conductivity of the insulators

This increased conductivity can be brought about by humidity, or by humidity in connection with conductivity producing agencies. This problem is discussed by J. A. MEDLEY[127]. In this instance, measurements were made on charging of wool fibres and filter paper passing over steel rollers. Tested were effects of relative humidity, of KCl solution up to 0.3 N yielding volume conductivity and surface agents, e.g. Aerosol OT in standard wool combing oil.

a. The volume Conductivity. The charging rate of an inhomogeneous material of conductivity k moving at a speed v transfers charge at a net rate I on metal roller contact. The charge density transferred is σ and X is the field responsible for conduction corresponding to the charge density. The value of σ depends on pressure on the rollers, the type of material, etc. Then for a sheet of thickness y

$$I_1 = v\sigma + ykX \tag{4.31}$$

and for a cylindrical fibre, (roving),

$$I_2 = 2\pi\sigma a v + \pi a^2 k X \tag{4.32}$$

where a is the fibre radius. These reduce to an empirical expression

$$I = v(\sigma + KX). \tag{4.33}$$

Theoretical maximum charging for a sheet rolled on one side is 10 esu/cm² as limited by *electrostatic back* discharge in air. In practice a charge of about 4.5 esu per cm² was achieved as a result of uneven charging. The material passing parallel to a plane sheet of metal 5 cm below it and of length 25 cm parallel to the sheet had its charge reduced to 3 esu by directing wool fibres more parallel to the roller surface and loss by leakage along the fibres. The constant K could be evaluated and compared for various measured conductivities of the wool for various agents used. The value of $K = y\left(\dfrac{k}{v}\right)$ for sheet and $K = \left(\dfrac{ak}{2v}\right)$ for cylindrical roving. To derive the exact value of K from a given degree of dissipation requires a series of laborious computations of σ and X at all points of the material. For rough computation assume that σ_A is the value of charge density at a point A beyond the rollers as limited by gaseous back discharge. The field X_A will be $4\pi\sigma_A \cos\theta_A$, where θ_A is the inclination of lines of force to the direction of motion. θ_A will be in excess of 90° because the lines of force strive to be normal to the cylinder as the fabric moves away. Then

$$I_A = v\sigma_A(1 + K\cos\theta). \tag{4.34}$$

For zero dissipation $I_A = v\sigma_A$ and for 50% dissipation

$$K_{\text{crit}} = -\frac{1}{8\pi\cos\theta}. \tag{4.35}$$

If θ_A lies between 95° and 120° K_{crit} must lie between 0.4 and 0.1. The observed values lie within this region. This is satisfactory agreement since the practical range of K values covers several powers of ten.

In studies using wetting agents for the surface these were applied either as a uniform thin film to the roller surfaces, (around 2×10^{-3} cm thick), or by wetting the fabric surface with it. This was accomplished through a conducting agency dissolved in the natural fat coming from the wool. These oiled rovings, or threads, *yielded a K less than 0.01*, even when there was a complete dissipation of charge, indicating that this was *not a body* conductivity. In this case with rollers oiled and rovings passing over them the dissipation was still good. Thus there was some relation between bulk conductivity of *the surface agency* and *dissipation*.

b. Surface conductivity. There is an alternative mechanism in dissipation; namely that the surface film acts like a leaky condenser for a short interval of time t after fibre to metal contact is broken*. Let D be the dielectric constant of the agent and ϱ its specific resistivity. The reduction in charge from q_0 to q by leakage from a condenser requires a time t given by $q/q_0 = e^{-t/RC}$, with R the resistance and C the capacity. The resistance will represent the specific resistance $\dfrac{\varrho\, d}{A} = R$ where A is the area of contact and d is the length of the resistance path. The capacity $C = \dfrac{A D}{4\pi d}$ where d is the plate separation which is the resistance path also. Thus

$$R C = \frac{AD}{4\pi d}\,\frac{\varrho\, d}{A} = \frac{D}{4\pi}\,\varrho. \tag{4.36}$$

Accordingly

$$\frac{q}{q_0} = e^{-n} = e^{-\frac{4\pi t}{D\varrho}}, \tag{4.37}$$

where $n = 4.6$ for $\dfrac{q}{q_0} = 100$ and 1 for $\dfrac{q}{q_0} = 0.37$. Thus $t = \dfrac{D}{4\pi}\,\varrho\, n$ in order to reduce the charge by a factor e^{-n}. Now the contacts are broken near the position of minimum clearance l between rollers. For 2 rollers of radius r each the time t is governed by the tangential movement of the fibres from them. Thus it is proper to set the distance of separation vt by means of the proportion, $\dfrac{vt}{2r} = \dfrac{l}{vt}$ such that $\dfrac{v^2 t^2}{2r} = l$ and $t = \sqrt{\dfrac{2r l}{v}}$. In consequence to reduce the charge by a factor

$$e^{-n},\ n \ge \frac{4\pi}{D\varrho v}\sqrt{2r l}. \tag{4.38}$$

If $n \sim 1$ electrification should appear and accumulate. Thus the condition

$$1 = \frac{4\pi}{D\varrho v}\sqrt{2r l} \tag{4.39}$$

is the rough condition for emergence of charging and is open to direct test.

The measurements confirm this if surface tension effects cause l to be somewhat greater than the thickness of the conductive agent when distributed uniformly over the rollers of radius r, or the 2.0×10^{-3} cm wool fibres. It is interesting to note for the volume process that charging fell sharply to zero when length conductance approached $\sim 10^{-11}$ ohm^{-1} cm and that with the oil the sharp decline in charge occurred when bulk conductivity of the oil was $\sim 5 \times 10^{-9}$ ohm^{-1} cm^{-1}.

Thus even in the mechanism of the discharge of static from wool one notes two different discharging processes one depending on bulk conductivity and one on the conductivity of surface films to be at work under different conditions.

c. Agencies aiding; Surface or volume conductivity. As P. S. H. HENRY[121] states in his study of charge dissipation the conductivity may not be adequate

---

* This action of the surface film reducing conductivity in consequence of back leakage is not unique to this type of charge separation. It has been discussed in cases of electrolytic type of electrification in the Helmholtz theory and in connection with other processes. In this instance, however, it is amenable to a different type of treatment.

in many charging processes. It is further pointed out that high humidity assists such processes. There are however conditions where this does not work for moisture may have to be excluded or else conducting agencies cannot be used. Agencies used in increasing surface and/or volume conductivity are listed as follows by J. C. FORREST[128].

Use of semi-conducting ceramic glazes. These glazes contain Fe_2O_3, ZnO, Cr_2O_3 etc. Stannic oxide films and stannous chloride films that are transparent and fairly stable with resistances of the order of a hundered ohms per cm² or less can readily be prepared.

Thin metallic films can readily be evaporated onto surfaces but do not adhere too well on usage.

Colloidal carbon in aqueous or other media has resistances from 10^3 to 10^6 ohms. Certain carbon containing paints and enamels are conducting. Carbon films can be deposited on surfaces at high temperature from gaseous hydro-carbon with resistivities from 100 to 10^4 ohm and considerable durability. Conducting rubber loaded with carbon black has a volume resistivity from 2.5 to 10^8 ohm cm.

Moisture films; Clean dry porcelain at 50% relative humidity has a resistivity of 100 000 $M\Omega$. When humidity exceeds 70% the decrease becomes marked and at 90% it is 1000 $M\Omega$. If surfaces are polluted by exposure to the atmosphere much lower values appear at higher humidity owing to the deposit of thin layers of salts.

Films of wetting agents can be used on normally water repellent surfaces. Thus "perspex" retained its charge of 3×10^{-9} coulomb/cm² for hours. When treated with I.C.I. Lissapol N and allowed to stand for 24 hours, its charge decayed in the order of a few seconds. This property lasted for several weeks. Shell Teepol 530 had a time constant of a minute for discharge of the same surface. These films may act in two ways as leakage agents and in reducing charging. These materials also require high humidity.

3. The use of electrostatic dischargers

a. By self discharge. Electrostatic discharge in air or ambient gas may be employed to cause the material to discharge itself to unobjectionable values. This mechanism has been used in connection with steel rollers in the wool industry to reduce charge in the fibres by J. A. MEDLEY[127]. The sheet or fibre material passed through a pair of steel rollers insulated but grounded. Roller pressure could be varied to change the charging rate. Various conductors were used near the material; $a \neq 18$ S.W.G. copper wire 2 cm from the point of contact and 5 mm below it, a plate 25 cm long parallel to the direction of motion 5 cm below the fibre, a small loop 4 mm diameter of No. 36 stainless steel surrounding the wool cord placed at 2 cm beyond the point of contact of the rollers. Currents to all three were measured. The studies were made at 24° C with 40% relative humidity. Various substances were tried, woolen, taffeta cloth and Whatman No. 4 filter paper. Materials were initially dried and allowed to come to equilibrium in the room. Careful cleaning was resorted to. Successive soakings in alcohol, Lissapol N solution and rinsing in running and finally distilled water gave reproducible results.

The copper rod dissipated more charge than the plane parallel electrode
The use of the loop practically completely discharged the roving cord, the charg
being less than 0.2 esu/cm^2, despite some small contact between steel roving an
loop. In these cases the very highly electrified fine wool fibres yielded coron
discharge, thus reducing the charge to low values.

The method does not however, remove charge completely, though it can b
much more effective if the gas pressure is reduced to a few mm. Then the poten
tial of the surfaces will be reduced to ~400 volts. If the surfaces are smooth
the use of a conductor with many fine points of platinized whiskers near the surfac
might help a great especially at lower gas pressures.

b. **Use of floating insulated systems.** If the rollers used by MEDLEY had beei
highly insulated, they would have at first yielded high charge, but as the material
continued to run over them, charge would have been low. Working with tape
of camera film. J. A. WELLS, in discussing MEDLEY's paper, indicated that metal
or even perspex rollers, charge to such a degree that no more charge can be trans
ferred from the roller to the film and after some two or three feet of film hac
rolled it thereafter remained uncharged. If the insulation on the rollers hac
a value of 10^{12} ohm, there was no charging and if it was 5×10^{11} ohm charging
was only 10% of that with a grounded roller. In criticism of this techniques a:
applied to the textile industry MEDLEY remarked that despite the lack of charge
the residual attraction between material and machinery remains, i.e., the fabric
is still bound to the roller indirectly. He also indicated the difficulty of maintain
ing insulation to a resistance of 10^{12} ohms in operating industrial machinery
Finally, he indicated that the charge accumulations in such large insulated system:
are high leading to dangerous shock hazards.

c. **Ions from electrostatic sources.** The surfaces may be discharged by one
of two types of devices using ionization in air produced by an electrical discharge
The most effective production of ions utilizes the ionization of one sign createc
by corona discharge from highly stressed wires, or points. The ionization createc
by a wire or a point with steady potential is of the sign of the charge on the point
Currents of the order of 10^{-5} to 10^{-4} amperes can be expected from points undei
the more desirable operating conditions. Charge generation to be dissipatec
varies over a wide range of values under different conditions. However, a high
charging rate would run to ~5×10^{-5} ampere for a m wide surface moving
1 m/sec. Thus a corona discharge generator with 10 points or a single wire ol
10^{-2} cm diameter 1 m long operating enough above threshold to give a continuous current, e.g., $10 \mu a$ per point placed a few cm above the surface would
neutralize the charge. The potential needed depends on current desired, point
radius and in a minor degree on air gap from points to surface. It will run in the
neighborhood of from 8000—12000 volts. By using a grid of uniform parallel
wires over and normal to the direction of motion of a moving surface *quite* uniform
charge distribution can be achieved. The points can operate on alternating potential, or by steady potential. There are two basic mechanisms to consider.

α) *By ions of both signs.* Discharge by creation of ions of both signs. This is
accomplished by the use of alternating potential, or as indicated by HENRY[128]
by a series of two sets of parallel points or wires of opposite polarity separated
by a plastic shield discharging to the surface.

The value of the bipolar deionizers lies in the fact that they will eliminate charging of either sign and reduce it to near zero. In many industrial processes, the conditions are not sufficiently controlled so that continuity of charge and sign of charge on supposedly the same surfaces remain constant or invariable. The use of a neutral plasma insures neutrality or near neutrality if enough ions and time of exposure remain. The bipolar alternating or steady current suffers from some charge dissipation—especially over larger distances from corona to surface through recombination of ions. Thus larger currents may be required.

β. *By monitored ions of one sign.* The other approach assuming uniformity of charge sign uses an elimination of one sign, negative wire corona for a positive surface and vice versa. Here the problem arises of so regulating the neutralizing current as to eliminate the charge without perhaps over or undercharging it. This situation can be controlled by using a probe exposed to the charged surface as it leaves the rollers which can monitor the discharge potential and current to yield a neutral surface. Possibly better still would be a probe to sense and to measure the residual charge on the surface. Drawbacks to the electrostatic schemes are as follows: They may be awkward or inconvenient to install in the proper place. The high tension presents a hazard to operating personnel. Relative to shock hazard some relief may be had. Commercially the TAKK Static Eliminator and another developed by the Chapman Electrical Neutralizer Co.[128] and the Simco Co. have a row of points capacitatively coupled to a highly insulated bus bar served by a transformer giving 8000 and 12000 volts. The capacitative coupling ensures that currents drawn if points are touched or shorted are not dangerous. The sparks drawn from one device are supposed to be too weak to ignite inflammable mixtures. Another English design by I. G. McDonald and J. B. Todd[128] uses two rows of points of opposite polarity with opposite D.C. potentials. It is designated as the "Shirley" Electric Static Eliminator. There is one other drawback which appears usually not to be mentioned. All electrical discharges in air produce ozone and nitric oxide, two *very irritating gases*, which even in low concentrations if inhaled for long periods, produce chronic respiratory irritation. Several means of eliminating such actions are possible such as supplying the eliminators with a constant slow flow of a chemically inert gas such as N_2 or He in which the discharge takes place or else providing adequate ventilation of the spaces where these are used.

d. The use of radioactive ionizing radiations. The use of radioactive materials for such purposes on an industrial scale has been made possible as a result of the production of bi-products of the uranium purification process and the reactor piles. Under some conditions, the radiations from purified Po devoid of RaD and other γ ray emittors can be safe. Here *monitoring* of the *discharging current* in too great a measure is *not* possible. Again great care must be used in protecting the atmosphere and personnel from the recoil radiation contamination if α particles are used and from γ rays if other radiations are used. To obtain suitable currents, strong sources are needed and this increases the hazards. Rigid State or Government inspection should be required where used.

e. The problems of elimination in dust and spray electrification. The third and perhaps most serious situation that has to be met in static generation is the one encountered in industry where the processing of various solids or liquids in dispersed form in which the dust, spray or cloud is highly charged and mixed with air, becomes an explosive hazard. Thus the handling of powdered sugar, flowers of sulfur and dispersal of ether, gasoline, crude oil coal dust, flour, Al or Mg powder etc. present serious hazards. Probably the real dangers lie in the nature of the explosive dust-air or ether-air mixtures. The static spark detonating such a mixture can be a pretty small and localized affair. In operating room explosions, it presumably comes from the manipulation of dry blankets and plastic sheets around the metal accessories on rubber tired vehicles. In this case the spray electrification of the ether is not the cause. Trouble can be avoided by use of conducting rubber, avoidance of plastics and other substances that create static and maintenance of a reasonably moist atmosphere.

In sulfur mills, sugar mills, in coal mines and storage bins and granaries, aluminium and magnesium plants, the situation is more acute. In certain hazardous operating bins static accumulation and sparks may be rendered innocuous by using CO_2, flue gases or N_2 gas which is non-explosive if sparks occur. In more voluminous or open spaces this is not possible. The grounding of all metal accumulating surfaces helps. Charged dusts can, however, accumulate on insulated surfaces. Care should be taken to avoid or prevent such local accumulations. No one knows in many instances how the igniting spark in the many fatal explosions took place.

Whether the *clouds of dust themselves ever build up potentials* such as to lead to sparks between oppositely charged clouds or to surfaces is debatable. That lightning strokes do occur between charged water clouds is true. But in the first instance, charge accumulations are very high in virtue of the high turbulence and more massive particles and to secondary mechanisms related to the ice phase and secondly the sparks occur at much lower fields in the presence of water drops than in the presence of insulating coal or grain dust. In this connection, metal dusts such as of Al or Mg would be among the most dangerous because of the ease of corona discharge from small points at more moderate potentials. Such dusts should always be handled in an oxygen free atmosphere.

Much hazard would be avoided in granaries and in coal storage where dust hazards exist if all the surfaces were rendered quasi-conducting and grounded through use of some cheap conducting surface that can be sprayed over all walls. The situation in coal mines is probably the most difficult to handle. However, opinion appears to ascribe most explosions and fires in mines to marsh gas and inadequate ventilation rather than dust explosion. Knowing the sources of hazard with stainless steel at hand, there should be little hazard in processing sugar by handling it in all metal containers, duly grounded.

f. Precipitation static in aircraft. One of the most troublesome and serious charging hazards is the problem of precipitation static in aircraft[13]. Considerable money and time have been spent on this problem by competent scientists without arriving at a solution. Aircraft in flight in dust, but more so in rain or hail, and especially in snow, acquire exceedingly high charges. Currents may run to 500 microamperes. Plane charges vary, being usually negative, but depend on surface and are capricious*. The charges become so great, that corona breaks

---

* In the absence of charge on planes, the exhaust gases cause charging depending on operating conditions notably connected with soot or solid particles ejected, see p. 205.

out all over the plane at points of high curvature, e.g., propeller blades, antenna, wing edges, etc. Corona currents, positive or negative, in air, when pretty well above thresholds, are electrically noisy. This is caused by the intermittent discharge in positive streamers and in negative Trichel pulses in which di/dt is very high. These produce so much static noise that communications and beam signals are wiped out—especially if discharges are adjacent to antenna. The sign of the charge on the plane as well as magnitude of charge vary over such wide limits that whatever device is used must be of such nature as to meet all conditions. The sign of charge on a plane, for example, has been known to change as a result of servicing and the wiping of the plane surfaces with greasy rags. It was found possible to discharge the plane by letting water run out of its trailing edges. The amount of water needed was prohibitively great. For slow charging rates, the charge can be kept down, using fine metal coated fibres. These fibres each yield a corona, but the current is so small and there are so many fibres, that the noise is low. However, such fibres fail to deliver the current when heavy charging occurs. Some discharging, around 12%, occurs from engine exhausts at highest fields*. Two independent workers[130] have indicated that larger corona points charged by small, high tension generators within the plane, discharging to a ring or cylinder, and placed in the slip stream of the plane near or on the tail structure with polarity appropriate to the charge on the plane can be caused to discharge the plane without serious noise at the antenna owing to the shielding and distance from the antenna. Owing to scientific politics, neither of these devices have been given a chance for a fair trial on planes in flight, even though in wind tunnels their performance has proven adequate. The device must at all costs, be given a fair trial since there appears to be no other solution and planes are still crashing owing to precipitation static.

D. Thunderstorm electrification[131]

In discussing static generation and prevention perhaps in view of what has been said earlier one might summarize the possible mechanisms active in a turbulent thunderstorm. Extensive meteorological studies of the thunderstorm indicate several facts about generation of static leading to lightning in thunder clouds.

1. Accumulation of charge and generation of lightning strokes are dependent on the formation of precipitation.

2. In all active thunderstorms, little charge is generated until precipitation appears above the freezing isotherm.

3. No generation occurs unless there is much turbulence and convection. The more violent the convection and turbulence, the more the electrification.

4. In the great majority of electrical storms studied, the main or negative charge of a cell resides in a region above some 2—3 km and extends to some 9 km. The base area of the cell is of the orders km in diameter. This charge resides on rain, or graupel pellets that will become rain.

5. Above the negative charge is the high domelike thunderhead going up to 12 km with positive charge largely on fine sublimed snow or ice crystals.

---

* See the Footnote p. 214.

6. In many storms, there is, or is unmasked near the throat of the advancing cloud, below the freezing isotherm a small but intense positive charge on larger raindrops.

a. Capture or fission theories. While theories dealing with the capture of ions moving in the earth's potential gradient can yield the right sign of charge segregation in clouds, turbulence is not needed and in fact disturbs the mechanism and the process is ineffective unless there are many more ions and fields are higher than with the normal earth's gradient. In a like measure mechanisms resulting from fusion or fission of cloud droplets in the field are again nonturbulent mechanisms and require initial high fields.

b. Spray charging. It is possible, but not certain, that the lower positive charge below the freezing isotherm could be generated by very turbulent wind action on large drops, by the Lenard spray mechanism carrying fine negative spray up to the main cell and leaving larger positive drops behind. The positive charge would also be accounted for by the process of unmasking recently disclosed by J. KUETTNER[132]. The spray mechanism can yield sufficient charge for such action as indicated by CHAPMAN's study in turbulent drafts.

c. Ice-ice impact charging. The largest separation takes place by the impact of the colder finer sublimed crystals of ice against the warmer heavier graupel pellets in the upper portions of the cloud cell above the freezing isotherm. In these impacts the colder small crystals of nearly pure ice carry upward positive charge on regelation of the melted impact liquid from the graupel pellet. The latter descends to the main cell with negative charge while the small ice crystal is carried to the dome by updrafts.

d. Separation of charge through freezing potentials. At and about the freezing isotherm impact between super cooled graupel pellets, or hailstones and raindrops results in freezing potentials with the negative rain spray being carried aloft and the colder pellet descending with positive charge.

With the three mechanisms b, c, and d outlined above active and with data on the magnitude of the charge separation by these processes in the laboratory together with knowledge of the amount and rate of charge generation actually occurring, it is possible to account for the phenomena observed to date quantitatively.

E. Protection against lightning stroke

In conclusion possibly a few words might be said about the lightning stroke hazard and how to reduce it.

While the details of the cloud to ground stroke mechanism are not completely known there is sufficient knowledge so that some points can be made.

It is probable that the stroke initiates from the cloud at such heights and under such conditions that the underlying terrain has in general very little influence on the point of impact[130, 131]. Presumably turbulence brings the small positive scud cloud charge at the cell base into sudden proximity with the large negatively charged cell base such that fields in excess of 10000 volts/cm are created. These fields shatter the raindrops drawing them into spindle shaped corona points. These points undergo corona discharge and disperse positive charge upward by streamer corona, leaving negative charge behind. Presumably at the point of highest field distortion perhaps a group of streamers develop into a spark channel which by the upward spread of the rapidly propagating streamers lowers negative charge into the positive cloud.

From the earth end of this distortion and its conducting channel there emerges a negatively charged pilot leader tip feeling its way earthward in view of the field distortion ahead of it. After this leader has advanced a certain distance some, (10—30 m), it has spent its energy and slows down. The inductive inertia of the discharge behind and the high fields plus the burst of ionization at the *upper end* of the pilot leader channel builds up a gradient which eventually sends a wave of ionization and luminosity coursing down the spent pilot channel at some 2×10^8 cm/sec. Arrived at the tip of the pilot leader, the tip is reenergized and resumes its advance. The pilot leader is itself invisible, but the brilliantflash of the potential wave that revives it can in some cases be photographed.

The pilot leader in this stepwise fashion forges toward the earth creating field distortion ahead of it. The width of the transient stepped luminosity is some 10 m while the intensely ionized core of the pilot leader channel may only be some 20 cm in diameter. The advance of the pilot leader is erratic presumably depending on local field distortion and chance ionizing events. In vigorous strokes the pilot leader will *branch downwards* and laterally leading to the well known lightning forks. As the leader and/or its branches approach the ground, the electrical field distortion ahead makes itself felt. The downward velocity of the tip is of the order of 10^7 cm/sec for the pilot leader, while the steps advance down the pilot leader at some 10^8 cm/sec. The time of advance from cloud to ground in a 3 km stroke to ground is thus 0.03 sec. The action of the tip field as it approaches the ground is to induce a highly localized positive charge. Depending on ground contour, conductivity, etc., the distortion may become so great within the last 10^{-4} sec of reaching the ground, that one, or more, upward positive streamers may leave the ground attempting to complete the channel. These are usually not observed, but one photograph of a stroke to a beach, or to ocean near the beach, is on record with two unsuccessful streamers and one successful one. The upward advance of such a streamer before junction is of the order of 10 m or less under usual circumstances. It is much greater if a well grounded and highly conducting body projects upward from the earth. Thus positive streamers were the rule from McEACHRON's antenna at the top of the Empire State Building. In fact, these streamers *invited strokes* from clouds which would otherwise not have discharged in the area. The junction of pilot leader and streamer, or ground, produces a very steep potential distortion, which propagates upward as a wave of luminosity and conductivity from the ground towards the cloud. With the preexisting density of ionization of $\sim 10^{13}$ ions/cm^3 and less in the stepped leader channel, this distortion progresses ionizing and exciting as it goes multiplying the carriers by a factor of 10^4. The speed of advance of this steep luminous potential front ranges from the order of 10^{10} cm/sec down to 5×10^8 cm/sec. This *return stroke* and its *subsequent current flow* constitutes the brilliant flash of the lightning stroke and makes the path of the 20 cm diameter core of the pilot leader and some of its branches highly conducting. Temperatures achieved in the channel may reach 20000° K instantaneously for $\sim 10^{-6}$ sec and the molecules are all completely ionized to the full gas density ion densities reaching 10^{17} per cm^3. The current to ground lasts a relatively short time ranging from 10^3 to 10^5, (average 3×10^4), amp for perhaps 10^{-4} to 10^{-3} sec. Then the channel decays in ionization and may sweep on with the surface wind until it is revitalized by a dart leader, *unstepped*, advancing $\sim 2-5 \times 10^8$ cm/sec down the old channel, discharging the next ~ 0.7 km of cloud cell above the lower 0.7 km discharged in the first stroke. Several successive strokes may follow at about 0.03 sec intervals until the cell is discharged.

The locality on the ground struck by one flash thus has some 20 coulombs of charge deposited on it in the order of 10^{-5} of a second on some square meter of terrain. Some strokes go up to 200 coulombs and others are 10 coulombs or less. If this quantity can rapidly be dissipated by ready access to the ground and with circuits having an effectivelly small time constant, it will do relatively little damage. If conducting paths are poor with localized potentials of millions of volts and 20 coulombs to dissipate some of the most amazing vagaries in discharge paths are chosen.

It is clear that until near the earth, (borne out by field distortion measurements near the earth's surface), the location of the ground target of a stroke depends on pure vagaries of the storm, wind velocities, turbulence, and the factors which govern the cell movements aloft.

As the pilot leader nears the earth local electrical conditions can play a major role. Obviously higher objects in the landscape will be more vulnerable. The more conducting they are the more vulnerable they are.

The old adage that lightning does not strike twice in the same place is correct from statistical considerations dealing with conditions aloft initiating the stroke. There may however be environmental geographical and meteorological factors causing convection and cell generation in certain favored localities—e.g., presence of shallow lakes. There are also terrain factors channelling such localized cells along paths set by wind patterns bound to topography. Under such conditions certain regions are more vulnerable than others. Thus certain high hilltops are such that cell passage over them is more frequent and ground conditions invite strokes at such points. There are definite hills and localities, notably high buildings, hilltops, etc., where strokes are very frequent. Except for such special features, however, the location of a stroke is a matter of pure chance.

The modern theory of the lightning rod or conductor is that while it *invites* strokes that would come in its neighborhood, it disposes of the charge most effectively. This it will only do if its content of copper, or conductivity, is adequate, if it is clearly exposed above the structure it protects and if it is well grounded. It should be as nearly devoid of bends or loops which have inductance as possible. Strokes in fact have been observed to jump across an air gap rather than go around a loop or too complicated a bend in a metal conductor. A larger structure would do well to have one or more conducting paths. It was considered in earlier days that[134] the sharp pointed end on a conductor would ward off a stroke by pouring a cloud of positive charge upward during the corona preceding the stroke to "neutralize" the charge descending. It is true that the rod will deliver considerable corona current as field intensity increases. However, this will never equal the 20 coulombs deposited in some 10^{-3} sec when fields become intense. Actually, it will probably start the positive upward streamer which may project the junction point of pilot leader and its structure with its heavy field distortion well up into the air and perhaps distribute current flow temporally so that the charge is more easily disposed of without damage.

In this connection, SCHONLAND[134] points out that for the Empire State Building and the Washington Memorial Obelisk have many discharges from them starting as upward moving streamers which continuously remove the charge on the cloud without the brilliant intermitent process of any kind. These yield average currents of 80 amperes lasting for 0.6 sec compared with the 5×10^4 amperes in the usual strokes to ground lasting 10^{-4} sec. For shorter lightning conductors, the strokes are catalysmic, but the higher the lightning rod, the higher the positive streamer from the rod to the approaching negative stepped leader. Those from ordinary houses lead to streamers some 15 or so meters long. Thus the function of the rod is to project the streamer and to carry safely to ground the some 10^5 ampere of current rather than have the charge deposited on a chimney there to seek its capricious route to ground.

HARDER and CLAYTON[133] give some interesting data on protection of power transmission lines. Here shielding is accomplished by elevating a grounded wire or mast above the line to be protected. In general, a grounded wire shields a wedge of space below it. Thus, for a conductor of 50 feet in height, a ground wire 90 feet high will shield roughly an angle of $41°$ to the ground. The character of the time consumed in the transmission of the return stroke and building up the conductivity of the long channel acts to limit the instantaneous current drain on the current arrestors or protective wires. Thus the footing resistances need not be prohibitively low, ~ 20 ohms.

In *average* lightning hazard territory, there are ten strokes to ground per square mile per year. A transmission line will encounter 1 direct stroke per mile per year. A small structure or substation will average one stroke in ten years. The number increases as more than the first power of the height of the building. The Cathedral of Learning in Pittsburgh, 585 feet high, is struck twice a year on the average. These average values are subject to the usual statistical fluctuations and vary widely with terrain and local conditions.

The number of thunderstorm days per year varies from less than 5 on the West Coast of the United States to 90 in Florida. Areas of *average* activity, such as in the mid-Atlantic states, have from 30—45 storms per year. The lightning arrestor currents, owing to diversion over several paths, line surges and perhaps chopping by flash-over are generally less than stroke currents. The short duration surges of high intensity produce large disruptive effects, but present small fire hazard. The long-tailed waves, with thousands of micro-seconds of current of ~ 200 amperes present great fire hazards, but do little shock damage.

The destructive nature of lightning strokes before the use of the lightning rod became common is recorded by SCHONLAND[134]. In London, the steeple and roof of St. Paul's Church were set on fire and destroyed in a thunderstorm in 1561, that of St. Brides was severely damaged in 1764, and that of St. Martins-in-the-Fields as late as 1842. The Campanile of St. Marks, in Venice, was twice completely destroyed and *seven* times severely damaged between 1388 and 1766. In sixteen years, between 1799 to 1815, lightning damaged 150 vessels of the British Navy. Nearly 100 lower masts of line-of-battle ships and frigates were destroyed. One ship in eight was set on fire in some part of the rigging or sails, about 70 seamen were killed and 130 seriously hurt. In ten cases, ships were completely disabled and forced to leave their stations at critical periods in the Napoleonic wars. Several ships were lost with all hands in violent thundertorms. The appearance of St. Elmo's fire at the tips of masts, a corona discharge, was common under thunderstorm conditions. It was taken by the sailors as a good omen, but today in wooden ships, will be taken as a danger signal for high fields aloft.

In consequence, as a general rule, in thunderstorm country, all structures, houses, barns, etc., especially those in exposed positions, e.g., hilltops, or towering above other structures, should have lightning rod protection.

Probably relatively safe are conducting metal structures such as metal aircraft hulls, automobiles, oil tanks and steel ships. These are Faraday cages and thus if conductivity is adequate discharge currents will pass over the skin.

Aircraft are repeatedly struck by lightning and in most cases damage is slight. It is probable that where it was severe, the records are obliterated for obvious reasons. There are possibly several reasons for the relatively slight damage. First the cloud-to-ground stroke requires heavy charges, currents, and fields to reach the earth. Aircraft are struck by the weaker intracloud strokes. They also are continually in corona from the wing tips and act as equalizers for smaller charge accumulations leading to less charge dissipation by the stroke.

Automobiles again seem in general to carry off the charge successfully. However, here there are danger spots, especially before rain begins when tires are insulating. The radio antenna is also a potential source of trouble and should be retracted. The all metal railroad trains, bound as they are to effectively grounded structures, are safe. The steel ships at sea also appear comparatively safe.

It would be assumed that oil tanks were particularly immune. However, the author personally witnessed a devastating fire set by a stroke to a gasoline storage tank. The tanks were on relatively dry ground in central California in summer. Either there were no lightning rods or they were defective. It is not impossible that the stroke was exceedingly heavy and possibly burnt or bored a hole through the top, or side, tearing a section of wall out and igniting the gasoline vapor.

How such a rivetted tank could be damaged so as to be set on fire is not difficult to discern. Exposed to sea air without proper paint protection, the rust can penetrate the lapped riveted seams so that a whole plate could effectively be insulated from the rest of the tank. Thus the spark to the plate would jump the rust insulated section, vaporize moisture between and create pressure, snapping the rivets. The tanks described above were apparently in that condition.

Obviously strokes near vent pipes, from which, owing to lowering barometric pressure gas fumes are exuding, without the essential protective screen invites trouble. Thus despite their metal nature and good grounding, the dangerous nature of the contents in case of a puncture by stroke, suggest well grounded lightning rods.

Concerning the individual caught outside in a thunderstorm, there are a few don'ts to be observed:

1. Do not stand up in open fields. If one cannot get to safe shelter, ignominiously lie prone on the ground.

2. Do not seek shelter under trees. Trees have roots that find water. They have conducting sap. They project upwards above the earth. The taller the tree, the more it invites a stroke. When trees are struck, they may not dissipate the charge. The human body is a better conductor than trees or rocks. In illustration of this, a group of Sierra mountaineers were caught near the top of a rocky hill-crest in a heavy storm. Just below the crest were some shallow, dry caves under overhanging rocks. Shelter was sought in the caves. Two of the party took off their packs and rested with moist shirts and backs against the cave wall. Others leaned against the packs. The rock above them was struck by a heavy stroke. The discharge went through and around the rock. Those with packs were jolted; the two lying against the rock were killed. The current took the low resistance path through their bodies for only a few feet.

By the same token, it is best to avoid proximity to wires and even wire fences in the open. Carrying tall metal objects invites trouble. Keep away from grounded wires shorter than oneseff.

3. Indoors, close windows. Keep away from chimneys, open fireplaces, telephones, radios, television, and light fixtures, and even avoid metal furniture, e.g. brass bedsteads, etc.

* * *

By rare coincidence, after this book was in print, the author, on October 20, 1957, met the single eye-witness to the explosion of the gas tank in California. Since electrical storms are rare in that area, there was no lightning rod. The large crude oil tank was provided with a vent pipe of some 10 cm diameter or more, which projected of the order of a meter above the top of the tank. The weather was exceptionally warm and humid; the tank was being filled from the bottom with crude oil at the time of the stroke; and was venting the explosive mixture of crude oil fumes and air, when the lightning was seen to strike the vent pipe. The resulting explosion blew off the whole top of the tank and very quickly enveloped the tank in flames of burning oil. It is gratifying to have such a simple explanation of an otherwise mystifying phenomenon.

References

Introduction

[1] LOEB, L. B.: Science, Lancaster, Pa. **102**, 573 (1945).
[2] HENRY, P. S. H.: Brit. J. Appl. Phys. Suppl. **2**, S. 56 (1953).

Chapter I

[3] LANDOLT BÖRNSTEINS Tabellen, p. 1019ff. Berlin: Springer 1923.
[4] BOWDEN, F. P., and W. R. THROSSEL: Proc. Roy. Soc. Lond., Ser A **209**, 297 (1951).
[5] KUNKEL, W. B.: J. Appl. Phys. **21**, 820 (1950).
[6] DODD, E. E.: J. Appl. Phys. **24**, 73 (1953).
[7] PETERSON, J. W.: J. Appl. Phys. **25**, 501 (1954).
[8] GILL, E. W. B., and G. F. ALFREY: Nature, Lond. **163**, 172 (1949).
[9] COOPER, W. F.: Brit. J. Appl. Phys. Suppl. **2**, S. 11 (1953).
[10] DOLEZALEK, F.: Chem. Ind. **36**, 33 (1913). — Nature Paris (Oct. 1) **1930**, 342—344.
[11] McKEOWN. S. S., and V. WAUK: Industr. Engg. Chem. **34**, Nr. 6.
[12] LENARD, P.: Ann. d. Phys. **46**, 584 (1892).

[13] GUNN, R., and ASSOCIATES: Proc. Inst. Radio Engrs. **34**, 156 P, 162 P, 167 P, 234, 241, 248 (1946).

[14] VIEWEG, H. F.: J. Phys. Chem. **30**, 865 (1928).

[15] RICHARDS, H. F.: Phys. Rev. **22**, 122 (1923).

[16] DAVY, H.: Ann. Phys. **28**, 161 (1808).

[17] KNOBLAUCH, O.: Z. phys. Chem. **39**, 225 (1901).

[18] HELMHOLTZ, H. L. F.: Ann. d. Phys. **7**, 337 (1879).

[19] PERRIN, J.: J. Chem. Phys. **2**, 607 (1904).

[20] SMOLUCHOWSKI, W.: Kolloid.-Z. **18**, 190 (1916).

[21] GOUY, M.: J. Phys. Radium **9**, 457 (1910).

[22] DEBYE, P., and E. HÜCKEL: Phys. Z. **24**, 185, 305, 575 (1923); **25**, 97, 204 (1924).

[23] HENRY, W. S. P.: Proc. Roy. Soc. Lond., Ser. A **133**, 106 (1931).

[24] TAYLOR, H. S., and S. GLASSTONE: Treatise of Physical Chemistry, 3rd Ed., Vol. 2, pp. 628 ff. New York: D. van Nostrand & Co. 1951.

[25] EVERSOLE, W. G., and P. H. LAHR: J. Chem. Phys. **9**, 530 (1941).

[26] GUNN, R., and J. E. DINGER: Terrest. Mag. a. Electr. **51**, 477 (1946).

[27] WORKMAN, E. J., and S. E. REYNOLDS: Phys. Rev. **78**, 254 (1950).

[28] LOEB, L. B., A. F. KIP and A. W. EINARSSON: J. Chem. Phys. **6**, 265 (1937).

[29] WORKMAN, E. J., and S. E. REYNOLDS: New Mex. Inst. Min. a. Technol., Thunderstorm Electr. Report Nr. 9, Final, Aug., 1955.

[30] HENNIKER, J. C.: J. Coll. Sci. **7**, 443 (1949).

[31] HENNIKER, J. C.: Rev. Mod. Phys. **21**, 322 (1949).

[32] SCHAEFER, V. J.: Phys. Rev. **77**, 721 (1950).

[33] GILL, E. W. B., and G. F. ALFREY: Brit. J. Appl. Phys. Suppl. **2**, S. 16 (1953).

Chapter II

[34] HELMHOLTZ, H. L. F.: Ann. Phys. **7**, 337 (1879).

[35] KELVIN, LORD: Phil. Mag. (V) **46**, 82 (1898).

[36] MILLIKAN, R. A.: Phys. Rev. **7**, 18 (1916).

[37] WILSON, H. A.: Phil. Trans. Roy. Soc. Lond., Ser. A **197**, 429 (1901); A **202**, 258 (1903). — Phil. Mag. **24**, 196 (1912).

[38] RICHARDSON, O. W.: Jb. Radio Aktivitat **1**, 302 (1904). — Phil. Mag. **23**, 601, 619 (1912); **24**, 740 (1912).

[39] HERRMAN, G., and P. S. WAGENER: The Oxide Coated Cathode, Vol. II. London: Chapman-Hall 1951.

[40] NICHOLS, M. H.: Phys. Rev. **57**, 297 (1940).

[41] LANGMUIR, I., and K. H. KINGDON: Phys. Rev. **23**, 112 (1924).

[42] BECKER, J. A.: Phys. Rev. **28**, 341 (1926).

[43] LANGMUIR, I., and J. P. TAYLOR: Phys. Rev. **44**, 423 (1933).

[44] BOER, J. H. DE: Electron Emission and Absorbtion Phenomena. Cambridge: Press 1935.

[45] LANGMUIR, I.: Phys. Rev. **22**, 357 (1923).

[46] LINFORD, L. B.: Rev. Mod. Phys. **5**, 47 (1933).

[47] FOWLER, R. H.: Phys. Rev. **38**, 45 (1931).

[48] ZISMAN, W. A.: Rev. Sci. Instrum. **3**, 367 (1932).

[49] COMPTON, K. T.: Phil. Mag. **23**, 574 (1912).

[50] VOLLRATH, R. E.: Phys. Rev. **42**, 298 (1932).

[51] KUNKEL, W. B.: J. Appl. Phys. **21**, 829 (1950).

[52] BOWDEN, F. P., and D. TABOR: Friction and Lubrication of Solids. Oxford: Clarendon Press 1954.

[53] HARPER, W. R.: Proc. Roy. Soc. Lond., Ser. A **205**, 83 (1951).

Chapter III

[54] ELSTER, J., and H. GEITEL: Wien. Ber. **94** (1890).

[55] FRUMKIN, A.: Z. Phys. Chem. **109**, 34 (1924); **111**, 190 (1924); **116**, 485 (1925).

[56] CHALMERS, J. A., and F. PASQUILL: Phil. Mag. **23**, 88 (1937).

[57] COEHN, A., and U. RAYDT: Ann. d. Phys. **30**, 777 (1909). — COEHN, A., and H. MOZER: Ann. d. Phys. **43**, 1048 (1914). — COEHN, A., and LOTZ: Phys. Z. **21**, 327 (1920).

[58] McTaggart, H. A.: Phil. Mag. **27** (297); **28**, 367 (1914); **44**, 386 (1922).

[59] Alty, T.: Proc. Roy. Soc. Lond., Ser. A **106**, 316 (1924); A **112**, 235 (1926); A **122**, 622 (1929).

[60] Chapman, Seville: Phys. Rev. **52**, 184 (1937); **54**, 520, 528 (1938).

[61] Harper, W. R.: Brit. J. Appl. Phys. Suppl. **2**, S 19 (1953).

[62] Coehn, A.: Wied. Ann. **54**, 217 (1898). — Coehn, A., and U. Raydt: Ann. d. Phys. **30**, 777 (1909).

[63] DeBroglie, M.: Ann. Chim. et Phys. **16**, 50 (1909).

[64] Blanchard, D. C.: Nature, Lond. **175**, 334 (1955).

[65] Büsse, W.: Ann. d. Phys. **76**, 495 (1925).

[66] Aselman, E.: Ann. d. Phys. **19**, 960 (1906).

[67] Lenard, P.: Ann. d. Phys. **47**, 463 (1915).

[68] Chapman, Seville: Thunderstorm Electricity, H. Byers, Editor, Chap. IX, p. 209 ff. Univ. of Chicago Press 1953.

[69] Obolensky, W.: Ann. d. Phys. **39**, 971 (1912).

[70] Blackwood, O.: Phys. Rev. **16**, 85 (1920).

[71] Erickson, H. A.: Phys. Rev. **17**, 400 (1921); **18**, 100 (1921); **19**, 275 (1922).

[72] McBain, J. W., and R. C. Swain: Proc. Roy. Soc. Lond., Ser. A **154**, 608 (1936).

[73] Johonnot, E. S.: Phil. Mag. **47**, 501 (1899). — Phys. Rev. **20**, 388 (1905).

[74] Bernal, J., and R. H. Fowler: J. Chem. Phys. **1**, 515 (1933).

[75] Munson, R. J., A. M. Tyndall, H. G. David and K. Hoselitz: Proc. Roy. Soc. Lond., Ser. A **172**, 28, 139 (1940); A **177**, 183 (1941).

[76] Vassails, G.: Thesis. Paris: Masson & Cie. 1948.

[77] Kunkel, W. B.: J. Appl. Phys. **21**, 833 (1950).

[78] Loeb, L. B.: Basic Processes of Gaseous Electronics, Chap. 1, Sec. 13, p. 171. Univ. Calif. Press 1955.

[79] Chapman, Seville: Physics (Jour. Appl. Phys.) **5**, 150 (1934).

[80] Dyk, C.: Phys. Rev. **31**, 913 (1928).

[81] Hopper, V. D., and T. H. Laby: Proc. Roy. Soc. Lond., Ser. A **178**, 243 (1941).

[82] Hansen, J. W.: Phys. Rev. **72**, 741 (1947). — Hansen, J. W., and W. B. Kunkel: Rev. Sci. Instrum. **31**, 308 (1950).

[82A] Whytlaw-Gray, R., and H. Whittaker: Proc. Leeds Phil. Soc. **1**, 97 (1926).

[83] Margenau, H., and G. Murphy: Mathematics of Physics and Chemistry, pp. 422—424. New York: D. van Nostrand Co. Inc. 1934.

[84] Natanson, G. L.: Zhur. Fiz. Khim. **23**, 304 (1949).

[85] Bateman, H.: Phil. Mag. **21**, 745 (1911).

[86] Leontovich, M.: C. R. Acad. Sci. USSR. **53**, 111 (1946).

[87] Natanson, G. L.: C. R. Acad. Sci. USSR. **53**, 115 (1946).

[88] Zeleny, J.: Phys. Rev. **16**, 102 (1920). — Macky, W. A.: Proc. Roy. Soc. Lond., Ser. A **133**, 565 (1931). — English, W. N.: Phys. Rev. **74**, 179 (1948).

[89] Woodcock, A. H.: Amer. Soc. Test. Mat. **50**, 1151 (1950). — J. Meteorology **9**, 200 (1952); **10**, 362 (1953).

[90] Woodcock, A. H., C. F. Kientzler, A. B. Arons and D. C. Blanchard: Nature, Lond. **172**, 1144 (1953). — Tellus **6** (1954).

[91] Blanchard, D. C.: J. Coll. Sci. **9**, 321 (1954).

[92] Stuhlman, O.: Physics (Jour. Appl. Phys.) **2**, 457 (1932).

[93] Blanchard, D. C.: Nature, Lond. **175**, 334 (1955).

Chapter IV

[94] Medley, J. A.: Brit. J. Appl. Phys. Suppl. **2**, S 29 (1953).

[95] Debeau, D.: Phys. Rev. **66**, 9 (1944).

[96] Gill, E. W. B.: Phys. Rev. **74**, 842 (1948).

[97] Peterson, J. W.: Phys. Rev. **76**, 1882 (1949).

[98] Kunkel, W. B.: J. Appl. Phys. **19**, 1056 (1948).

[99] Kunkel, W. B.: J. Appl. Phys. **19**, 1053 (1948).

[100] Guest, P. A.: Static Electricity in Nature and Industry, U.S. Bureau Mines Bull. 36 B, Washington, D.C., 1933.

[101] RUDGE, W. A. O.: Phil. Mag. **24**, 852 (1912); **25**, 481 (1913).

[102] BEYERSDORFER, P.: Z. Ver. dtsch. Zuckerind. **72**, 475 (1922).

[103] BLACTIN, S. C., and H. ROBINSON: Safety Mines Research Paper 71, London 1931.

[104] WALTHER, R., and W. FRANKE: Braunkohle **28**, 789 (1929).

[105] BÖNING, P.: Z. techn. Phys. **8**, 385 (1927).

[106] HOAG, J. B.: Ann. Petrol. Inst. Bull. **9**, 181 (1928).

[107] THOMAS, D. G. A.: Brit. J. Appl. Phys. Suppl. **2**, S 55 (1953).

[108] PETERSON, J. W.: J. Appl. Phys. **25**, 907 (1954).

[109] WAGNER, P. E.: J. Appl. Phys. **27**, 1301 (1956).

[110] FINCH, G. I.: Nature, Lond. **138**, 1010 (1936).

[111] LEISE, K. H.: Z. Physik. **124**, 258 (1948).

[112] HARPER, W. R.: Proc. Roy. Soc. Lond., Ser. A **231**, 388 (1955).

[113] LOEB, L. B.: Thunderstorm Electricity, H. Byers Editor, Chap. VII, p. 150. Univ. Chicago Press 1953.

[114] SIMPSON, G. C., and F. J. SCRASE: Proc. Roy. Soc. Lond., Ser. A **177**, 281 (1937).

[115] PEARCE, D. C., and W. B. CURRIE: Canad. J. Res. **27**, 1 (1949).

[116] CHALMERS, J. A. J.: Atmos. a. Terr. Phys. **2**, 337 (1952).

[117] NORINDER, H., and R. SISKNA: Tellus **5**, 260 (1953). -- Ark. Geofysik **3**, 59 (1954).

[118] THOMPSON, J.: Proc. Roy. Soc. **11**, 473 (1861).

[119] HENRY, P. S. H.: Brit. J. Appl. Phys. Suppl. **2**, S 31 (1953).

[120] SHAW, P. E., and C. S. JEX: Proc. Roy. Soc. Lond., Ser. A **111**, 339 (1926).

[121] HENRY, P. S. H.: Brit. J. Appl. Phys. Suppl. **2**, S 9 (1953).

[122] PETERSON, J. W.: Phys. Rev. **76**, 1882 (1949).

[123] RINDE, C. H.: Masters Thesis in Physics. Univ. California, Berkeley 1926.

[124] LOEB, L. B.: Encyclopedia Britanica, XVI Edit. Chicago, Conduction Electricity in Gases, Sect. 14. 1948.

[125] LOEB, L. B.: Encyclopedia of Physics, Vol. 22, Sect. VII, pp. 506 ff. Berlin: Springer 1956.

[126] BULGIN, D.: Brit. J. Appl. Phys. Suppl. **2**, S 83 (1953).

[127] MEDLEY, J. A.: Brit. J. Appl. Phys. Suppl. **2**, S 23 (1953).

[128] FORREST, J. C.: Brit. J. Appl. Phys. Suppl. **2**, S 78 (1953).

[129] HENRY, P. S. H.: Brit. J. Appl. Phys. Suppl. **2**, S 78. (1953).

[130] CHAPMAN, SEVILLE: Cornell Aeronautical Laboratory, Report C.A.L. 68, March, 1956 also BENNETT, W. H.: Electrical Engineering, Oct. 1948.

[131] LOEB, L. B.: Thunderstorms and Lightning Strokes, Modern Physics for the Enginner, Edit. by L. N. Ridenour, Chap. 13, p. 331. New York: McGraw-Hill 1954.

[132] Papers by LOEB, L. B., GUNN, ROSS, KUETTNER: J. at. 250 Centennial Celebration in honor of BENJAMIN FRANKLIN, Amer: Acad. Arts and Sci., Cambridge, Mass., Jan., 1956 (to be published under title Atmospheric Explorations JOHN WILEY and Sons, NewYork 1958).

[133] Thunderstorm Electricity, H. Byers, Editor, loc. cit., Chap. XVI, p. 335.

[134] SCHONLAND, B. F. J.: Proc. Roy. Soc. Lond., Ser. A **235**, 433 (1956).

[135] BÖNING, P.: Other early papers. Arch. Elektrochem. **1928** and on. — A book entitled Elektrische Isolierstoffe, In Verhalten auf Grund der Ionenadsorption an inneren Grenzflächen. Braunschweig: Vieweg & Sohn 1938. Kolloid-Z. **92**, 136, 1940; **94**, 31 (1941). — Über das Auftreten elektrischer Ladungen bei Strömen isolierender Flüssigkeiten durch Faserstoffe. Diss. Braunschweig 1926. Staubelektrizität Z. Techn. Phys. **8**, 385 (1927). — Theorie der Aufladungserscheinungen an Staub, Papier und Spinnstoffen. Elektrotechn. Z., Ausg. A, **73**, H. 20 (1952). — Die universelle Bedeutung der Ionenadsorption in Isolierstoffen. Z. angew. Phys. **8**, 516—520 (1956).

Author Index

Subject Index

15*